Scientific Computation

L. C. Berselli   T. Iliescu   W. J. Layton

Mathematics of Large Eddy Simulation of Turbulent Flows

With 32 Figures

 Springer

Dr. Luigi C. Berselli
University of Pisa
Department of Applied Mathematics
"U. Dini"
Via Bonanno 25/b
I-56126 Pisa, Italy
e-mail: berselli@dma.unipi.it

Dr. William J. Layton
University of Pittsburgh
Department of Mathematics
Thackeray Hall 301
Pittsburgh, PA 15260, USA
e-mail: wjl@pitt.edu

Dr. Traian Iliescu
Virginia Polytechnic Institute
and State University
Department of Mathematics
456 McBryde Hall
Blacksburg, VA 24061, USA
e-mail: iliescu@math.vt.edu

Library of Congress Control Number: 2005930495

ISSN 1434-8322
ISBN-10 3-540-26316-0   Springer Berlin Heidelberg New York
ISBN-13 978-3-540-26316-6   Springer Berlin Heidelberg New York

Springer is a part of Springer Science+Business Media
springeronline.com

© Springer-Verlag Berlin Heidelberg 2006
Printed in Germany

The use of general descriptive names, registered names, trademarks, etc. in this publication does not imply, even in the absence of a specific statement, that such names are exempt from the relevant protective laws and regulations and therefore free for general use.

Typesetting: Data conversion by LE-TEX Jelonek, Schmidt & Vöckler GbR, Leipzig, Germany
Cover design: *design & production* GmbH, Heidelberg

Printed on acid-free paper   55/3141/YL   5 4 3 2 1 0

To Lucia, Raffaella, and Annette

Preface

Turbulence is ubiquitous in nature and central to many applications important to our life. (It is also a ridiculously fascinating phenomenon.) Obtaining an accurate prediction of turbulent flow is a central difficulty in such diverse problems as global change estimation, improving the energy efficiency of engines, controlling dispersal of contaminants and designing biomedical devices. It is absolutely fundamental to understanding physical processes of geophysics, combustion, forces of fluids upon elastic bodies, drag, lift and mixing. Decisions that affect our life must be made daily based on predictions of turbulent flows.

Direct numerical simulation of turbulent flows is not feasible for the foreseeable future in many of these applications. Even for those flows for which it is currently feasible, it is filled with uncertainties due to the sensitivity of the flow to factors such as incomplete initial conditions, body forces, and surface roughness. It is also expensive and time consuming–far too time consuming to use as a design tool. Storing, manipulating and post-processing the mountain of uncertain data that results from a DNS to extract *that which is needed* from the flow is also expensive, time consuming, and uncertain.

The most promising and successful methodology for doing these simulations of *that which matters* in turbulent flows is large eddy simulation or LES. LES seeks to calculate the large, energetic structures (the large eddies) in a turbulent flow. The aim of LES is to do this with complexity-independent of the Reynolds number and dependent only on the resolution sought. The approach of LES, developed over the last 35 years, is to filter the Navier–Stokes equations, insert a closure approximation (yielding an LES model), supply boundary conditions (called a Near Wall Model in LES), discretize appropriately and perform a simulation. The first three key challenges of LES are thus: *Do the solutions of the chosen model accurately reflect true flow averages? Do the numerical solutions generated by the chosen discretization, reflect solutions of the model? And, With the chosen model and method, how is simulation to be performed in a time and cost effective manner?* Although all three questions are considered herein, we have focused mostly on the first, i.e. the mathematical development of the LES models themselves. The second and third questions concerning numerical analysis and computational simulation of LES models are essential. However, the numerical analysis of LES should not begin by assuming a model is a correct mathematical realization of the intended physical phenomenon (in other words, that the model is well posed). To do so would be to build on a foundation of optimism. Numerical analysis of LES models with sound mathematical foundations is an exciting challenge for the next stage of the LES adventure.

One important approach to unlocking the mysteries of turbulence is by computational studies of key, building block turbulent flows (as proposed by von Neumann). The great success of LES in economical and accurate descriptions of many building block turbulent flows has sparked its explosive growth. Its development into a predictive tool, useful for control and design in complex geometries, is clearly the next step, and possibly within reach in the near future. This development will require much more experience with practical LES methods. It will also require fundamental mathematical contributions to understanding "How", "Why", and "When" an approach to LES can work and "What" is the expected accuracy of the combination of filter, model, discretization and solver.

The extension of LES from application to fully developed turbulence to include transition and wall effects and then to the delicate problems of control and design is clearly the next step in the development of large eddy simulation. Progress is already being made by careful experimentation. Even as "[The universe] is written in mathematical language" (Galileo), the Navier–Stokes equations are the language of fluid dynamics. Enhancing the universality of LES requires making a direct connection between LES models and the (often mathematically formidable) Navier–Stokes equations. One theme of this book is the connection between LES models and the Navier–Stokes equations rather than the phenomenology of turbulence. Mathematical development will complement numerical experimentation and make LES more general, universal, robust and predictive.

We have written this book in the hope it will be *useful* for LES practitioners interested in understanding how mathematical development of LES models can illuminate models and increase their usefulness, for applied mathematicians interested in the area and especially for PhD students in computational mathematics trying to make their first contribution. One of the themes we emphasize is that mathematical understanding, physical insight and computational experience are the three foundations of LES! Throughout, we try to present the first steps of a theory as simply as possible, consistent with correctness and relevance, and no simpler. We have tried, in this balancing act, to find the right level of detail, accuracy and mathematical rigor.

This book collects some of the fundamental ideas and results scattered throughout the LES literature and embeds them in a homogeneous and rigorous mathematical framework. We also try to isolate and focus on the mathematical principles shared by apparently distinct methodologies in LES and show their essential role in robust and universal modeling. In part I we review basic facets of on the Navier–Stokes equations; in parts II and III we highlight some promising models for LES, giving details of the mathematical foundation, derivation and analysis. In part IV we present some of the difficult challenges introduced by solid boundaries; part V presents a syllabus for numerical validation and testing in LES.

We are all too aware of the tremendous breadth, depth and scope of the area of LES and of the great limitations of our own experience and understanding. Some of these gaps are filled in other excellent books on LES. In particular, we have learned a lot ourselves from the books of Geurts [131], John [175], Pope [258], and Sagaut [267]. We have tried to complement the treatment of LES in these excellent books by developing mathematical tools, methods, and results for LES . Thus, many of the same topics are often treated herein but with the magnifying glass of mathematical analysis. This treatment yields new perspectives, ideas, language and illuminates many open research problems.

We offer this book in the hope that it will be *useful* to those who will help develop the field of LES and fill in many of the gaps we have left behind herein.

It is a pleasure to acknowledge the help of many people in writing this book. We thank Pierre Sagaut for giving us the initial impulse in the project and for many detailed and helpful comments along the way. We owe our friend and colleague Paolo Galdi a lot as well for many exciting and illuminating conversations on fluid flow phenomena. Our first meeting came through one such interaction with Paolo. We also thank Volker John, who throughout our LES adventure has been part of our day to day "battles".

Our understanding of LES has advanced through working with friends and collaborators Mihai Anitescu, Jeff Borggaard, Adrian Dunca, Songul Kaya, Roger Lewandowski, and Niyazi Sahin.

The preparation of this manuscript has benefited from the financial support of the National Science Foundation, the Air Force office of Scientific Research, and Ministero dell'Istruzione, dell'Università e della Ricerca.

Pisa, Italy   *Luigi C. Berselli*
Blacksburg, USA   *Traian Iliescu*
Pittsburgh, USA   *William J. Layton*
April, 2005

Index of Acronyms

Contents

Part II Eddy Viscosity Models

Part III Advanced Models

Part IV Boundary Conditions

Part I

Introduction

1

Introduction

Large Eddy Simulation, LES, is about approximating local, spatial averages of turbulent flows. Thus, LES seeks to predict the dynamics (the motion) of the organized structures in the flow (the eddies) which are larger than some user-chosen length scale δ. Properly, LES was born in 1970 with a remarkable paper by Deardorff [87] in which the question of closure, boundary conditions and accuracy of approximation are studied via computational experiments. Since then, LES has undergone explosive development as a computational technology. Such a rapid development has, naturally, raised many questions in LES, some of which are essentially mathematical in nature. Many of these mathematical issues in LES are important for advancing practical computations. Many are also important for broadening the usefulness of LES from a research methodology to a design tool and increasing its universality beyond fully developed turbulence to the heterogeneous mix of laminar, transitional, and fully developed turbulence typically found in practical flows.

The great challenge of simulating turbulence is that equations describing averages of flow quantities cannot be obtained directly from the physics of fluids. On the other hand, the equations for the pointwise flow quantities are well known, but intractable to solution and sensitive to small perturbations and uncertainties in problem data. These pointwise equations for velocity and pressure in an incompressible, viscous, Newtonian fluid are the Navier–Stokes equations (abbreviated NSE) for the velocity $\mathbf{u}(\mathbf{x}, t) = u_j(x_1, x_2, x_3, t)$, $(j = 1, 2, 3)$ and pressure $p(\mathbf{x}, t) = p(x_1, x_2, x_3, t)$ given by

$$\mathbf{u}_t + \mathbf{u} \cdot \nabla \mathbf{u} - \nu \Delta \mathbf{u} + \nabla p = \mathbf{f}, \quad \text{in } \Omega \times (0, T), \tag{1.1}$$

$$\nabla \cdot \mathbf{u} = 0, \quad \text{in } \Omega \times (0, T), \tag{1.2}$$

where $\nu = \mu/\rho$ is the kinematic viscosity, $\mathbf{f}$ is the body force, and $\Omega \subset \mathbb{R}^d$ $(d = 2 \text{ or } 3)$ is the bounded flow domain with a sufficiently regular boundary $\partial \Omega$. The NSE are supplemented by the initial condition and the usual pressure normalization condition

$$\mathbf{u}(\mathbf{x}, 0) = \mathbf{u}_0(\mathbf{x}), \text{ for } \mathbf{x} \in \Omega \qquad \text{and} \qquad \int_\Omega p(\mathbf{x}, t) \, d\mathbf{x} = 0, \qquad (1.3)$$

and appropriate boundary conditions, such as the no-slip condition,

$$\mathbf{u} = \mathbf{0} \quad \text{on} \quad \partial\Omega.$$

When it is useful to uncouple the difficulties arising from the equations of motion from those connected with interaction of the fluid with the boundary, it is usual to study (1.1), (1.2), and (1.3) on $\Omega = (0, 2\pi)^d$, under periodic boundary conditions (instead of the no-slip condition) with zero mean imposed upon the velocity and all data[1]

$$\begin{cases} \mathbf{u}(\mathbf{x} + 2\pi\mathbf{e}_j, t) = \mathbf{u}(\mathbf{x}, t), \quad \text{and} \quad \int_\Omega \mathbf{u}(\mathbf{x}, t) \, d\mathbf{x} = \mathbf{0}, \\ \\ \text{where } \int_\Omega \mathbf{u}_0(\mathbf{x}) \, d\mathbf{x} = \mathbf{0}, \quad \text{and} \quad \int_\Omega \mathbf{f}(\mathbf{x}, t) \, d\mathbf{x} = \mathbf{0}, \text{ for } 0 \leq t \leq T. \end{cases}$$

We observe that, due to the divergence-free constraint, the nonlinear term $\mathbf{u} \cdot \nabla \mathbf{u}$ can be written in two equivalent ways:

$$\mathbf{u} \cdot \nabla \mathbf{u} = \sum_{j=1}^3 u_j \frac{\partial u_i}{\partial x_j} \qquad \text{or} \qquad \nabla \cdot (\mathbf{u}\,\mathbf{u}) = \sum_{j=1}^3 \frac{\partial}{\partial x_j} (u_i u_j).$$

The NSE follow directly from conservation of mass, conservation of linear momentum and a linear stress–strain relation, see Sect. 2.2 for further details. One fundamental property of the Navier–Stokes equations that is a direct connection between the physics of fluid motion and its mathematical description is the *energy inequality*, proved by J. Leray in his 1934 paper [213]. Here $\| \cdot \|$ denotes the usual $L^2(\Omega)$-norm of a vector field $\|\mathbf{u}\| = (\int_\Omega |\mathbf{u}(\mathbf{x})|^2 \, d\mathbf{x})^{\frac{1}{2}}$ and $(\cdot, \cdot)$ the associated $L^2(\Omega)$ inner product.

Theorem 1.1. *For each divergence-free initial datum $\mathbf{u}_0$ (under either periodic or no-slip boundary conditions) there exist weak[2] solutions in the sense of Leray and Hopf. All of them satisfy, for $t > 0$, the following energy inequality*

$$\frac{1}{2}\|\mathbf{u}(t)\|^2 + \int_0^t \nu\|\nabla\mathbf{u}(\tau)\|^2 \, d\tau \leq \frac{1}{2}\|\mathbf{u}_0\|^2 + \int_0^t (\mathbf{f}(\tau), \mathbf{u}(\tau)) \, d\tau.$$

If $\mathbf{u}$ is a strong solution then the energy inequality holds with inequality replaced by equality.

---

[1] Here $\mathbf{e}_j$ $j = 1, \ldots, d$ are the canonical basis functions in $\mathbb{R}^d$.

[2] We will present later the exact definitions of strong and weak solutions referred to in this theorem.

The above energy inequality is the direct link between the physics of fluid motion and the abstract theory of the NSE. In fact, each term has a direct physical interpretation (see also Sect. 2.4.3):

$$\text{kinetic energy } k(t) = \frac{1}{2}\|\mathbf{u}(t)\|^2,$$

$$\text{energy dissipation rate } \varepsilon(t) = \frac{\nu}{|\Omega|}\|\nabla\mathbf{u}(t)\|^2,$$

$$\text{power input } P(t) = (\mathbf{f}(t), \mathbf{u}(t)),$$

and, as we will see extensively in Part II and Part III, an energy inequality will be the core for most of the analytical existence theorems for LES models. In particular, in Chap. 2 we will review the main results on weak and strong solutions for the NSE to give the reader at least the flavor of the mathematical difficulties involved in the study of the NSE. At the same time, we try to give a reasonable number of details, in such a way that the reader can understand and practice some of the basic tools in the analysis of nonlinear partial differential equations.

The NSE, supplemented by appropriate boundary-initial-conditions quite likely include all information about turbulence. The wide separation between the largest and smallest scales of turbulence (Chap. 2 and onward) creates problems however. Extracting that information reliably (meaning, performing a direct numerical simulation (DNS) in which the mesh is chosen fine enough to resolve the smallest persistent eddy) is not feasible for many important flows.

The key control parameter in the Navier–Stokes equations is the (non-dimensional) Reynolds [262] number Re defined by

$$Re := \frac{UL}{\nu}, \qquad U = \text{characteristic velocity, } L = \text{characteristic length,}$$

$$\nu = \text{kinematic viscosity.}$$

$$(1.4)$$

When Re is large, the flow typically increases in complexity and in the range of solution scales that persist. For large enough Re, the flow becomes turbulent. Since turbulent flows are the typical case in nature, predicting turbulent flows is an important challenge in scientific and engineering applications. Indeed, design decisions that impact on our lives are made daily based upon doubtful simulations of turbulent flow. Obviously, design and control must be preceded by description and understanding, and reliable prediction requires a synthesis of both theory and experiment as foreseen by Sir Francis Bacon in 1620 in *Novum Organum*:

Nature, to be commanded, must be obeyed.

Kolmogorov's 1941 theory of homogeneous, isotropic turbulence (which is described in Chap. 2 in more detail) predicts that small scales exist down to

$O(Re^{-3/4})$, where $Re > 0$ is the Reynolds number, see (1.4). Thus, in order to capture them on a mesh, we need a mesh size $h \approx Re^{-3/4}$, and consequently (in 3D) $N = Re^{9/4}$ mesh points. To give the flavor of the overall computational cost, here are some representative Reynolds numbers

- model airplane (characteristic length 1 m, characteristic velocity 1 m/s)
  $Re \approx 7 \cdot 10^4$
  requiring $N \approx 8 \cdot 10^{10}$ mesh points per time-step for a DNS
- cars (characteristic velocity 3 m/s)
  $Re \approx 6 \cdot 10^5$
  requiring $N \approx 10^{13}$ mesh points per time-step for a DNS
- airplanes (characteristic velocity 30 m/s)
  $Re \approx 2 \cdot 10^7$
  requiring $N \approx 2 \cdot 10^{16}$ mesh points per time-step for a DNS
- atmospheric flows
  $Re \approx 10^{20}$
  requiring $N \approx 10^{45}$ mesh points per time-step for a DNS

Even though DNS is obviously unsuitable for many numerical simulations of turbulent flows, it can be useful to validate turbulence models. Moreover, even if DNS were feasible for turbulent flows, a major hurdle would be defining precise initial and boundary conditions. At high Reynolds numbers the flow is unstable. Thus, even small boundary perturbations may excite the already existing small scales. This results in unphysical noise being introduced in the system, and in the random character of the flow. Indeed, as observed in Aldama [7], the uncontrollable nature of the boundary conditions (in terms of wall roughness, wall vibration, differential heating or cooling, *etc.*) forces the analyst to characterize them as "random forcings" which, consequently, produce random responses. In such settings, calculating average values of flow quantities makes more sense than point values. See Sect. 2.6 for further details.

Further, that information, if extractable comprises a data set so large that sifting through it to calculate the quantities needed for flow simulation is a computational challenge by itself. Often these quantities are flow statistics or averages. Thus, the clear practical solution to both aspects which has evolved is to try to calculate directly the sought averages. Further, there is considerable evidence, which comes from analyzing data from observations of turbulent flows in nature, that the large scales in turbulence are not chaotic but deterministic, while the sensitivity, randomness, and chaotic dynamics is restricted to the small scales. Thus, it is usual to seek not to predict the pointwise (molecular) couple velocity–pressure $(\mathbf{u}, p)$, but rather suitable averages of it, $(\overline{\mathbf{u}}, \overline{p})$.

1.1 Characteristics of Turbulence

In 1949, John von Neumann wrote in one of his reports, privately circulated for many years (see [117]):

These considerations justify the view that a considerable mathematical effort toward a detailed understanding of the mechanism of turbulence is called for. The entire experience with the subject indicates that the purely analytical approach is beset with difficulties, which at this moment are still prohibitive. The reason for this is probably as was indicated above: That our intuitive relationship to the subject is still too loose – not having succeeded at anything like deep mathematical penetration in any part of the subject, we are still quite disoriented as to the relevant factors, and as to the proper analytical machinery to be used.

Under these conditions there might be some hope to 'break the deadlock' by extensive, well-planned, computational efforts. It must be admitted that the problems in question are too vast to be solved by a direct computational attack, that is, by an outright calculation of a representative family of special cases. There are, however, strong indications that one could name certain strategic points in this complex, where relevant information must be obtained by direct calculations. If this is properly done, and then the operation is repeated on the basis of broader information then becoming available, etc., there is a reasonable chance of effecting real penetrations in this complex of problems and gradually developing a useful, intuitive relationship to it. This should, in the end, make an attack with analytical methods, that is truly more mathematical, possible.

Since we are still far from a mathematically rigorous understanding of turbulence, the physical markers of turbulence in experiments are important.

However, this path is by no means easy. This is apparent when we try to define turbulence. It is usual to describe turbulence by listing its characteristic features. (For a detailed presentation, the reader is referred to Lesieur [214], Frisch [117], Pope [258], and Hinze [151].)

- Turbulent flows are *irregular*. Because of irregularity, the deterministic approach to turbulence becomes impractical, in that it appears intractable to describe the turbulent motion in all details as a function of time and space coordinates. However, it is believed possible to indicate average (with respect to space and time) values of velocity and pressure.
- Turbulent flows are *diffusive*. This causes rapid mixing and increased rates of momentum, heat and mass transfer. Turbulent flows are able to mix transported quantities much more rapidly than if only molecular diffusion processes were involved. For example, if a passive scalar is being transported by the flow, a certain amount of mixing will occur due to molecular diffusion. In a turbulent flow, the same sort of mixing is observed, but in a much greater amount than predicted by molecular diffusion. From the practical viewpoint, diffusivity is very important: the engineer, for instance, is concerned with the knowledge of turbulent heat diffusion coefficients, or the turbulent drag (depending on turbulent momentum diffusion in the flow).

- Turbulent flows are *rotational*. For a large class of flows, turbulence arises due to the presence of boundaries or obstacles, which create vorticity inside a flow which was initially irrotational. Turbulence is thus associated with vorticity, and it is impossible to imagine a turbulent irrotational flow.
- Turbulent flows occur at *high Reynolds numbers*. Turbulence often arises as a cascade of instabilities of laminar flows as the Reynolds number increases.
- Turbulent flows are *dissipative*. Viscosity effects will result in the conversion of kinetic energy of the flow into heat. If there is no external source of energy to make up for this kinetic energy loss, the turbulent motion will decay (see [117]).
- Turbulence is a *continuum* phenomenon. As noticed in [151], even the smallest scales occurring in a turbulent flow are ordinarily far larger than any molecular length scale.
- Turbulence is a feature of *fluid flows*, and not of fluids. If the Reynolds number is high enough, most of the dynamics of turbulence is the same in all fluids (liquids or gases). The main characteristics of turbulent flows are not controlled by the molecular properties of the particular fluid.

1.2 What are Useful Averages?

We have seen that it is usual to seek to predict suitable averages of velocity and pressure. Several different, useful averages play important roles in our presentation. Many more averaging operations are used in practice. At this point, there is no clear consensus on which averaging operation is most promising and, naturally, there's a lot of experimentation with different possibilities.

Conventional turbulence models approximate time averages of flow quantities, such as

$$\langle \mathbf{u}\rangle(\mathbf{x}) := \lim_{T\to\infty} \frac{1}{T} \int_0^T \mathbf{u}(\mathbf{x},t)\, dt, \qquad \langle p\rangle(\mathbf{x}) := \lim_{T\to\infty} \frac{1}{T} \int_0^T p(\mathbf{x},t)\, dt.$$

Due to the pioneering work [262] of Osborne Reynolds, these are also known as Reynolds averages. There are also flows for which the central features of turbulence are inherently dynamic. For these flows time averaging will completely erase the features one seeks to predict, which are retained by using instead a local, spatial average. LES seeks to approximate these local, spatial averages of the flow variables.

Mesh cell averaging is natural for finite difference calculations on structured meshes. Thus, the simplest example is *averaging over a mesh cell* (for example a box about $\mathbf{x} = (x_1, x_2, x_3)$ with equal sides of length δ):

$$\overline{\mathbf{u}}(\mathbf{x},t) = \frac{1}{\delta^3} \int_{x_1-\frac{\delta}{2}}^{x_1+\frac{\delta}{2}} \int_{x_2-\frac{\delta}{2}}^{x_2+\frac{\delta}{2}} \int_{x_3-\frac{\delta}{2}}^{x_3+\frac{\delta}{2}} \mathbf{u}(y_1, y_2, y_3, t)\, dy_1 dy_2 dy_3 \qquad (1.5)$$

However natural[3], this definition has many disadvantages. The resulting model cannot be rotation invariant. The model's solution will be very sensitive to the mesh orientation, so convergence will be hard to assess. The model also will not be smoothing and hence the property that the averages be deterministic in nature might fail. A better approach is to define the averages by convolution with a smooth function that is rotationally symmetric. Thus, let $g(\mathbf{x})$ be a filter kernel that is smooth, rotationally symmetric, and satisfying

$$0 \le g(\mathbf{x}) \le 1, \qquad g(\mathbf{0}) = 1, \qquad \int_{\mathbb{R}^d} g(\mathbf{x}) \, d\mathbf{x} = 1. \qquad (1.6)$$

Pick the length scale $\delta > 0$ of the eddies that are sought and define:

$$g_\delta(\mathbf{x}) := \frac{1}{\delta^d} g\left(\frac{\mathbf{x}}{\delta}\right).$$

Then, the LES average velocity $\overline{\mathbf{u}}$ and (turbulent) fluctuation $\mathbf{u}'$ are defined by

$$\overline{\mathbf{u}}(\mathbf{x}, t) = (g_\delta * \mathbf{u})(\mathbf{x}, t) := \int_{\mathbb{R}^d} g_\delta(\mathbf{x} - \mathbf{x}')\mathbf{u}(\mathbf{x}', t) \, d\mathbf{x}', \quad \text{and } \mathbf{u}' = \mathbf{u} - \overline{\mathbf{u}}. \ (1.7)$$

It is interesting to note that, while this decomposition into means and fluctuations was developed by Reynolds, it was advanced much earlier by, for example, da Vinci in 1510 in his description of vortices trailing a blunt body (as translated by Piomelli, http://www.glue.umd.edu/~ugo)

> "Observe the motion of the water surface, which resembles that of hair, that has two motions: One due to the weight of the shaft the other to the shape of the curls; thus water has two eddying motions, one part of which is due to the principal current, the other to the random and reverse motion."
> L. da Vinci, *Codice Atlantico*, 1510.

This definition overcomes many of the disadvantages of averaging over a mesh cell. Further, $\mathbf{u}' \to \mathbf{0}$ as $\delta \to 0$, so the closure problem is essentially that of estimating the effects of small quantities on large quantities, and thus hopeful.

To make this precise it is necessary to introduce some notation (*e.g.* Adams [4], Dautray and Lions [84], or Galdi [121], and for further details see also Chap. 2).

Definition 1.2. *The $L^2(\Omega)$ norm, denoted $\|\,.\,\|$, is*

$$\|\mathbf{u}\| := \left[\int_\Omega |\mathbf{u}|^2 \, d\mathbf{x}\right]^{1/2}.$$

---

[3] The original definition of O. Reynolds used exactly (1.5), which relies on space average within a small volume (this is (4) on p. 134 of the original paper [262]). He also considered the time average over a sliding time window (p. 135 of the original paper). Therefore, the original Reynolds operator is the LES box filter or its temporal counterpart!

The H^k-norm, denoted $\| \cdot \|_{H^k}$, is

$$\|\mathbf{u}\|_{H^k} := \left[\sum_{|\alpha| \leq k} \left\| \frac{\partial^{|\alpha|}}{\partial x_1^{\alpha_1} \dots \partial x_d^{\alpha_d}} \mathbf{u} \right\|^2 \right]^{1/2}$$

and $H^k(\Omega)$ denotes the closure of the infinitely smooth functions in $\| \cdot \|_k$.

For a constant averaging radius δ a lot is known about filtering, some of which is summarized next (see also Sect. 2.4.5). One common filter we will treat is the Gaussian. In three dimensions it is

$$g_\delta(\mathbf{x}) = \left(\frac{\gamma}{\pi}\right)^{\frac{3}{2}} \frac{1}{\delta^3} e^{-\gamma \frac{|\mathbf{x}|^2}{\delta^2}},$$

and typically $\gamma = 6$. Here we summarize the main properties of this filter and we will use them in the sequel.

Theorem 1.3. *Let δ be constant (not varying with position $\mathbf{x}$). Then,*

(a) If $\mathbf{u} \in L^2(\Omega)$ and $\mathbf{u}$ is extended by zero off Ω to compute $\bar{\mathbf{u}}$, then $\bar{\mathbf{u}} \to \mathbf{u}$ as $\delta \to 0$, i.e. $\|\mathbf{u} - \bar{\mathbf{u}}\| \to 0$.
(b) If $\mathbf{u} \in L^2(\Omega)$ and $\nabla \mathbf{u} \in L^2(\Omega)$ with $\mathbf{u} = \mathbf{0}$ on $\partial\Omega$ and extended by zero off Ω to compute $\bar{\mathbf{u}}$, then

$$\|\nabla(\mathbf{u} - \bar{\mathbf{u}})\| \to 0 \ \ as \ \delta \to 0.$$

(c) If the velocity field $\mathbf{u}$ has bounded kinetic energy then so does $\bar{\mathbf{u}}$:

$$\frac{1}{2} \int_\Omega |\bar{\mathbf{u}}|^2 \, d\mathbf{x} \leq \frac{C}{2} \int_\Omega |\mathbf{u}|^2 \, d\mathbf{x},$$

where the constant C is independent of δ.
(d) In the absence of boundaries (e.g. in the whole space or under periodic boundary conditions), filtering and differentiation commute:

$$\frac{\partial^{|\alpha|}}{\partial x_1^{\alpha_1} \dots \partial x_d^{\alpha_d}} \bar{\mathbf{u}} = \overline{\left(\frac{\partial^{|\alpha|}}{\partial x_1^{\alpha_1} \dots \partial x_d^{\alpha_d}} \mathbf{u} \right)} \qquad \forall \, \alpha \in \mathbb{N}^d.$$

(e) In the absence of boundaries (under periodic boundary conditions), for smooth $\mathbf{u}$, $\mathbf{u} = \bar{\mathbf{u}} + O(\delta^2)$. Specifically, we have

$$\|\mathbf{u} - \bar{\mathbf{u}}\| \leq C\delta^2 \|\mathbf{u}\|_{H^2}, \quad for \ \mathbf{u} \in H^2(\Omega). \tag{1.8}$$

Remarks on the Proof: We sketch the proof since it can be done by using the well-known technique of the Fourier transform. By definition, the Fourier transform of ϕ is

$$\hat{\phi}(\mathbf{k}, t) := \int_{\mathbb{R}^d} \phi(\mathbf{x}, t) \, e^{-i\mathbf{k} \cdot \mathbf{x}} \, d\mathbf{x},$$

where $\mathbf{k}$ represents the wavenumber vector. As a notation convention, from now on we will denote the Fourier transform of ϕ by either $\widehat{\phi}$, or $\mathcal{F}(\phi)$. Parts (a)–(d) are standard results for averaging by convolution. Part (c) is known as Young's inequality. Part (e) can be proved several different ways (see the books [117, 152, 158]). For example, by using basic properties of Fourier transforms:

$$\|\mathbf{u} - \overline{\mathbf{u}}\|^2 = \|\widehat{\mathbf{u} - \overline{\mathbf{u}}}\|^2 = \|(1 - \widehat{g}_\delta)(\mathbf{k})\widehat{\mathbf{u}}(\mathbf{k})\|^2$$
$$= \left(\int_{|\mathbf{k}| \geq \pi/\delta} + \int_{|\mathbf{k}| \leq \pi/\delta} \right) |1 - \widehat{g}_\delta(\mathbf{k})|^2 \, |\widehat{\mathbf{u}}(\mathbf{k})|^2 dk, \qquad (1.9)$$

we observe (this is one of the main tools when using Gaussian filters) that the Fourier transform of the Gaussian is again a Gaussian:

$$\widehat{g}_\delta(\mathbf{k}) = e^{-\dfrac{\delta^2}{4\gamma}(k_1^2 + k_2^2 + k_3^2)}.$$

In the sequel C will denote possibly different constants, not depending on δ and $\mathbf{u}$. On $0 \leq |\mathbf{k}| \leq \pi/\delta$, Taylor series expansion shows that

$$|1 - \widehat{g}_\delta(\mathbf{k})|^2 \leq C\delta^4 |\mathbf{k}|^4, \qquad \text{for } 0 \leq |\mathbf{k}| \leq \frac{\pi}{\delta}, \qquad (1.10)$$

while on $|\mathbf{k}| \geq \pi/\delta$ it holds that

$$|1 - \widehat{g}_\delta(\mathbf{k})|^2 \leq 2^2 \leq C(1 + |\mathbf{k}|^2)^{-2}(1 + |\mathbf{k}|^2)^2 \leq C(1 + \pi^2 \delta^{-2})^{-2} \, (1 + |\mathbf{k}|^2)^2.$$

Thus,

$$|1 - \widehat{g}_\delta(\mathbf{k})|^2 \leq C \, \delta^4 (1 + |\mathbf{k}|^2)^2, \qquad \text{for } |\mathbf{k}| \geq \frac{\pi}{\delta}. \qquad (1.11)$$

Combining (1.10) and (1.11) in (1.9) gives

$$\|\mathbf{u} - \overline{\mathbf{u}}\|^2 \leq C \, \delta^4 \int (1 + |\mathbf{k}|^2)^2 \, |\widehat{\mathbf{u}}(\mathbf{k})|^2 \, dk.$$

We note that, again by Plancherel's theorem,

$$\int (1 + |\mathbf{k}|^2)^2 |\widehat{\mathbf{u}}(\mathbf{k})|^2 dk \leq C \|\mathbf{u}\|_{H^2}^2$$

so that (1.8) follows, since the latter is an equivalent definition of the space H^2, see [84].

It is interesting to note that the fact the averaging over space can also slow down time variability of a flow was described by W. Wordsworth:

"Yon foaming flood seems motionless as ice;
Its dizzy turbulence eludes the eye,
Frozen by distance."
W. Wordsworth, 1770-1850, from *Address to Kilchurn Castle*.

Naturally, definition (1.7) for $\overline{\mathbf{u}}$ only makes sense if $\mathbf{u}(\mathbf{x}, t)$ can be extended off the flow domain Ω. For example, for Ω a box with periodic boundary conditions, a periodic extension of $\mathbf{u}$ suffices. If Ω is a box again and $\mathbf{u}$ vanishes on the boundary then $\mathbf{u}$ can be extended oddly off Ω. For more general domains, finding such an agreeable extension of $\mathbf{u}$ off Ω (which, through the equations of motion, determines the extension of the body force $\mathbf{f}$ needed to compute $\overline{\mathbf{f}}$) is not possible, see also Chap. 9.

Commonly used Spatial Filters

Many filter kernels are used. A good survey of the spatial filters commonly used in LES is given in Aldama [7], Coletti [67], and in the recent book by Sagaut [267]. Here we recall the most widely used.

Let $\phi(\mathbf{x}, t)$ be an instantaneous flow variable (velocity or pressure) in the NSE, and g denote an averaging kernel (1.6), with $g(\mathbf{x}) \rightarrow 0$ rapidly as $|\mathbf{x}| \rightarrow \infty$. The corresponding filtered flow variable is defined by convolution:

$$\overline{\phi}(\mathbf{x}, t) := \int_{\mathbb{R}^d} g(\mathbf{x} - \mathbf{x}')\, \phi(\mathbf{x}', t)\, d\mathbf{x}'. \tag{1.12}$$

The effect of the filtering operation becomes clear by taking the Fourier transform of expression (1.12). By the convolution theorem (roughly speaking the Fourier transform converts convolution into product), we get

$$\widehat{\overline{\phi}}(\mathbf{k}, t) = \widehat{g}(\mathbf{k})\, \widehat{\phi}(\mathbf{k}, t).$$

Thus, if $\widehat{g} = 0$, for $|\mathbf{k}_i| > \mathbf{k}_c$, $1 \leq i \leq d$, where $\mathbf{k}_c$ is a "cut-off" wavenumber, all the high wavenumber components of ϕ are filtered out by convolving ϕ with g. In 1958 Holloway [154] denoted a filter with these characteristics an "Ideal Low Pass Filter." However, if $\widehat{g}$ falls off rapidly (exponentially, say), an effective cut-off wavenumber can also be defined.

In addition to the ideal low pass filter, most commonly box filters and Gaussian filters have been used [267, 258]. The box filter (also known as "moving average" or "top hat filter") is commonly used in practice for experimental or field data.

- **Ideal Low Pass Filter**

$$g(\mathbf{x}) := \prod_{j=1}^{d} \frac{\sin \frac{2\pi x_j}{\delta}}{\pi x_j} \tag{1.13}$$

$$\widehat{g}(\mathbf{k}) = \begin{cases} 1 & \text{if } |k_j| \leq \frac{2\pi}{\delta}, \quad \forall\, 1 \leq j \leq d, \\ 0 & \text{otherwise.} \end{cases} \tag{1.14}$$

- **Box Filter**

$$g(\mathbf{x}) := \begin{cases} \dfrac{1}{\delta^3} & \text{if } |x_j| \leq \frac{\delta}{2}, \quad \forall\, 1 \leq j \leq d, \\ 0 & \text{otherwise.} \end{cases} \tag{1.15}$$

$$\widehat{g}(\mathbf{k}) = \prod_{j=1}^{d} \frac{\sin \frac{\delta k_j}{2}}{\frac{\delta k_j}{2}} \tag{1.16}$$

- **Gaussian Filter**

$$g(\mathbf{x}) := \left(\frac{\gamma}{\pi}\right)^{3/2} \frac{1}{\delta^3} e^{-\dfrac{\gamma\,|\mathbf{x}|^2}{\delta^2}} \tag{1.17}$$

$$\widehat{g}(\mathbf{k}) = e^{-\dfrac{\delta^2\,|\mathbf{k}|^2}{4\gamma}} \tag{1.18}$$

In formulas (1.13)–(1.18), δ represents the radius of the spatial filter g, and γ is a shape parameter often chosen to have the value $\gamma = 6$. For the ideal low pass filter a clear cut-off wavenumber, equal to $2\pi/\delta$ can be defined. In contrast, the Fourier transform of the box filter is a damped sinusoid and thus, spurious "amplitude reversals" are produced by its use in the Fourier space. Finally, the Fourier transform of the Gaussian filter is also a Gaussian and decays very rapidly. In fact, for all practical purposes, it is essentially contained in the range $\left[-\frac{2\pi}{\delta}, \frac{2\pi}{\delta}\right]$.

Differential Filters. An alternative well-known class of filters is that of differential filters, that were proposed in two pioneering papers by Germano [126, 127], see also Chap. 9. By using a differential filter, he obtained an LES model similar to the Rational LES model we will present in Chap. 7. Although differential filters are very appealing, they have been less used in practice than the three filters defined above. On the other hand, there is a strong argument that differential filters are a correct extension of filtering by convolution to bounded domains. Thus, we believe that they will become more central to LES as it develops.

We also observe that the Gaussian is the heat kernel. Thus, a natural extension of filtering from $\mathbb{R}^3$ to bounded domains is *via* a differential filter. In this case the average $\overline{\mathbf{u}}$ is the solution of

$$-\delta^2 \triangle \overline{\mathbf{u}} + \overline{\mathbf{u}} + \nabla\lambda = \mathbf{u}(\mathbf{x}, t), \text{ and } \nabla \cdot \overline{\mathbf{u}} = 0, \text{ in } \Omega,$$

subject to appropriate conditions on the boundary. For example, if Ω is a bounded domain with the no-slip condition $\mathbf{u} = 0$ on the boundary, we impose $\overline{\mathbf{u}} = 0$ on $\partial\Omega$ as the boundary condition for the above.

Any reasonable, local, spatial filter has two key properties:

(i) $\overline{\mathbf{u}} \to \mathbf{u}$ in $L^2(\Omega)$ as $\delta \to 0$;
(ii) $\|\overline{\mathbf{u}}\| \leq C\|\mathbf{u}\|$, uniformly in δ.

Remark 1.4. In the above presentation, we have assumed that δ, the filter radius, is constant. Often the filter radius is allowed to vary in space: $\delta := \delta(\mathbf{x})$. In Part IV, we will discuss in more detail the reasons for this choice and the possible consequences.

1.3 Conventional Turbulence Models

Time averaging, used in conventional turbulence models CTM (such as the k-ϵ model) was introduced by O. Reynolds [262]. It commutes with differentiation; thus, averaging the NSE gives an equilibrium problem for this flow average $\langle \mathbf{u} \rangle(\mathbf{x})$.

$$-\frac{1}{Re}\triangle\langle\mathbf{u}\rangle + \nabla \cdot \langle\mathbf{u}\,\mathbf{u}\rangle + \nabla\langle p\rangle = \langle\mathbf{f}\rangle, \quad \text{and} \quad \nabla \cdot \langle\mathbf{u}\rangle = 0, \quad \text{in } \Omega. \quad (1.19)$$

This problem is affected by the closure problem. The closure problem arises in (1.19), since $\langle\mathbf{u}\,\mathbf{u}\rangle \neq \langle\mathbf{u}\rangle\langle\mathbf{u}\rangle$. Thus, some closure model is needed. Since the closure problem occurs in a very similar way in LES, it is useful to look at it briefly here, for time averaging. The fluctuations about the average $\langle\mathbf{u}\rangle$, $\mathbf{u}'(\mathbf{x}, t)$ are defined by $\mathbf{u}'(\mathbf{x}, t) := \mathbf{u}(\mathbf{x}, t) - \langle\mathbf{u}\rangle(\mathbf{x})$. Time averaging has many convenient mathematical properties. For example, if $\mathbf{u} = \mathbf{0}$ on the boundary and if the boundary does not itself move, then $\langle\mathbf{u}\rangle = \mathbf{0}$ on the same boundary. Thus, for time averaging, correct boundary conditions are known. Any other boundary condition imposed is for economy or convenience, not necessity. Other important properties include: $\langle\mathbf{u}'\rangle = \mathbf{0}$, $\langle\langle\mathbf{u}\rangle\rangle = \langle\mathbf{u}\rangle$, and $\langle\langle\mathbf{u}\rangle\,\mathbf{v}\rangle = \langle\mathbf{u}\rangle\,\langle\mathbf{v}\rangle$, see Mohammadi and Pironneau [239]. Since $\mathbf{u} = \langle\mathbf{u}\rangle + \mathbf{u}'$ we can expand the nonlinear term using these properties:

$$\langle\mathbf{u}\,\mathbf{u}\rangle = \langle\langle\mathbf{u}\rangle\langle\mathbf{u}\rangle\rangle + \langle\langle\mathbf{u}\rangle\mathbf{u}' + \mathbf{u}'\langle\mathbf{u}\rangle\rangle + \langle\mathbf{u}'\mathbf{u}'\rangle = \langle\mathbf{u}\rangle\langle\mathbf{u}\rangle + \langle\mathbf{u}'\mathbf{u}'\rangle.$$

Thus, the time–averaged Navier–Stokes equations are

$$-\frac{1}{Re}\triangle\langle\mathbf{u}\rangle + \nabla \cdot \langle\mathbf{u}\rangle\langle\mathbf{u}\rangle + \nabla \cdot \langle\mathbf{u}'\mathbf{u}'\rangle + \nabla\langle p\rangle = \langle\mathbf{f}\rangle, \text{ and } \nabla \cdot \langle\mathbf{u}\rangle = \mathbf{0}, \text{ in } \Omega.$$

If we think of the average $\langle\mathbf{u}\rangle$ as being the observable and the fluctuation $\mathbf{u}'$ as being the unknowable, the closure problem can be restated pessimistically as follows:

"model the mean action of the unknowable upon the observable," that is, replace $\langle \mathbf{u}'\mathbf{u}' \rangle$ by terms only involving $\langle \mathbf{u} \rangle$. If the closure problem for CTM can be solved, then it gives economical prediction of flow statistics, meaning time averages. Closure is one of the main challenges of conventional turbulence modeling: find models which give (incremental) more reliable predictions of flow statistics for various flow configurations.

As an example of a common closure, let's briefly consider eddy viscosity models, which we will consider in more detail in Chaps. 3 and 4. Turbulent flow has long been observed to have stronger mixing and energy dissipation properties than laminar flows (see, for example, the experimental laws of fully developed turbulence discussed in Chap. 5 of Frisch [117]). This and other considerations led Boussinesq [43] to postulate that

"turbulent fluctuations are dissipative in the mean,"

now known as the Boussinesq assumption or eddy viscosity hypothesis, Chap. 3. Mathematically, this corresponds to the model

$$\nabla \cdot \langle \mathbf{u}'\,\mathbf{u}' \rangle \approx -\nabla \cdot (\nu_T \nabla^s \langle \mathbf{u} \rangle) + \text{ terms incorporated into the pressure.}$$

Here ∇^s denotes the symmetric part of the gradient tensor,

$$(\nabla^s \mathbf{v})_{ij} := \frac{1}{2}(v_{i,x_j} + v_{j,x_i})$$

and ν_T is the unknown eddy viscosity coefficient. Dimensional analysis suggests that the form of the turbulent viscosity coefficient ν_T should be given by the Prandtl–Kolmogorov relation:

$$\nu_T = Constant\; l\; \langle \sqrt{k'} \rangle,$$

where

$$l = l(\mathbf{x}, t) : \text{ local length scale of turbulent fluctuations,} \tag{1.20}$$

$$k' = \frac{1}{2}|\mathbf{u}'(\mathbf{x}, t)|^2 : \text{ kinetic energy of turbulent fluctuations.}$$

Assuming the eddy viscosity hypothesis, which is itself at best an analogy rather than a systematic approximation, the closure problem of CTM revolves around the almost equally hard problem of predicting k' and l.

The basic difficulty with conventional turbulence modeling is that $\langle \mathbf{u} \rangle$ and $\mathbf{u}'$ are, typically, both $O(1)$ and thus so are k' and (arguably) l. Thus, very accurate models are needed to produce accurate statistics. This often changes conventional turbulence modeling into a problem of model calibration, which means data fitting many undetermined model parameters to specific flow settings. Challenges for conventional turbulence models include how to produce reliable data for their calibration and how to simulate essentially dynamic flow behavior (known as URANS modeling). Contributions to these questions and others also are being made by LES.

1.4 Large Eddy Simulation

LES is connected to a natural computational idea: when a computational mesh is so coarse that the problem data and solution sought fluctuates significantly inside each mesh cell, it is only reasonable to replace the problem data by mesh cell averages of that data and for the approximate solution to represent a mesh cell average of the true solution. This observation was made by L. F. Richardson in his 1922 book [263]! Mathematically, if δ is the mesh cell width, then we should seek to approximate not the pointwise fluid velocity $\mathbf{u}(\mathbf{x}, t)$ but rather some mesh cell average $\overline{\mathbf{u}}(\mathbf{x}, t)$, the simplest of which is given by (1.5).

These cell averages are just convolution of the velocity $\mathbf{u}$ with the filter function $\delta^{-3} g(\mathbf{x}/\delta)$. For example, for the simplest cell average, (1.5), $g(\mathbf{x})$ is given by

$$g(\mathbf{x}) = \begin{cases} 1, & \text{if all } |x_j| \leq \frac{1}{2} \\ 0, & \text{otherwise.} \end{cases}$$

As noted above, many other filters, $g(\mathbf{x})$, are useful and important as well, such as Gaussian and differential filters.

Then this is the idea of LES in a nutshell: pick a useful filter $g(\mathbf{x})$ and define $\overline{\mathbf{u}}(\mathbf{x}, t) := (g_\delta * \mathbf{u})(\mathbf{x}, t)$ by (1.7). Derive appropriate equations for $\overline{\mathbf{u}}$ by filtering the NSE. Solve the closure problem; impose accurate boundary conditions for $\overline{\mathbf{u}}$. Then discretize the resulting continuum model and solve it!

Generally, such an averaging suppresses any fluctuations in $\mathbf{u}$ below $O(\delta)$ and preserves those on scales larger than $O(\delta)$. Averaging the NSE with this $g_\delta(\mathbf{x})$ reveals that under periodic boundary conditions (after some calculations and simplifications) $\overline{\mathbf{u}}$ satisfies $\overline{\mathbf{u}}(\mathbf{x}, 0) = \overline{\mathbf{u}}_0(\mathbf{x})$ and

$$\overline{\mathbf{u}}_t - \nu \triangle \overline{\mathbf{u}} + \overline{\mathbf{u}} \cdot \nabla \overline{\mathbf{u}} + \nabla \overline{p} + \nabla \cdot (\overline{\mathbf{u}\,\mathbf{u}} - \overline{\mathbf{u}}\,\overline{\mathbf{u}}) = \overline{\mathbf{f}}, \text{ in } \Omega \times (0, T) \quad (1.21)$$

$$\nabla \cdot \overline{\mathbf{u}} = 0, \text{ in } \Omega \times (0, T). \quad (1.22)$$

This system is often called the Space Filtered Navier–Stokes Equation (SFNSE). Again the closure problem arises since $\overline{\mathbf{u}\,\mathbf{u}} \neq \overline{\mathbf{u}}\,\overline{\mathbf{u}}$. With an appropriate closure model for $\overline{\mathbf{u}\,\mathbf{u}} - \overline{\mathbf{u}}\,\overline{\mathbf{u}}$, apparently (1.21) and (1.22) can be supplemented by boundary conditions then discretized and solved to give an approximation of $\overline{\mathbf{u}}$.

Complex models are thus coupled with complex discretization and solution algorithms which contain implicit and grid-dependent stabilization. These can swamp the subtle effects the model is attempting to simulate. If the qualitative predictions of the simulation are grid dependent, the question arises: does the continuum LES model have a solution which would then be mesh independent or is the code trying to hit a target that moves as the mesh width $h \to 0$? Thus, the fundamental mathematical questions of existence, uniqueness and stability of a continuum LES model have direct bearing on interpreting results of simulations. Unfortunately, these mathematical problems and others introduced by boundaries are also nontrivial.

1.5 Problems with Boundaries

This derivation of the SFNSE takes advantage of the fact that convolution and differentiation commute. In fact,

in the absence of boundaries $\quad g * \left(\dfrac{\partial \mathbf{u}}{\partial x_i} \right) = \dfrac{\partial}{\partial x_i}(g * \mathbf{u}), \quad$ for $i = 1, \ldots, d$

i.e. for $\Omega = \mathbb{R}^3$ or for Ω a box with periodic boundary conditions imposed on its boundary. The "first difficult issue" with boundaries is then associated with the "very first step" in the derivation of the SFNSE and in phrases like *away from walls, in the absence of boundaries*, and *we first focus on the interior equations*. Briefly, averaging/convolution and differentiation do not commute when boundaries are present and this introduces an extra term, the boundary commutation error term $A_\delta(\mathbf{u}, p)$, into the correctly derived SFNSE. Let the stress tensor be denoted by $\boldsymbol{\sigma}(\mathbf{u}, p) := -p\,\mathbb{I} + 2\nu\nabla^s\mathbf{u}$. In Chap. 9 this boundary commutator error term is calculated as

$$A_\delta(\mathbf{u}, p) = \int_{\partial\Omega} g_\delta(\mathbf{x} - \mathbf{s})\,\boldsymbol{\sigma}(\mathbf{u}, p)(\mathbf{s}) \cdot \mathbf{n}(\mathbf{s})\, dS(\mathbf{s}),$$

where $\mathbf{n}$ is the outward unit normal vector of $\partial\Omega$.

A careful analysis of the equations in Chap. 9, which follows [101], shows that $\|A_\delta(\mathbf{u}, p)\| \to 0$, as $\delta \to 0$, if and only if $\boldsymbol{\sigma}\cdot\mathbf{n} \equiv \mathbf{0}$, on $\partial\Omega$. The expression $\boldsymbol{\sigma}(\mathbf{u}, p)\cdot\mathbf{n}$ is the force that the unknown, underlying turbulent flow $(\mathbf{u}, p)$ exerts on the boundary. Thus, the term vanishes only if all variables can be extended across the boundary so that there's no net pointwise force on the boundary. One inescapable conclusion of this result is that within the usual constant averaging radius filtering approach to LES, a model of the commutation error term must be included for turbulent flow in which boundaries are important! So far, this term seems intractable, although encouraging attempts were made in [83, 39]. Lack of good models for $A_\delta(\mathbf{u}, p)$ might be one contributing reason LES experiences difficulties with near wall turbulence.

Developing effective computational models of the boundary commutation error term $A_\delta(\mathbf{u}, p)$ is thus an important open problem in LES. Another, complementary research challenge is to develop more – fully alternative – approaches, such as using a variable averaging radius $\delta = \delta(\mathbf{x}) \to 0$ as $\mathbf{x} \to \partial\,\Omega$, as developed by Vasilyev, Lund, and Moin [304], and differential filters, Germano [127, 126]. In both these cases, commutation error terms appear that are more uniformly distributed through the domain instead of piling up near $\partial\Omega$.

Very often, LES models have difficulty predicting turbulence generated by interactions of a (mostly laminar) flow with a (usually complex) boundary. Thus, the issue of finding boundary conditions for flow averages that are both accurate and well posed is an important one. With constant averaging radius, the problem of finding accurate boundary conditions for $\overline{\mathbf{u}}$ is also unavoidable.

In LES, such conditions are known as *near wall models*. The difficulty in near wall modeling is that $\overline{\mathbf{u}}$ on the boundary depends nonlocally on $\mathbf{u}$ near the boundary. Thus, simply imposing $\overline{\mathbf{u}} = \mathbf{0}$ on the boundary has two negative consequences:

(i) it degrades the overall accuracy of the model,
(ii) it introduces artificial boundary layers near the boundary that are smaller than $O(\delta)$.

Many ideas about using wall laws in conventional turbulence models for efficiency have been imported into near wall modeling in LES. The approach we study in Chap. 10 is to decompose $\mathbf{u} = \mathbf{0}$ on $\partial\Omega$ into its two component parts:

no-penetration: $\mathbf{u} \cdot \mathbf{n} = 0$ and   *no-slip:* $\mathbf{u} \cdot \boldsymbol{\tau}_j = 0$, on the boundary,

where $\boldsymbol{\tau}_j$ denotes unit tangent vectors.

Motivated by the work of Maxwell in 1879 [234] we consider near wall models for $\overline{\mathbf{u}}$ retaining no-penetration but replacing no-slip by a slip-with-friction condition:

No penetration of large eddies: $\overline{\mathbf{u}} \cdot \mathbf{n} = 0,$
Slip-with-friction along the boundary: $\beta\,\overline{\mathbf{u}} \cdot \boldsymbol{\tau}_j + \mathbf{n} \cdot \boldsymbol{\sigma}(\overline{\mathbf{u}}, \overline{p}) \cdot \boldsymbol{\tau}_j = 0.$

For $\beta \geq 0$, these boundary conditions lead to well-posed problems. The simplest example of such a β was derived by J.C. Maxwell [234], using the kinetic theory of gases. If we identify the LES microlength scale with δ (for a gas it is a mean free path), then Maxwell's analysis suggests

$$\beta \approx \frac{L Re^{-1}}{\delta}. \tag{1.23}$$

In Chap. 10, we show how the friction parameter $\beta = \beta(\overline{\mathbf{u}}, \delta, Re)$ can be constructed using boundary layer theory. These constructions give near wall models with the correct double asymptotics in Re and δ, see Sect. 10.4. However, boundary layer theory is less accurate for complex geometries, and it does not apply to turbulent flows with time-dependent boundary conditions, such as those in a control setting. We will present an alternative set of boundary conditions for these types of flows in Chap. 10.

Because of the twin difficulties of commutator error and near wall modeling, we take the reductionist approach: the closure problem will be treated for periodic boundary conditions in Parts I, II, and III. This uncouples the modeling and model validation problems from the problems of boundaries. Then, the question of boundaries will be separately considered in Part IV.

1.6 The Interior Closure Problem in LES

In contrast to conventional turbulence models, LES retains all the dynamics of the large scales. Like in conventional turbulence models, the closure problem arises because the average of the product is not the product of the averages. As before, define fluctuations $\mathbf{u}' = \overline{\mathbf{u}} - \mathbf{u}$ so we can write $\mathbf{u} = \overline{\mathbf{u}} + \mathbf{u}'$. In spatial filtering $\overline{\mathbf{u}'} \neq \mathbf{0}$ and, in general, $\overline{\overline{\mathbf{u}}} \neq \overline{\mathbf{u}}$ so that the nonlinear term retains all four addends

$$\overline{\mathbf{u}\,\mathbf{u}} = \overline{\overline{\mathbf{u}}\,\overline{\mathbf{u}}} + \overline{\overline{\mathbf{u}}\mathbf{u}'} + \overline{\mathbf{u}'\overline{\mathbf{u}}} + \overline{\mathbf{u}'\mathbf{u}'},$$

or equivalently,

$$\overline{\mathbf{u}\,\mathbf{u}} - \overline{\mathbf{u}}\,\overline{\mathbf{u}} = (\overline{\overline{\mathbf{u}}\,\overline{\mathbf{u}}} - \overline{\mathbf{u}}\,\overline{\mathbf{u}}) + \overline{\overline{\mathbf{u}}\mathbf{u}'} + \overline{\mathbf{u}'\overline{\mathbf{u}}} + \overline{\mathbf{u}'\mathbf{u}'}.$$

Compared to conventional turbulence models, the closure problem in LES is more difficult in that more interactions must be modeled than in the former. The critical reason for optimism in LES closure modeling is that, generally, δ is small and getting smaller as computers improve[4] and feasible meshes get finer. Thus, in many flows, *the portion of the flow that must be modeled,* $\mathbf{u}'$ *is small relative to the portion that is calculated,* $\overline{\mathbf{u}}$. Generally, a crude model with a large percentage error of a small variable is, in absolute terms, more accurate than a complex and highly tuned model of a large variable. As a result, models in LES tend to be both simple and accurate and overall computational cost tends not to be much greater than doing an (unreliable, under-refined) solution of the NSE on the same mesh!

This was the LES idea of deriving equations for space averaged variables mentioned by Richardson [263] in 1922! It can also be argued that the first use of this idea for mathematical understanding of the Navier–Stokes equations was by J. Leray [213] in the 1930s. Indeed, if we make the simple closure substitution

$$\overline{\mathbf{u}\,\mathbf{u}} \approx \overline{\overline{\mathbf{u}}\,\overline{\mathbf{u}}}$$

then the SFNSE become a closed system for a velocity and pressure, $(\mathbf{w}, q)$, which (hopefully) approximate $(\overline{\mathbf{u}}, \overline{p})$, given by

$$\mathbf{w}_t + \overline{\mathbf{w}} \cdot \nabla \mathbf{w} - \nu \Delta \mathbf{w} + \nabla q = \overline{\mathbf{f}} \quad \text{in } \Omega \times (0, T),$$
$$\nabla \cdot \mathbf{w} = 0, \quad \text{in } \Omega \times (0, T).$$

Leray developed (for the Cauchy problem) the mathematical properties of the system which are quite favorable, see Chaps. 2 and 8. By considering the behavior of $\mathbf{w}$ as $\delta \to 0$, he recovered a solution of the NSE. Missing from his treatment are two central issues in LES:

(i) *What is the accuracy of the approximation* $\mathbf{w} \approx \overline{\mathbf{u}}$?
(ii) *How accurately do statistics calculated from* $\mathbf{w}$ *represent the same statistics from* $\mathbf{u}$ *or* $\overline{\mathbf{u}}$?

For recent work revisiting Leray's model see Cheskidov *et al.* [59].

---

[4] For geophysical flows, however, computers are not yet powerful enough for $\mathbf{u}'$ to be small in absolute terms and the situation might not be as optimistic.

Remark 1.5 (The NS-α Model). Among the many interesting related topics not covered herein there is the NS-α model. The NS-α model is an interesting recent model derived in Camassa and Holm [53] by averaging a Lagrangian rather than Eulerian formulation of the Euler equations. It is an appealing model because it is supported by rigorous mathematical analysis, [53], Foiaş, Holm, and Titi [110, 111], and Marsden and Shkoller [232]. Interestingly, it has recently been shown by Guermond, Oden, and Prudhomme [145] that the NS-α model also comes about as a correction which restores frame invariance to the above Leray regularization of the NSE.

1.7 Eddy Viscosity Closure Models in LES

Here we briefly anticipate some facts regarding eddy viscosity models (developed fully in Part II). If the flow domain Ω is a box and boundary conditions are periodic, the boundary commutation error vanishes (so we will drop it for the moment). If all the remaining nonclosed terms are lumped together and the eddy viscosity hypothesis is postulated then we can write

$$\nabla \cdot (\overline{\mathbf{u}\mathbf{u}} - \overline{\mathbf{u}}\,\overline{\mathbf{u}}) \simeq -\nabla \cdot (\nu_T \nabla^s \overline{\mathbf{u}}) + \text{terms incorporated into the pressure,}$$

where, as in CTM, the form of ν_T is given by dimensional analysis to be $\nu_T = Constant \cdot l \cdot \sqrt{k'}$. In LES the length scale associated with fluctuations is known, $l = \delta$. Further, because (generically) $\overline{\overline{\mathbf{u}}} \neq \overline{\mathbf{u}}$, an estimate of k' can also be given by extrapolation from resolved to unresolved scales. As $\mathbf{u}' = \mathbf{u} - \overline{\mathbf{u}}$, this implies $\overline{\mathbf{u}'} = \overline{\mathbf{u}} - \overline{\overline{\mathbf{u}}}$ and we can derive the estimate

$$\sqrt{k'} = \sqrt{\frac{1}{2}\overline{|\mathbf{u} - \overline{\mathbf{u}}|^2}} \approx \sqrt{\frac{1}{2}|\overline{\mathbf{u}} - \overline{\overline{\mathbf{u}}}|^2},$$

so that we obtain an easily calculable expression for the turbulent viscosity coefficient:

$$\nu_T := \mu_0 \delta |\overline{\mathbf{u}} - \overline{\overline{\mathbf{u}}}|.$$

The value of the constant μ_0 can be fitted to homogeneous isotopic turbulence – Sect. 3.2 – and is around 0.17. Actually, any expression which is dimensionally consistent with this is possible. (This is also why the averaging can be moved around to within the accuracy of the expression.) Thus, in LES there are at least three natural turbulent viscosity coefficients:

$$\nu_T := \mu_0 \delta |\overline{\mathbf{u}} - \overline{\overline{\mathbf{u}}}|, \tag{1.24}$$

$$\nu_T := \mu_1 \delta^2 |\nabla^s (\overline{\mathbf{u}} - \overline{\overline{\mathbf{u}}})|, \tag{1.25}$$

$$\nu_T := \mu_2 \delta^3 |\Delta(\overline{\mathbf{u}} - \overline{\overline{\mathbf{u}}})|. \tag{1.26}$$

If a differential filter is used then $-\delta^2 \Delta \overline{\mathbf{u}} + \overline{\mathbf{u}} = \mathbf{u}$, so that $\mathbf{u} - \overline{\mathbf{u}} = -\delta^2 \Delta \overline{\mathbf{u}}$. This gives a fourth expression for the turbulent viscosity coefficient which is both directly connected to the idea of turbulent diffusion and which is computationally agreeable:

$$\nu_T := \mu_3 \delta^3 |\overline{\overline{\Delta \mathbf{u}}}|. \tag{1.27}$$

The eddy viscosity LES model is then, in the periodic case,

$$\mathbf{w}_t + \mathbf{w} \cdot \nabla \mathbf{w} - \nu \Delta \mathbf{w} - \nabla \cdot (\nu_T \nabla^s \mathbf{w}) + \nabla q = \overline{\mathbf{f}} \quad \text{in } \Omega \times (0, T) \tag{1.28}$$
$$\nabla \cdot \mathbf{w} = 0, \quad \text{in } \Omega \times (0, T), \tag{1.29}$$

where the eddy viscosity coefficient ν_T is given either by (1.24), (1.25), (1.26), or (1.27). All of them are computationally agreeable and, so far, seem to produce good results by the standards expected of eddy viscosity models. All however, give a system whose highest order term is a nonmonotone non-linearity. With the first one (1.24), the nonlinearity also has an unbounded coefficient and, as a consequence, the mathematical theory is not highly developed: all that is known is that a distributional solution exists, see Layton and Lewandowski [208] and Chap. 4. When ν_T is given by the second relation (1.25), the model is close enough to the Smagorinsky model [277] that the analysis of Ladyžhenskaya [195] and Du and Gunzburger [95] should be extendable to the model. For (1.26), nothing is known. Interestingly, the model (1.27) using the Gaussian Laplacian is very regular. Because this ν_T is bounded, a first step [170] at a complete theory has been possible, Chap. 4.

A value for the constant μ_j, can be estimated by following a calculation of Lilly [219] matching the models time averaged energy dissipation rate

$$\langle \epsilon_{model} \rangle := \lim_{T \to \infty} \sup \frac{1}{T} \int_0^T \frac{1}{|\Omega|} \int_\Omega [\nu + \nu_T(\mathbf{w})] |\nabla^s \mathbf{w}|^2 \, d\mathbf{x} dt$$

to that of the Navier–Stokes equations

$$\langle \epsilon \rangle := \lim_{T \to \infty} \sup \frac{1}{T} \int_0^T \frac{1}{|\Omega|} \int_\Omega \nu |\nabla^s \mathbf{u}|^2 \, d\mathbf{x} dt,$$

for the case of fully developed, homogeneous isotopic turbulence, Chap. 3. This setting also gives an indication of the successful uses of eddy viscosity models: they can give good prediction of time averaged statistics of fully developed turbulence. They have more difficulties when integrated over long time intervals, for problems with delicate energy balance, for transitional flows and for predicting the dynamics of coherent eddies rather than their statistics. Often eddy viscosity models, which are reliable for fully developed turbulence, fail in transitional flows in which the turbulence must develop. One speculation as to a source of some of these difficulties is that eddy viscosity should be limited to modeling the actions of turbulent fluctuations on the mean flow, *i.e.*, the $\nabla \cdot (\overline{\mathbf{u}'\mathbf{u}'})$ term. This means that other, non-diffusive models are needed for the first two terms in (1.6). Eddy viscosity models also fail to predict *backscatter*, the inverse transfer of energy from small eddies to the large ones. Backscatter is an important feature of the subfilter-scale stress tensor $\tau = \overline{\mathbf{u}\mathbf{u}} - \overline{\mathbf{u}}\,\overline{\mathbf{u}}$, and should be included in the LES model. However, the mathematical theory associated with the backscatter is very challenging. A detailed description of the phenomenon of backscatter and numerical illustrations are presented in Chap. 12.

1.8 Closure Models Based on Systematic Approximation

Since in the case of LES the nonlinear term retains four terms,

$$\overline{\mathbf{u}\,\mathbf{u}} = \overline{\overline{\mathbf{u}}\,\overline{\mathbf{u}}} + \overline{\overline{\mathbf{u}}\mathbf{u}'} + \overline{\mathbf{u}'\overline{\mathbf{u}}} + \overline{\mathbf{u}'\mathbf{u}'}$$

one way to generate closure models is to find a method of either representing $\mathbf{u}$ in terms of $\overline{\mathbf{u}}$ (for example $\mathbf{u} \approx O(\overline{\mathbf{u}})$) or $\mathbf{u}'$ in terms of $\overline{\mathbf{u}}$. Both are equivalent formulations of the problem of deconvolution. Here we summarize the systematic approximation, that we will present in Chap. 7, together with recent existence results for the corresponding models.

With a deconvolution approximation,

$$\mathbf{u} \approx O(\overline{\mathbf{u}})$$

the closure problem can be solved by $\overline{\mathbf{u}\mathbf{u}} \approx \overline{O(\overline{\mathbf{u}})O(\overline{\mathbf{u}})}$. Unfortunately, the deconvolution problem is ill-posed. Since it is also a fundamental question of image processing [236], many approximations and regularizations have been developed for it. The necessary requirements for a deconvolution approximation to be useful in LES are:

(i) *For smooth* $\mathbf{u}$, $\|\mathbf{u} - O(\overline{\mathbf{u}})\| \to 0$, *rapidly as* $\delta \to 0$.
This is a mathematical statement of the requirement that models equations for the large scales be very close to the Navier–Stokes equations.

(ii) *When used as a closure model* $\overline{\mathbf{u}}\,\overline{\mathbf{u}} \approx \overline{O(\overline{\mathbf{u}})O(\overline{\mathbf{u}})}$, *the resulting continuum model for the large scales* $\mathbf{w} \approx \overline{\mathbf{u}}$,

$$\mathbf{w}_t + \nabla \cdot \overline{O(\mathbf{w})O(\mathbf{w})} - \nu \Delta \mathbf{w} + \nabla q = \overline{\mathbf{f}}, \text{ and } \nabla \cdot \mathbf{w} = 0,$$

is well posed.

(iii) *Statistics computed from solving the above continuum LES model are close to those obtained from the Navier–Stokes equations.*
This can be considered to be a condition of accuracy on the small scales.

How can such deconvolution operators be generated? One method is by asymptotics, either in physical or wavenumber space. For example, let $g_\delta(\mathbf{x})$ be the Gaussian filter and $\overline{\mathbf{u}} = g_\delta * \mathbf{u}$, so $\widehat{\overline{\mathbf{u}}}(\mathbf{k}) = \widehat{g_\delta}(\mathbf{k})\widehat{\mathbf{u}}(\mathbf{k})$. Thus, $\widehat{\mathbf{u}}(\mathbf{k}) = \widehat{g_\delta}(\mathbf{k})^{-1}\widehat{\overline{\mathbf{u}}}(\mathbf{k})$. If $\widehat{g_\delta}(\mathbf{k})^{-1}$ is expanded in a Taylor series in δ we obtain $\widehat{g_\delta}(\mathbf{k})^{-1} = 1 - \frac{\delta^2}{4\gamma}|\mathbf{k}|^2 + O(\delta^4)$. Using this approximation and then inverting the Fourier transform gives a deconvolution operator (see Chap. 7)

$$\mathbf{u}(\mathbf{x}) \approx \mathcal{F}^{-1}\left[\left(1 - \frac{\delta^2}{4\gamma}|\mathbf{k}|^2\right)\widehat{\overline{\mathbf{u}}}(\mathbf{k})\right],$$

where $\mathcal{F}^{-1}$ denotes the inverse Fourier transform.

Using this deconvolution approximation and collecting terms gives, after simplification, the *gradient model* [212, 65]

$$\overline{\mathbf{u}\mathbf{u}} - \overline{\mathbf{u}}\,\overline{\mathbf{u}} \approx \frac{\delta^2}{2\gamma}\nabla\overline{\mathbf{u}}\,\nabla\overline{\mathbf{u}}^T, \tag{1.30}$$

where

$$(\nabla\overline{\mathbf{u}}\,\nabla\overline{\mathbf{u}}^T)_{i,j} = \sum_{l=1}^{d}\frac{\partial\overline{u}_i}{\partial\mathbf{x}_l}\frac{\partial\overline{u}_j}{\partial\mathbf{x}_l}.$$

One way to test a model, Jimenez [172], is through *a priori* testing, see Chaps. 7 and 12: perform a direct numerical simulation, obtain a velocity field $\mathbf{u}$, then compute the model's consistency of approximation $\|\overline{\mathbf{u}\mathbf{u}} - \overline{O(\overline{\mathbf{u}})O(\overline{\mathbf{u}})}\|$. The gradient model performs well in these types of tests. However, stability problems have been reported for it consistently and it has recently been shown in [169] that the gradient model fails the above condition (ii): the kinetic energy of the model can blow up in finite time. Thus, improvements in the asymptotic derivation of the model are considered.

The next critical improvement on the gradient model considered in Chap. 7 is to replace Taylor series asymptotics by Padé asymptotics. Sub-diagonal Padé approximations are attractive because they preserve the attenuation of high frequencies in the Gaussian filter. The (0,1)-Padé approximation to the Gaussian is given by

$$\widehat{g}_\delta(\mathbf{k}) := \frac{1}{1 + \frac{\delta^2}{4\gamma}|\mathbf{k}|^2} + O(\delta^4).$$

The same procedure as before (take the Fourier transform, replace $\widehat{g}_\delta$ by its (0,1)-Padé approximation, then take the inverse Fourier transform) gives the deconvolution approximation

$$\mathbf{u} := \left(-\frac{\delta^2}{4\gamma}\Delta + \mathbb{I}\right)\overline{\mathbf{u}} + O(\delta^4).$$

This deconvolution approximation leads to the Rational LES model [122], Chap. 7,

$$\mathbf{w}_t + \mathbf{w}\cdot\nabla\mathbf{w} - \nu\Delta\mathbf{w} + \frac{\delta^2}{2\gamma}\nabla\cdot\left[\left(-\frac{\delta^2}{4\gamma}\Delta + \mathbb{I}\right)^{-1}(\nabla\mathbf{w}\nabla\mathbf{w}^T)\right] + \nabla q = \overline{\mathbf{f}},$$

$$\nabla\cdot\mathbf{w} = 0.$$

The Rational LES model has been shown to give good performances in numerical tests, Chap. 7 and Part V, especially when combined with an eddy viscosity model of the neglected $O(\delta^4)$, $\nabla \cdot \overline{\mathbf{u}'\mathbf{u}'}$ term. The Rational LES model seems to be a step along a good path to develop accurate and stable LES models. At this point, it does not seem to be the final step and the mathematical theory of the rational model is not complete, yet. The model's derivation and theoretical foundation are presented in Chap. 7. To capture $\overline{\mathbf{u}'\mathbf{u}'}$, the term neglected in the Rational LES model, a higher-order subfilter-scale model [33] and its supporting mathematical analysis are also presented in Chap. 7.

The next approach to deconvolution we consider is by extrapolation from resolved to unresolved scales. In other words, any model which can be thought of as being a scale-similarity model. Chapter 8 begins with an introduction to some common scale-similarity models and examples of extensions of them for which mathematical development is possible. Next we consider a very promising family of such deconvolution models pioneered by Stolz and Adams [285], and Stolz, Adams, and Kleiser [289, 290]. The first two examples of these Stolz–Adams scale similarity/deconvolution models are:

(1) *constant extrapolation from resolved to unresolved scales*

$$\mathbf{u} \approx \overline{\mathbf{u}} + O(\delta^2), \qquad \text{giving} \quad \overline{\mathbf{u}\,\mathbf{u}} \approx \overline{\overline{\mathbf{u}}\,\overline{\mathbf{u}}} + O(\delta^2), \text{ and} \qquad (1.31)$$

(2) *linear extrapolation from resolved to unresolved scales*

$$\mathbf{u} \approx 2\overline{\mathbf{u}} - \overline{\overline{\mathbf{u}}}, \quad \text{giving} \quad \overline{\mathbf{u}\,\mathbf{u}} \approx \overline{(2\overline{\mathbf{u}} - \overline{\overline{\mathbf{u}}})(2\overline{\mathbf{u}} - \overline{\overline{\mathbf{u}}})} + O(\delta^4).$$

The mathematical theory of the whole family of deconvolution models has recently been completed in Layton and Lewandowski [208, 210, 209] and Dunca and Epshteyn [98]. We present this new theory in Chap. 8. The development of these models is an outgrowth of recognition of their kinetic energy balance. To be more precise, consider the LES model arising from (1.31) given by

$$\mathbf{w}_t + \nabla \cdot \overline{\mathbf{w}\,\mathbf{w}} - \nu \Delta \mathbf{w} + \nabla q = \overline{\mathbf{f}}, \qquad (1.32)$$

$$\nabla \cdot \mathbf{w} = 0. \qquad (1.33)$$

Supposing that the averaging operator is the differential filter $\overline{\phi} := (-\delta^2 \triangle + \mathbb{I})^{-1}\phi$, it can be proved that any weak solution to the above model (1.32) and (1.33), under periodic boundary conditions, satisfies the energy inequality:

$$k_{LES}(t) + |\Omega| \int_0^t \epsilon_{LES}(\tau)\,d\tau \leq k_{LES}(0) + \int_0^t P_{LES}(\tau)\,d\tau,$$

where

$$k_{LES}(t) := \frac{1}{2}\left[\|\mathbf{w}(t)\|^2 + \delta^2\|\nabla\mathbf{w}(t)\|^2\right],$$

$$P_{LES}(t) := \int_\Omega \mathbf{f} \cdot \mathbf{w}\,dx,$$

$$\epsilon_{LES}(t) := \frac{\nu}{2|\Omega|}\left[\|\nabla\mathbf{w}(t)\|^2 + \delta^2\|\Delta\mathbf{w}(t)\|^2\right].$$

For further details see Sect. 8.5. The above energy inequality is a very strong regularity result shared by weak solutions, strong solutions (if they exist) and the usual Galerkin approximations of weak solutions. Based on this observation, standard mathematical techniques will allow us to conclude an existence result for the model, Chap. 8.

These scale similarity models have a higher state of mathematical development than most LES models. Nevertheless, there are still important questions left open with these deconvolution models such as how to obtain a globally stable approximation when using the model coupled with appropriate wall laws.

1.9 Mixed Models

In practical problems, with the idea of using the "good" properties of each model (stability for eddy viscosity and accuracy for models derived by systematic approximation), combinations of different models are used: the resulting models are called mixed models. In numerical tests on three-dimensional turbulent flows, almost invariably mixed models are used. These models generally arise by taking a combination of a chosen LES model with an eddy viscosity model. There are (at least) three reasons for using mixed models:

(1) An eddy viscosity term is added *ad hoc* to a model of high formal accuracy because calculations with the model alone show instabilities. In this scenario, the eddy viscosity terms must be large enough to stabilize the other modeling terms.

(2) In going from a continuum model to a discretization of it, some build up of kinetic energy is observed around the cut-off length scale. This can be corrected by mesh refinement at constant filter width or by adding an eddy viscosity term calibrated to the regions and scales at which this build up occurs. The latter, being cheaper, is usually selected.

(3) Accurate LES models must be based upon some truncated asymptotic expansion of the space filtered Navier–Stokes equation's nonclosed term. Often, an eddy viscosity term is a sensible addition to the model to incorporate physical effects of the neglected terms. For example, in the expansion of the subfilter scale stress tensor

$$\overline{\mathbf{u}\,\mathbf{u}} - \overline{\mathbf{u}}\,\overline{\mathbf{u}} = \overline{\overline{\mathbf{u}}\,\overline{\mathbf{u}}} - \overline{\mathbf{u}}\,\overline{\mathbf{u}} + \overline{\overline{\mathbf{u}}\,\mathbf{u}'} + \overline{\mathbf{u}'\,\overline{\mathbf{u}}} + \overline{\mathbf{u}'\,\mathbf{u}'},$$

the last Reynolds stress term $\overline{\mathbf{u}'\,\mathbf{u}'}$ is formally $O(\delta^4)$. Often it is formally negligible and yet it is thought to describe an important physical process best captured by an $O(\delta^4)$ eddy viscosity term. Thus it is sensible to combine dispersive models of the first two terms on the right-hand side with an eddy viscosity model for the Reynolds stress term. Selection of the combination of LES model plus eddy viscosity model is often done by the normal approach in our field (model-solve-look-model-solve-look...). A better understanding of the individual components of the mixed model is necessary to develop better combinations of models. How best to combine different models in one simulation is clearly an important research problem!

1.10 Numerical Validation and Testing in LES

Although the focus of this book is on the mathematical theory of LES, it should be emphasized that, as its name implies, LES is a *computational* approach! Thus, the development and analysis of an LES model is not complete until the model has been validated and tested in numerical simulations.

Any LES test can be decomposed into a sequence of steps summarized below:

- **Step 1** Choose the numerical method

- **Step 2** Choose the test problem

- **Step 3** Run the numerical simulation

- **Step 4** Interpret the results

All these steps are essential and strongly interdependent, although not equally developed.

Step 1 is essential for the numerical validation and testing of the LES model. Finite differences and (pseudo) spectral methods are the traditional numerical methods used in LES validation and testing. The main reason is their high-order accuracy, which is believed to be important in the numerical simulation of turbulent flows. The finite element method, appropriate for complex geometries, is less developed as a tool in the numerical simulation and testing of LES.

Numerical analysis is an essential LES component: for example, many important decisions, such as the relationship between the grid size and the filter size, are made based on heuristics instead of a sound numerical analysis. With so many open mathematical questions at the core of the theory of LES, however, the time is not yet ripe for a universal numerical analysis of LES. Chap. 11 presents some numerical analysis issues related to LES. In particular it gives some ideas about Hughes' [160] Variational Multiscale Method (VMM). The VMM is an exciting recent development in which the actual discretization acts, in effect, as a sort of expert system to pick and adjust the closure model. Admittedly, this chapter ends with more questions than answers. On the other hand, it gives some background for the simulations given in Chap. 12.

Steps 2–4 are considerably more developed than *Step 1*.

Step 2 offers a wide variety of choices for the test problem in LES. Probably two of the most popular test problems are

1. homogeneous, isotropic turbulence
2. channel flow.

The first one is representative of the class of unbounded turbulence (turbulence away from solid boundaries), and usually employs (pseudo) spectral numerical schemes. The second is one of the most popular test problems for wall-bounded turbulence.

Some other popular choices are forced isotropic turbulence, jets (unbounded flows), pipe flow, flat plate flow, lid-driven cavity, and backward-facing step (wall-bounded flows). Each test problem has its own characteristics/important features that need to be captured by an LES model. The validation of LES models should include as many such test problems as possible: the more test problems successfully run, the better the LES model. This is an important point since, in general, LES models tend to run successfully on some tests, and poorly on others.

Step 3 illustrates the close relationship among *Steps 1–4*: depending on the test problem chosen in *Step 2*, one needs to specify different boundary conditions and initial conditions; depending on the important features/characteristics of the test problem that need to be collected, monitored, and interpreted in *Step 4*, different flow quantities need to be collected and stored. These quantities are mainly statistics for statistically steady state flows and pointwise values for time-dependent flows.

We also mention a few practical issues associated with *Step 3*. First, the LES runs are usually computationally intensive: a turbulent channel flow LES run can take a couple of days on a 32 processor machine. The generation of the initial conditions can be several times more expensive. Secondly, the storage of the output data could be a challenge: a generic flow field file could be several Mbytes – if one needs to store thousands of such files for each LES run (to generate a movie, for example), storage becomes critical.

Thus, before starting any LES validation and testing, one needs to make sure that the computational resources are available.

Step 4 is another critical step in the numerical validation and testing of LES. First, one needs to make sure that the monitored quantities correspond to the important features of the flow considered. Secondly, care needs to be taken when comparing several LES models: a sound validation requires not only the test problem to be the same, but also the entire computational setting (such as, numerical method, initial conditions, boundary conditions, machine architecture, *etc.*) It is also recommended that an extensive (DNS) database associated with the test problem be available. This is generally true for many of the most popular LES test problems (*e.g.*, for channel flows [242]) and provide a reliable benchmark for the numerical validation.

Remark 1.6. The numerical approach described in *Step 1–4* is usually referred to as *"a posteriori"* testing, implying the fact that the LES model is effectively tested in an actual numerical simulation. This is in contrast with *"a priori"* testing, where results from a fine DNS are filtered and then used to compute the LES approximation τ^{LES} (for example, we recall that $\tau^{LES} = -\nu_T(\nabla^s \overline{\mathbf{u}}) \, \nabla^s \overline{\mathbf{u}}$ for eddy viscosity models) to the "true" subfilter-scale stress tensor

$$\tau = \overline{\mathbf{u}\mathbf{u}} - \overline{\mathbf{u}}\ \overline{\mathbf{u}}.$$

The closer τ^{LES} to τ, the better the LES model. It should be stressed that the *"a posteriori"* testing is the final means of validating and testing an LES model, the *"a priori"* testing representing just a step in this process. We also need to mention that there exist LES models that perform very well in *"a priori"* tests, while performing poorly in *"a posteriori"* tests (classical scale-similarity models are such an example). This is probably related to the complex interplay between the continuum LES modeling and the numerical method used in the discretization process.

Chapter 12 represents an introduction to the numerical validation and testing of LES models. Most of Chap. 12 centers around the turbulent channel flow, one of the most popular test problems for LES. We explain in detail the computational setting, the generation of initial conditions, and the way we collect statistics. We put a special emphasis on *backscatter* (the inverse transfer of energy from small scales to large scales), an important feature in LES.

In our careful numerical exploration, we focus on the LES models introduced in the previous chapters. We permanently relate our numerical findings to the mathematical results in the earlier chapters. Thus, Chap. 12 represents not only an introduction to the numerical validation and testing in LES, but also the perfect illustration of the intrinsic connection among mathematics, physics, and numerics in LES.

2

The Navier–Stokes Equations

2.1 An Introduction to the NSE

The history of the development of the NSE is replete with the names of the great natural philosophers, beginning with Archimedes (287–212 BC). In Book I of the first treatise on mathematical fluid mechanics, *On Floating Bodies*, Archimedes lays down the basic principles of hydrostatics:

> Any solid lighter than a fluid will, if placed in the fluid, be so far immersed that the weight of the solid will be equal to the weight of the fluid displaced. (Proposition 5).

Book II, a collection of mathematical gems, deals with the application of euclidean geometry to the determination of positions of rest and stability of bodies floating in a fluid.

After the discovery of calculus, important contributions to the field of fluid mechanics came from D. Bernoulli (1700–1782) and his masterpiece *Hydrodynamica*. Another fundamental contribution to fluid mechanics is that of L. Euler, who was a student of J. Bernoulli and worked together with D. Bernoulli in St. Petersburg. Euler published several major pieces now collected in volume 11-12-13 of his *Opera Omnia* (including, for example, *Principes généraux du mouvement des fluids,* Hist. Acad. Berlin 1755), deriving the main formulas for the continuity equation, the Laplace velocity potential equation, and the *Euler equations* for the motion of an ideal incompressible fluid. In 1752 he wrote:

> However sublime are the researches on fluids which we owe to Messrs Bernoulli, Clairaut and d'Alembert, they flow so naturally from my two general formulæ that one cannot sufficiently admire this accord of their profound meditations with the simplicity of the principles from which I have drawn my two equations ...

Together with a similar assumption made by Euler for ideal fluids, the fundamental discovery of A.-L. Cauchy (1827) is the *stress principle*. This principle (translation by C. Truesdell) states that "upon any imagined closed surface $\mathcal{S}$ there exists a distribution of *stress vectors* whose resultant and moment are equivalent to those of the actual forces of material continuity exerted by the material outside $\mathcal{S}$ upon that inside"

> This principle has the simplicity of genius. Its profound originality can be grasped only when one realizes that a whole century of brilliant geometers had treated very special elastic problems in very complicated and sometimes incorrect ways without ever hitting upon the basic idea, which immediately became the foundation of the mechanics of distributed matter
> (C. Truesdell, 1953)

C.L.M.H. Navier (1785–1836), in the paper *Mémoire sur les lois du mouvement des fluides* (1823), derived the (as we call today) Navier–Stokes equations of a viscous fluid, despite not fully understanding the physics of the situation which he was modeling. He did not understand shear stress in a fluid, but rather he based his work on modifying Euler's equations to take into account forces between the molecules in the fluid. Although his reasoning is not acceptable today:

> The irony is that although Navier had no conception of shear stress and did not set out to obtain equations that would describe motion involving friction, he nevertheless arrived at the proper form for such equations. (Anderson, 1997).

The first rigorous derivation of the Navier–Stokes equations was obtained by G.G. Stokes (1819–1903). Under the advice of W. Hopkins, Stokes began to undertake research into hydrodynamics and in the 1845 paper *On the theories of the internal friction of fluids in motion* he derived the "Navier–Stokes" equations in a satisfactory way.[1]

Today it is widely accepted that the Navier–Stokes equations provide a very accurate description of most flows of almost all liquid and gases. The basic variables are:

ρ : density, $\mathbf{u} = (u_1, u_2, u_3)$: fluid velocity,

p : pressure, $\boldsymbol{\sigma}$: stress tensor associated with viscous forces,

$\mathbf{f}$: external (body) forces/unit volume.

As we will see in detail in Sect. 2.2, the NSE are simply a mathematical realization of conservation of mass,

$$\rho_t + \nabla \cdot (\rho\,\mathbf{u}) = 0,$$

conservation of linear momentum,

---

[1] As we have seen Stokes was not the first to obtain the equation. Navier, Poisson, and Saint-Venant had already started the analysis of the problem.

$$\rho\left(\mathbf{u}_t + \mathbf{u} \cdot \nabla \mathbf{u}\right) - \nabla \cdot \boldsymbol{\sigma} = \mathbf{f},$$

and a linear stress–strain relation

$$\boldsymbol{\sigma} = \mu(\nabla^s \mathbf{u}) + \left(\xi - \frac{2\mu}{3}\right)(\nabla \cdot \mathbf{u})\,\mathbb{I},$$

where μ and ξ are material parameters known as the first and second viscosities, while

$$(\nabla^s \mathbf{u})_{ij} := \frac{1}{2}\left(\frac{\partial u_i}{\partial x_j} + \frac{\partial u_j}{\partial x_i}\right), \qquad i, j = 1, \dots, d$$

is the deformation tensor. The derivation of such equations requires some deep physical assumptions to simplify the formulas and, as we have seen, these derive from the intuition and genius of the past centuries.

The mathematical structure of the NSE is best understood for incompressible fluids. Setting $\rho \equiv \rho_0 = \text{constant}$ and nondimensionalizing the resulting equations, yields the system we will study herein: the incompressible Navier–Stokes equations (in nondimensional form):

$$\mathbf{u}_t + \mathbf{u} \cdot \nabla \mathbf{u} - \frac{1}{Re}\Delta \mathbf{u} + \nabla p = \mathbf{f} \quad \text{in } \Omega \times (0, T), \tag{2.1}$$

$$\nabla \cdot \mathbf{u} = 0 \quad \text{in } \Omega \times (0, T), \tag{2.2}$$

where the Reynolds number $Re > 0$ is given by

$$Re = \frac{UL}{\mu/\rho_0} = \frac{\text{characteristic velocity} \times \text{characteristic length}}{\text{kinematic viscosity}}.$$

It is worthwhile for theorists to see a few representative values of Re.

Table 2.1. Representative values of Re

cm. sphere moving 1 cm/s in water	$Re \doteq 100,$
subcompact car	$Re \doteq 6 \times 10^5,$
small airplane	$Re \doteq 2 \times 10^7,$
competitive swimmer	$Re \doteq 1 \times 10^6,$
geophysical flows	$Re \doteq 10^{20}$ and higher.

The NSE (2.1) and (2.2) are assumed to hold in the flow domain (hereafter Ω) over some time interval $0 < t \le T$, and are supplemented by an initial velocity

$$\mathbf{u}(\mathbf{x}, 0) = \mathbf{u}_0(\mathbf{x}) \quad \mathbf{x} \in \Omega,$$

and appropriate boundary conditions. We will use mainly the *no-slip* boundary conditions

$$\mathbf{u}(\mathbf{x}, t) = \mathbf{0}, \quad \mathbf{x} \in \partial\Omega, \ t \in [0, T],$$

appropriate for internal flow. In several cases analytical and computational studies are done with periodic boundary conditions (an "easy case" that uncouples the equations from the boundaries):

$$\text{(periodic b.c.s)} \quad \mathbf{u}(\mathbf{x} + 2\pi\mathbf{e}_i, t) = \mathbf{u}(\mathbf{x}, t), \quad \Omega = (0, 2\pi)^3, \qquad (2.3)$$

where $\mathbf{e}_j$ are the canonical basis functions in $\mathbb{R}^d$ and (for technical reasons) subject to a zero mean over $(0, 2\pi)^3$ on the solution $\mathbf{u}(\mathbf{x}, t)$ and on all problem data.

2.2 Derivation of the NSE

The Navier–Stokes equations are a continuum model for the motion of a fluid. There are various ways to develop the Navier–Stokes equations. For example, the Boltzmann equation describes the motion of molecules in a rarefied gas. The Navier–Stokes equations can follow by taking spatial averages of the Boltzmann equation. They can likewise arise from the kinetic theory of gases. They have even been derived from quantum mechanics by a suitable averaging procedure.

The approach we are taking is the more classical approach of continuum mechanics in which all the flow variables:

$$\text{density } \rho, \text{ velocity } \mathbf{u}, \text{ pressure } p, \cdots$$

are assumed to be continuous functions of space and time from the beginning. This approach can be made completely axiomatic; see, for example, Serrin's [275] beautiful article. We will give a middle path which is axiomatic "enough", but which is compact and still retains a connection to the physical ideas.

Conservation of Mass
The equation describing conservation of mass is called the continuity equation. If mass is conserved, the rate of change of mass in a volume V must equal the net mass flux across ∂V:

$$\frac{d}{dt} \int_V \rho \, d\mathbf{x} = - \int_{\partial V} (\rho\,\mathbf{u}) \cdot \mathbf{n} \, dS,$$

where $\mathbf{n}$ denotes the outward normal unit vector to ∂V. The divergence theorem thus implies

$$\int_V \rho_t + \nabla \cdot (\rho\,\mathbf{u}) \, d\mathbf{x} = 0.$$

If all the variables are continuous, shrinking V to a point gives:

$$\rho_t + \nabla \cdot (\rho\,\mathbf{u}) = 0,$$

which is the first equation of mathematical fluid dynamics. If the fluid is homogeneous and incompressible,

$$\rho(\mathbf{x}, t) \equiv \rho_0$$

then conservation of mass reduces to

$$\nabla \cdot \mathbf{u}(\mathbf{x}, t) = 0,$$

which is a constraint on the fluid velocity, $\mathbf{u}$.

Conservation of Momentum

Conservation of momentum states that the rate of change of linear momentum must equal the net forces acting on a fluid particle, or

$$\text{force} = \text{mass} \times \text{acceleration}.$$

Let us consider a fluid particle. If it is at $(\mathbf{x}, t)$ (position $\mathbf{x}$ at times t) then at time $t + \Delta t$ it has flowed to (up to the accuracy of the linear approximation)

$$(\mathbf{x} + \mathbf{u}(\mathbf{x}, t)\Delta t, t + \Delta t).$$

Its acceleration is therefore:

$$\mathbf{a} = \lim_{\Delta t \to 0} \frac{\mathbf{u}(\mathbf{x} + \mathbf{u}(\mathbf{x}, t)\Delta t, t + \Delta t) - \mathbf{u}(\mathbf{x}, t)}{\Delta t}$$

$$= \mathbf{u}_t + \sum_{j=1}^{d} u_j \frac{\partial u_i}{\partial x_j} = \mathbf{u}_t + \mathbf{u} \cdot \nabla \mathbf{u}.$$

Thus, the mass $\times$ acceleration in a volume V,

$$\int_V \rho \left(\mathbf{u}_t + \mathbf{u} \cdot \nabla \mathbf{u} \right) d\mathbf{x},$$

must be balanced by external (body) forces and internal forces.

External forces include gravity, buoyancy, and electromagnetic forces (in liquid metals). These are collected in a body force term which has accumulated net force on the volume V given by

$$\int_V \mathbf{f} \, d\mathbf{x}.$$

Internal forces are the forces that a fluid exerts on itself and include pressure and the viscous drag that a fluid element exerts on the adjacent fluid. The internal forces of a fluid are *contact* forces: they act on the surface of the fluid element V. If $\mathbf{t}$ denotes this internal force vector, then the net contribution of the internal forces on V is

$$\int_{\partial V} \mathbf{t}(\mathbf{s}) \, dS.$$

Modeling these internal forces correctly is critical to predicting the fluid motion correctly. We will look at these internal forces more carefully next.

Stress and Strain in a Newtonian Fluid

The internal forces in a fluid are the key to "fluidity" and to the difference among solids, liquids, and gases. They also differentiate among different fluids.

The idea of Cauchy is that on any (imaginary) plane there is a net force that depends (geometrically) only on the orientation of that plane.

If this is true, we must have

$$\mathbf{t} = \mathbf{t(n)}, \quad \mathbf{n} = \text{normal vector to an imaginary plane.}$$

The exact dependence of $\mathbf{t}$ upon $\mathbf{n}$ can be determined rigorously by using other accepted principles of continuum mechanics. We shall summarize this below.

Theorem 2.1. *If linear momentum is conserved, then the stress forces must be in local equilibrium (i.e. (A1) holds).*

Assumption. The stress forces are in local equilibrium, *i.e.*

(A1)
$$\lim_{\substack{\text{as } V \text{ shrinks} \\ \text{to a point}}} \frac{1}{\text{surface area } (V)} \int_{\partial V} \mathbf{t(n)}\, dS = 0.$$

Theorem 2.2. *If (A1) holds, then $\mathbf{t}$ is a linear function of $\mathbf{n}$. Thus, there is a 3×3 matrix (a tensor $\boldsymbol{\sigma}$) with*

$$\mathbf{t(n)} = \mathbf{n} \cdot \boldsymbol{\sigma}, \quad \boldsymbol{\sigma} = \boldsymbol{\sigma}(\mathbf{x}, t).$$

With this *stress tensor* $\boldsymbol{\sigma}$ we can write the equation for conservation of linear momentum as follows. For a (spatially) fixed volume V

$$\int_V \rho\, (\mathbf{u}_t + \mathbf{u} \cdot \nabla \mathbf{u})\, dx = \int_{\partial V} \mathbf{t(n)}\, dS + \int_V \mathbf{f}\, dx$$

$$= \int_{\partial V} \mathbf{n} \cdot \boldsymbol{\sigma}\, dS + \int_V \mathbf{f}\, dx = \int_V (\nabla \cdot \boldsymbol{\sigma} + \mathbf{f})\, dx.$$

Shrinking V to a point gives

$$\rho\, (\mathbf{u}_t + \mathbf{u} \cdot \nabla \mathbf{u}) = \nabla \cdot \boldsymbol{\sigma} + \mathbf{f} \quad \text{in } \Omega,$$

which is the momentum equation.

Theorem 2.3. *Angular momentum is conserved if and only if $\boldsymbol{\sigma}$ is symmetric: $\sigma_{ij} = \sigma_{ji}$.*

Remark 2.4. Cauchy proved $\Rightarrow$ and Boltzmann $\Leftarrow$.

More about Internal Forces

A fluid has several types of internal forces:

• **Pressure Forces:** = normal forces. These act on a surface purely normal to that surface:

$$\text{pressure forces } = -p\,\mathbb{I}\,\mathbf{n}, \qquad p(\mathbf{x}, t) \text{ is the "dynamic pressure"}.$$

One basic postulate due to Cauchy is that a fluid at rest cannot support tangential stresses. Thus, only pressure forces can exist for a fluid at rest.

• **Viscous Forces:** The nonpressure part of the stress tensor is called the viscous stress tensor and is given by

$$V := \boldsymbol{\sigma} - p\,\mathbb{I}.$$

Remark 2.5. The simplest fluid model is that of a *perfect fluid*. A perfect fluid is incompressible and without internal viscous forces. Its motion is governed by the *Euler equations*:

$$\mathbf{u}_t + \mathbf{u}\cdot\nabla\mathbf{u} + \nabla\left(\frac{p}{\rho_0}\right) = \frac{\mathbf{f}}{\rho_0} \quad \text{in } \Omega \times (0, T)$$

$$\nabla\cdot\mathbf{u} = 0 \quad \text{in } \Omega \times (0, T).$$

Our system of equations is not closed until $\boldsymbol{\sigma}$ is related to the deformation tensor, $\nabla^s\mathbf{u}$. The simplest relation is a linear law (analog to Hooke's law) between stress and strain (force and deformation).

Assumption: Let $\nabla^s\mathbf{u} = \frac{1}{2}(\nabla\mathbf{u} + \nabla\mathbf{u}^T)$. Then,

$$\boldsymbol{\sigma} = 2\mu\,\nabla^s\mathbf{u} + \left(\xi - \frac{2\mu}{3}\right)(\nabla\cdot\mathbf{u})\,\mathbb{I},$$

where μ and ξ are the first and second viscosities of the fluid. The physical parameter μ is called the dynamic or shear viscosity.

More about $\boldsymbol{\sigma}$

It is good to keep in mind in these cases that a linear stress–strain relation is only a linear approximation about $\nabla^s\mathbf{u} = \mathbf{0}$ in a more general and nonlinear relation for a real fluid. The first scientist to postulate a linear stress–strain relation was Newton! For this reason, a fluid satisfying this assumption is called a "Newtonian fluid".

More general relations for $\boldsymbol{\sigma} = \boldsymbol{\sigma}(\nabla^s\mathbf{u})$ exist and are appropriate for fluids with larger stresses. We shall not go into detail about these fluids herein other than to state that they are of great practical importance and are not yet completely understood.

We thus have the NSE:

$$\rho_t + \nabla \cdot (\rho \, \mathbf{u}) = 0 \quad \text{in } \Omega \times (0, T)$$

$$\rho(\mathbf{u}_t + \mathbf{u} \cdot \nabla \mathbf{u}) + \nabla p - \nabla \cdot \left[2\mu \nabla^s \mathbf{u} + \left(\xi - \frac{2\mu}{3} \right)(\nabla \cdot \mathbf{u}) \, \mathbb{I} \right] = \mathbf{f} \quad \text{in } \Omega \times (0, T).$$

If the fluid is incompressible and μ is constant, these reduce to

$$\nabla \cdot \mathbf{u} = 0 \quad \text{in } \Omega \times (0, T) \tag{2.4}$$

$$\mathbf{u}_t + \mathbf{u} \cdot \nabla \mathbf{u} + \nabla \left(\frac{p}{\rho_0} \right) - \frac{\mu}{\rho_0} \Delta \mathbf{u} = \frac{1}{\rho_0} \mathbf{f} \quad \text{in } \Omega \times (0, T). \tag{2.5}$$

The pressure p is simply redefined to be p/ρ_0. The parameter μ is the fluid's viscosity coefficient and

$$\frac{\mu}{\rho_0} =: \nu = \quad \text{the kinematic viscosity.}$$

Further Remarks
The derivation of the equation of motion we have given is completely non-rigorous, but we think that can be more interesting than a formally accurate one. Furthermore, we have no space for the discussion of all the hypotheses hidden in the derivation of the Euler and Navier–Stokes equations. Complete details can be found in several references. For the sake of completeness we must cite Lamb's book [198] that even if the first edition dates back to 1879, while the last version is dated 1932, is a reference that is still up-to-date. Another famous paper is that of Serrin [274], that contains a deep discussion of several physical and variational problems related to the fluid dynamics. The reader may also consult some more recent references. Among the others we recall the well-known books by Batchelor [17], Landau and Lifshitz [199], and the graduate text by Chorin and Marsden [64].

2.3 Boundary Conditions

Consider the viscous, incompressible NSE in a bounded domain $\Omega \subset \mathbb{R}^3$. Boundary conditions must be imposed on $\partial \Omega$ to have a completely specified problem. Let $\Gamma \subset \partial \Omega$ be a solid wall. The first boundary condition is easy:

$$\text{no penetration} \quad \Leftrightarrow \quad \mathbf{u} \cdot \mathbf{n} = 0, \text{ on } \Gamma.$$

The tangential component is more complex. Navier proposed the following slip with friction condition:

$$\mathbf{u} \cdot \boldsymbol{\tau}_j + \beta \mathbf{n} \cdot \nabla^s \mathbf{u} \cdot \boldsymbol{\tau}_j = 0, \quad j = 1, 2,$$

where $\boldsymbol{\tau}_1$ and $\boldsymbol{\tau}_2$ are orthogonal unit vectors, tangent to $\partial \Omega$.

The NSE can also be derived from the kinetic theory of gases and it gives exactly this condition where

$$\beta \sim \frac{\text{mean free passes of molecules}}{\text{macroscopic length}}.$$

Thus, where the stresses are $O(1)$, the no-slip condition

$$\mathbf{u} \cdot \boldsymbol{\tau}_j = 0, \text{ on } \Gamma, \quad j = 1, 2,$$

is used and agrees well with experiments; if the boundary Γ is moving, this is modified to read

$$\mathbf{u} = \mathbf{g} \text{ on } \Gamma, \quad \mathbf{g} \text{ is the velocity of } \Gamma.$$

Infinite stresses arise where boundary velocities are incompatible. For example, if a piston pushes a fluid down a tube, the point where the piston and tube meet has a discontinuity in the boundary condition. Physically, one would expect leakage or the fluid to slip there. Thus, excluding leakage and infinite stresses in the physical model requires imposing a slip with friction condition near the contact point. A similar (but less dramatic) example occurs in flow over an object with sharp corners protruding into the fluid.

Of course, a liquid completely enclosed by stationary solid walls is rightly considered an "easy" case. Yet it is still hard enough that analytical and computational studies are done with periodic boundary conditions (2.3), to uncouple the equations from the boundaries.

This describes only the conditions at a solid, smooth, fixed, and nonporous wall. Many other conditions are important in practical flow problems. Often, the most vexing problems in flow simulations are associated with inflow and outflow boundary conditions, neither of which are considered herein.

2.4 A Few Results on the Mathematics of the NSE

The modern theory of the NSE (actually of a wide[2] area of partial differential equations, PDE in the sequel) began with the work of J. Leray [213]. The Leray theory (see for instance Galdi [121]) begins with the most concrete and physically meaningful possible point: the global energy inequality. From that the most abstract and (even today) mathematically complete theory of the NSE is directly constructed.

It is impossible to compress the basic theory of NSE into a single chapter. In this section we try to present some of the ideas underlying the mathematical analysis of the NSE. We hope to interest the reader (also those that are not

---

[2] The Leray–Schauder theory began from studies on incompressible fluids done by Leray in the early 1930s, in his doctoral thesis.

mathematicians) in this field. All the details can be found in the extensive bibliography that is cited throughout the section.

The study of the NSE opened a really challenging field. In fact the basic problem of global existence and uniqueness of solutions still resists the effort of mathematicians! The mathematical research around this topic is very intense and the reader can find about 4000 papers having the words Navier–Stokes in the title, about 200 *per* year over recent decades (AMS MathSciNet Source, http://www.ams.org/mathscinet).

2.4.1 Notation and Function Spaces

In this section we introduce the basic function spaces needed for the mathematical theory of the NSE and we give the definition of weak and strong solutions. We also recall some existence results, sketching some proofs. The reader may note that in the seminal paper by Leray [213] weak solutions were called *turbulent solutions*, since, in principle, they are not regular and the name was given in the attempt that such solutions may describe the chaotic behavior of turbulent flows.

We try to keep the book self-contained and at a level of mathematical depth understandable to a wide audience. The reader can find an excellent survey of the mathematics needed in the applied analysis of PDE in the series of books by Dautray and Lions [84]. In particular, see volume 5 for evolution problems.

In the sequel, we will need the classical L^p-spaces. We will not distinguish between scalar, vector, or tensor valued functions.

Given an open bounded[3] set $\Omega \subset \mathbb{R}^d$ ($d = 2, 3$) we say that a function $f : \Omega \to \mathbb{R}^n$ (with $n \in \mathbb{N}$) belongs to $L^p(\Omega)$, for $1 \le p \le \infty$, if f is measurable (with respect to the Lebesgue measure) and if the norm

$$\|f\|_{L^p} = \begin{cases} \left[\int_\Omega |f(\mathbf{x})|^p \, d\mathbf{x} \right]^{1/p} & \text{if } 1 \le p < \infty \\[2ex] \operatorname*{ess\,sup}_{\mathbf{x} \in \Omega} |f(\mathbf{x})| & \text{if } p = \infty \end{cases}$$

is finite. The spaces $(L^p(\Omega), \|\cdot\|_{L^p})$ are Banach spaces and we recall the *Hölder inequality*: if $f \in L^p(\Omega)$ and $g \in L^{p'}(\Omega)$, with $1/p + 1/p' = 1, 1 \le p, p' \le \infty$, then

$$\left| \int_\Omega f \, g \, d\mathbf{x} \right| \le \|f\|_{L^p} \|g\|_{L^{p'}}.$$

---

[3] Throughout the book we will consider smooth bounded open sets. There are problems having a physical meaning in which the domain may not be bounded (*e.g.* a channel or an exterior domain). In these cases the mathematical theory becomes more complicated since some properties, especially of divergence-free functions, may vary a lot. For full details see Galdi [120].

In the sequel, since the case $p = 2$ corresponds to the Hilbert space $L^2(\Omega)$, which is the most important for our applications, we will use the following notation:

$$\|f\| := \left[\int_\Omega |f(\mathbf{x})|^2\, d\mathbf{x}\right]^{1/2}.$$

The Hilbert space $(L^2(\Omega), \|\,.\,\|)$ is very important since it is the widest function space on which the *kinetic energy* is well-defined. It is a Hilbert space with the natural scalar product

$$(u, v) = \int_\Omega u\,v\,d\mathbf{x}.$$

Sobolev Spaces

In the variational formulation of mathematical physics problems, we shall encounter very often *Sobolev spaces*. In a first step it will be necessary to introduce at least the spaces $H^1(\Omega)$ and $H_0^1(\Omega)$. The space $H^1(\Omega)$ is the subspace of $L^2(\Omega)$ consisting of (equivalence classes of) functions with first-order distributional derivatives in $L^2(\Omega)$. The space $C_0^\infty(\Omega)$ will denote the infinitely differentiable functions on Ω with compact support.

Definition 2.6. *The Sobolev space $H^1(\Omega)$ is defined by*

$$H^1(\Omega) := \left\{ \begin{array}{l} u \in L^2(\Omega): \quad \text{there exist } g_i \in L^2(\Omega),\ i = 1, \ldots, d \text{ such that} \\[2mm] \displaystyle\int_\Omega u\,\frac{\partial\phi}{\partial x_i}\,d\mathbf{x} = -\int_\Omega g_i\,\phi\,d\mathbf{x}, \quad \forall\,\phi \in C_0^\infty(\Omega) \end{array} \right\}.$$

Given $u \in H^1(\Omega)$, we denote

$$\frac{\partial u}{\partial x_i} = g_i \quad \text{and} \quad \nabla u = (g_1, \ldots, g_d) = \left(\frac{\partial u}{\partial x_1}, \ldots, \frac{\partial u}{\partial x_d}\right). \qquad (2.6)$$

Remark 2.7. In (2.6) the function $g_i \in L^2(\Omega)$ represents the *weak derivative*, with respect to x_i, of the function u. Weak derivatives are defined through an integration by parts of the product with smooth functions. This definition is meaningful since for smooth functions weak derivatives coincide with the usual ones. The interested reader can find extensive investigations of Sobolev spaces in the book by Adams [4].

The space $H^1(\Omega)$ is a Hilbert space, equipped with the scalar product

$$(u, v)_{H^1(\Omega)} := \int_\Omega u\,v\,d\mathbf{x} + \int_\Omega \nabla u\,\nabla v\,d\mathbf{x},$$

and the corresponding norm

$$\|u\|_{H^1(\Omega)} := \left[\|u\|^2 + \|\nabla u\|^2\right]^{1/2}.$$

For a function u belonging to $H^1(\Omega)$, it is not possible (if $d > 1$) to define the pointwise values, but it makes sense to define the value of u on the boundary. We consider at least those functions that are vanishing on the boundary $\partial\Omega$.

Definition 2.8. *The Sobolev space $H_0^1(\Omega)$ is the closure of $C_0^\infty(\Omega)$ with respect to the norm $\|.\|_{H^1}$.*

The space $H_0^1(\Omega)$ represents the subspace of $H^1(\Omega)$ of functions vanishing on the boundary. These functions vanish in the *traces sense*, *i.e.* in the sense of $H^{1/2}(\partial\Omega)$. Without entering into details, we refer again to [4] for the introduction and properties of fractional Sobolev spaces. To use these space, the reader should be at least familiar with the fact that $H_0^1(\Omega)$ is the space of functions in $H^1(\Omega)$ that vanish on the boundary, in a generalized sense. Again, $u \in H_0^1(\Omega)$ means that $u \in H^1(\Omega)$ and $u_{|\partial\Omega} = 0$, provided u is smooth.

The functions belonging to $H_0^1(\Omega)$ satisfy the following property:

Lemma 2.9 (Poincaré inequality). *Let Ω be a bounded[4] subset of $\mathbb{R}^d$. Then there exists a positive constant C_P (depending on Ω) such that*

$$\|u\| \le C_P \|\nabla u\| \qquad \forall\, u \in H_0^1(\Omega).$$

Consequently, $\|\nabla u\|$ is a norm on $H_0^1(\Omega)$ equivalent to $\|.\|_{H^1}$. Furthermore, $\int_\Omega \nabla u \, \nabla v \, d\mathbf{x}$ is equivalent in $H_0^1(\Omega)$ to the scalar product $(u, v)_{H^1}$.

Function Spaces in Hydrodynamics

In the mathematical theory of incompressible fluids there is a need to consider functions that are divergence-free. A possible way to treat this feature is to include this constraint directly in the function spaces. In this respect it is well known (starting from the work of Helmholtz [150] in electromagnetism and a more recent analysis initiated by Weyl [313]) that any vector field $\mathbf{w} : \mathbb{R}^3 \to \mathbb{R}^3$ (that is decaying to zero sufficiently fast) can be uniquely decomposed as the sum of a "gradient" and of a "curl"

$$\mathbf{w} = \nabla\phi + \nabla \times \boldsymbol{\psi}.$$

This expression shows how to write a function as a gradient and a divergence-free part.

In the case of a smooth, bounded, and simply connected domain Ω we can define (the subscript "σ" stands for solenoidal)

$$L_\sigma^2 := \left\{ \mathbf{u} \in [L^2(\Omega)]^d : \nabla \cdot \mathbf{u} = 0 \text{ and } \mathbf{u} \cdot \mathbf{n} = 0 \text{ on } \partial\Omega \right\}. \qquad (2.7)$$

---

[4] Note that it is enough to require the domain Ω to be bounded at least in one direction, *i.e.* that Ω may be included in a "strip". In the case of periodic functions the inequality works too, provided we consider functions with vanishing mean value, *i.e.* $\int_\Omega u \, d\mathbf{x} = 0$.

The space $[L^2(\Omega)]^d$ is decomposed as the following direct sum:

$$[L^2(\Omega)]^d = L^2_\sigma \oplus G, \tag{2.8}$$

where G is the "subspace of gradients" (provided Ω is smooth and bounded):

$$G := \left\{ \mathbf{u} \in [L^2(\Omega)]^d : \mathbf{u} = \nabla p, \; p \in H^1(\Omega) \right\}.$$

The orthogonal projection operator $P : [L^2(\Omega)]^d \to L^2_\sigma$ is often called the *Leray projection* operator. We observe that functions in the definition (2.7) belong to $L^2(\Omega)$, so the divergence-free constraint is defined in a weak sense:

$$\nabla \cdot \mathbf{u} = 0 \qquad \text{means} \qquad \int_\Omega \mathbf{u} \cdot \nabla \Phi \, d\mathbf{x} = 0 \quad \forall \Phi \in C_0^\infty(\Omega).$$

The fact that $\mathbf{u} \cdot \mathbf{n} = 0$, where $\mathbf{n}$ denotes the exterior normal to $\partial\Omega$, has to be intended in the very weak sense of $H^{-1/2}(\partial\Omega)$, the topological dual of $H^{1/2}(\partial\Omega)$. Again, the reader not familiar with these spaces can better understand these properties if we recall that L^2_σ is the closure, with respect to the norm of $[L^2(\Omega)]^d$, of

$$\mathcal{V} := \left\{ \mathbf{v} \in [C_0^\infty(\Omega)]^d : \nabla \cdot \mathbf{v} = 0 \right\},$$

and passing to the limit, only the constraint on the normal part of $\mathbf{v}$ is kept. This is due to the fact that the L^2-norm is not strong enough to control the value of $\mathbf{v}$ on $\partial\Omega$. We refer the reader interested in full details to Ladyžhenskaya [197], Girault and Raviart [137], and Temam [295]. See also Galdi [120] for further details on the Helmholtz decomposition in L^p-spaces and unbounded domains. Note that the definition of differential operators through multiplication by smooth functions and integration by parts is one of the basic tools in the modern analysis of PDE.

Likewise, we can define the following space:

$$H^1_{0,\sigma} := \left\{ \mathbf{u} \in [H^1_0(\Omega)]^d : \nabla \cdot \mathbf{u} = 0 \right\},$$

embedded with the norm of $H^1_0(\Omega)$: that is $\|\mathbf{u}\|_{H^1_{0,\sigma}} = \|\nabla \mathbf{u}\|$. The space $H^1_{0,\sigma}$ is the closure of $\mathcal{V}$ with respect to the norm of $[H^1_0(\Omega)]^d$ (this property may fail or may be unknown for unbounded or non-smooth domains).

As usual in the study of evolution problems we may consider a function $f : \Omega \times [0, T] \to \mathbb{R}^d$ as

$$f : \; t \to f(t, \mathbf{x}),$$

i.e. as a function of time into a suitable Hilbert space $(X, \|\cdot\|_X)$. We define $L^p(0, T; X)$ as the linear space of strongly measurable functions $f : (0, T) \to X$ such that the functional

$$\|f\|_{L^p(0,T;X)} = \begin{cases} \left[\displaystyle\int_0^T \|f(\tau)\|_X^p \, d\tau \right]^{1/p} & \text{if } 1 \le p < +\infty \\[2mm] \operatorname*{ess\,sup}_{0<\tau<T} \|f(\tau)\|_X & \text{if } p = +\infty \end{cases}$$

is finite. In our case X will be either L^2_σ or $H^1_{0,\sigma}$.

Remark 2.10. For the numerical approximation we shall need completely different function space settings. In fact, it is very challenging to explicitly construct finite dimensional subspaces of L^2_σ or of $H^1_{0,\sigma}$ [146]. In the numerical approximation it is very common to resort to the so called *mixed-formulation*, where one uses spaces that are not divergence-free and imposes the constraint in an approximate way. The reader is referred to Gunzburger [146] for an excellent introduction to the finite element method for incompressible flows, and to Girault and Raviart [137] for an exquisite mathematical presentation.

2.4.2 Weak Solutions in the Sense of Leray–Hopf

In the study of the NSE it is necessary to introduce a suitable concept of solution. In generic situations it is hopeless to find smooth solutions, in such a way that all the space-time derivatives appearing in (2.1) exist in the usual classical sense. As we will see in the sequel, with this more general definition of solution it is possible to prove existence (but not uniqueness); see the work of Leray [213] for the Cauchy problem and Hopf [155] for the initial-boundary-value problem.

Definition 2.11 (Leray–Hopf weak solutions). *We say that a measurable function* $\mathbf{u} : \Omega \times [0, T] \to \mathbb{R}^d$ *is a weak solution to the NSE (2.1) and (2.2) if*

1. $\mathbf{u} \in L^\infty(0, T; L^2_\sigma) \cap L^2(0, T; H^1_{0,\sigma})$;
2. $\mathbf{u}$ *satisfies (2.1)–(2.2) in the weak sense, i.e. for each* $\phi \in C^\infty_0(\Omega \times [0, T))$, *with* $\nabla \cdot \phi = 0$, *the following identity holds:*

$$\int_0^\infty \int_\Omega \left[\mathbf{u}\, \phi_t - \frac{1}{Re} \nabla \mathbf{u} \nabla \phi - \mathbf{u} \cdot \nabla \mathbf{u}\, \phi \right] dx\, dt$$

$$= - \int_0^\infty \int_\Omega \mathbf{f}\, \phi\, dx\, dt - \int_\Omega \mathbf{u}_0\, \phi(0)\, dx; \tag{2.9}$$

3. the "energy inequality" is satisfied for $t \in [0, T]$:

$$\frac{1}{2} \|\mathbf{u}(t)\|^2 + \frac{1}{Re} \int_0^t \|\nabla \mathbf{u}(\tau)\|^2\, d\tau \le \frac{1}{2} \|\mathbf{u}_0\|^2 + \int_0^t \int_\Omega \mathbf{f}(\mathbf{x}, \tau) \mathbf{u}(\mathbf{x}, \tau)\, dx\, d\tau. \tag{2.10}$$

Weak Formulation

The above identity (2.9) is obtained by multiplying the NSE by a smooth ϕ and performing suitable integrations by parts in space-time variables. In particular, note that the pressure disappears, thanks to the following equality:

$$\int_\Omega \nabla p\, \phi\, d\mathbf{x} = \int_{\partial\Omega} p\, \phi \cdot \mathbf{n}\, d\sigma - \int_\Omega p\, \nabla \cdot \phi\, d\mathbf{x}, \qquad \forall\, t \in [0, T],$$

which is another way to restate the decomposition (2.8). The first integral on the right-hand side vanishes due to the fact that ϕ is zero on $\partial\Omega$; the second one vanishes since $\nabla \cdot \phi = 0$. The reader may note that in the definition of weak solutions there is no requirement for the time derivative of $\mathbf{u}$; furthermore, there are at most space derivatives of the first order. This is the basic idea behind the weak formulation of PDE: define a wider class (weak solutions) of functions that are solutions, by means of an *integral formulation.* After having proved the existence of more general solutions (this is generally simpler), the problem is then to show that these *weak solutions* are unique and, provided they are smooth, are also classical solutions to the original problem.

Remark 2.12. In the definition of weak solutions the pressure disappears. It is always possible to associate to each weak solution a corresponding pressure field, otherwise the weak solution concept will not be meaningful; unfortunately this requires rather sophisticated mathematical tools, based on the Helmholtz decomposition (2.8). For the introduction of the pressure field, we refer to some monographs, see for instance Ladyžhenskaya [197], Temam [295], and Galdi [121].

Remark 2.13. The introduction of weak solutions is based on the philosophical idea that "looking for solutions in a bigger set, it is easier to find them". The irregularity of turbulent flows also suggests that solutions to the NSE may be not very regular. In spite of these observation and the difficulty of finding explicit solutions for systems of PDE, the reader can find in Berker [26], a survey paper in the *Handbuch der Physik,* several (about 400 pages!) exact solutions, that may help to understand the basic features of incompressible flows and for benchmarking numerical experiments.

2.4.3 The Energy Balance

In this section we use the dimensional form (2.4) and (2.5) of the NSE, since we will deal with some physical quantities. However, in the rest of the book we will use essentially the nondimensional form (2.1) and (2.2).

If $(\mathbf{u}, p)$ are classical solutions to the NSE, subject to either no-slip or periodic boundary conditions, then multiplying (2.4) and (2.5) by p and $\mathbf{u}$, respectively, integrating over Ω, and applying the divergence theorem one immediately shows that

$$\int_\Omega \mathbf{u}_t \cdot \mathbf{u} + \nu \nabla \mathbf{u} : \nabla \mathbf{u} - \mathbf{f} \cdot \mathbf{u} \, d\mathbf{x} = 0. \tag{2.11}$$

In particular, note that (in the periodic-case the boundary integral vanishes) the nonlinear term disappears since:

$$\int_\Omega \mathbf{u} \cdot \nabla \mathbf{u} \, \mathbf{u} \, d\mathbf{x} = \int_\Omega \mathbf{u} \cdot \nabla \frac{|\mathbf{u}|^2}{2} \, d\mathbf{x} = \int_{\partial\Omega} \mathbf{u} \cdot \mathbf{n} \frac{|\mathbf{u}|^2}{2} \, d\mathbf{x} - \int_\Omega \frac{|\mathbf{u}|^2}{2} \nabla \cdot \mathbf{u} \, d\mathbf{x} = 0.$$

Integrating over time (2.11) gives the energy *equality:*

$$k(t) + |\Omega| \int_0^t \epsilon(\tau)\,d\tau = k(0) + \int_0^t P(\tau)\,d\tau \tag{2.12}$$

where

$$k(t) := \text{ kinetic energy at time } t := \frac{1}{2}\int_\Omega |\mathbf{u}|^2(t)\,d\mathbf{x} = \frac{1}{2}\|\mathbf{u}\|^2,$$

$$\epsilon(t) := \text{ energy dissipation rate} := \frac{\nu}{|\Omega|}\int_\Omega |\nabla\mathbf{u}|^2(t)\,d\mathbf{x} = \frac{\nu}{|\Omega|}\|\nabla\mathbf{u}\|^2,$$

$$P(t) := \text{ power input through force} - \text{flow interaction } := \int_\Omega \mathbf{f}\cdot\mathbf{u}\,d\mathbf{x}.$$

The energy *equality* (2.12) holds for classical solutions (which may not exist). Weak solutions satisfy – in principle – only the energy *inequality* (2.10), since the above calculations are "formal" if performed on weak solutions; in particular, the integral $\int_0^t \int_\Omega \mathbf{u}_t\,\mathbf{u}\,d\mathbf{x}d\tau$ is not well-defined due to the lack of regularity of $\mathbf{u}_t$. As we will see soon the energy equality and inequality are the basic tools in the proof of existence of weak solutions. In fact, by using these results it is possible to get a powerful *a priori* estimate.

With this tool Leray was able to prove the following result, if $\Omega = \mathbb{R}^3$. For a smooth bounded $\Omega \subset \mathbb{R}^3$, see Hopf [155].

Theorem 2.14 (J. Leray (1934), E. Hopf (1951)). *Consider* $\mathbf{u}_0$ *and* $\mathbf{f}$ *with*

$$\mathbf{u}_0 \in L_\sigma^2 \quad \text{and} \quad \mathbf{f} \in L^2(0, T; L_\sigma^2).$$

Then, there exists at least one weak solution to the NSE on $[0, T]$. *Weak solutions satisfy the energy inequality* (2.10) *that, in a bounded domain, can be rewritten in a dimensional form as*

$$k(t) + |\Omega| \int_0^t \epsilon(t')\,dt' \le k(0) + \int_0^t P(t')\,dt', \quad \forall t \in [0, T]. \tag{2.13}$$

Uniqueness of weak solutions is still not known. (It is a Clay-prize problem with a million dollar prize offered.) Uniqueness appears to be connected to the time regularity of the energy dissipation rate. It is known, for example, that all weak solutions satisfy

$$\int_0^T \epsilon(t')\,dt' < \infty, \tag{2.14}$$

while weak solutions are unique if, *e.g.*

$$\int_0^T \epsilon^2(t')\,dt' < \infty. \tag{2.15}$$

In fact, Leray conjectured connection between turbulence and breakdown of uniqueness in weak solutions to the NSE. In particular, conjecturing that perhaps $\epsilon(t)$ has singularities which are integrable but not square integrable: (2.14) holds but (2.15) might fail. This conjecture is still an open question and it is still unknown if equality or inequality holds in (2.13); see Duchon and Robert [97], and Galdi [121] for a very clear elaboration of this theory.

As successful as the Leray theory has been, it has taken many years to begin to establish a connection between it and the Kolmogorov (physical) theory of homogeneous, isotropic turbulence. The status of this connection is well presented in [112] so we shall skip to the essential elements of Kolmogorov's theory (often called the "K-41" theory) needed in this exposition. For more details see the paper by Kolmogorov [191] and the clear exposition in [117, 214, 258].

Consider the NSE under periodic boundary conditions. Let $\mathcal{F}(\mathbf{u}) = \widehat{\mathbf{u}}$ denote the Fourier transform of the velocity field with dual variable $\mathbf{k}$ with $k := |\mathbf{k}| = (k_1^2 + k_2^2 + k_3^2)^{1/2}$. Define

$$E(k,t) := \frac{1}{2} \int_{|\mathbf{k}|=k} |\widehat{\mathbf{u}}(\mathbf{k})|^2 \, d\mathbf{k}, \quad \text{and} \quad E(k) := \lim_{T\to\infty} \frac{1}{T} \int_0^T E(\mathbf{k},t) \, dt.$$

Data from many different turbulent flows (see Fig. 7.4 in Frisch [117]) reveal a universal pattern. Plotting the data on $(\log(k), \log E(k))$ axes, the universal pattern is a $k^{-5/3}$ decay in $E(k)$ through a wide range of wavenumbers known as the inertial range. By combining Richardson's [263] idea of an energy cas-

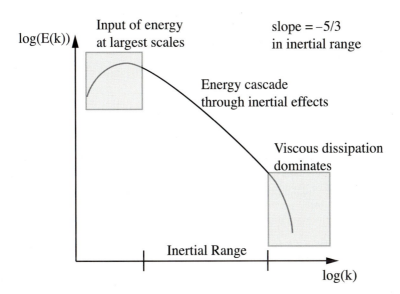

Fig. 2.1. A depiction of the observed energy cascade

cade in turbulent flows with audacious physical guesswork and dimensional analysis, Kolmogorov was able to give a clear explanation of Fig. 2.1. We recall Richardson's famous verse on big whirls and lesser whirls

"Big whirls have little whirls what feed on their velocity, little whirls have smaller whirls, and so on to viscosity" (L.F. Richardson)

was inspired by J. Swift's description of a cascade of poets:

"So, nat'ralists observe a flea
Hath smaller fleas that on him prey;
And these have smaller yet to bite 'em.
And so ad infinitum.
Thus, every poet, in his kind,
Is bit by him that comes behind." (J. Swift)

and by L. da Vinci's descriptions of turbulent flows as composed of an area with energy input at the large scales, an area of interactions and an area of decay into small scales:

"where the turbulence of water is generated,
where the turbulence of water maintains for long,
where the turbulence of water comes to rest." (L. da Vinci)

Kolmogorov began his analysis with the assumption that, roughly speaking, far enough away from walls, after a long enough time, and for high enough Reynolds numbers

time averages of turbulent quantities depend only on one number, the time-averaged energy dissipation rate:

$$\langle \epsilon \rangle := \lim_{T \to \infty} \frac{1}{T} \int_0^T \epsilon(t)\, dt.$$

Two remarkable consequences were that:

(1) the smallest persistent eddy in a turbulent flow is of diameter $O(Re^{-3/4})$;
(2) $E(k)$ must take the universal form

$$E(k) = \alpha \, \langle \epsilon \rangle^{2/3} \, k^{-5/3}, \quad \alpha \cong 1.4,$$

with $\langle \epsilon \rangle$ the only parameter changing from one turbulent flow to another.

The first estimate of $O(Re^{-3/4})$ accounts for the often quoted requirement of $O(Re^{9/4})$ grid points in space for the direct numerical simulations of a turbulent flow. Considering the magnitudes of representative Reynolds numbers (Table 2.1), it also explains the 1949 assessment of turbulence of von Neumann:

"It must be admitted that the problems are too vast to be solved by a direct computational attack." (J. von Neumann, 1949),

which is still true today and provides the motivation for the development of LES!

2.4.4 Existence of Weak Solutions

In this section, which requires a little more mathematical background, we sketch the existence proof for weak solutions, following essentially the approach of Hopf. The very interesting idea of Leray is also recalled at the end of the section. This section requires, at least, knowledge of the basic results of linear functional analysis; see for instance the first chapters in Brezis [45].

The existence of weak solutions will be given by using the Faedo–Galerkin method introduced by Faedo [104] and Galerkin [123]. The main idea of this method is to approximate the natural Hilbert/Banach space V in which the solution $\mathbf{u}$ lives by a sequence of finite dimensional spaces $\{V_m\}_{m\geq 0}$, $V_m \subset V$. In this way the original problem can be reduced to a family of algebraic systems (for elliptic problems) or to a family of ordinary differential equations – ODE (for parabolic problems) for an unknown $\mathbf{u}_m \in V_m$. Then, if it is possible to prove suitable estimates independent of m, it is also possible to pass to the limit as $m \to \infty$ and, if we are lucky, $\mathbf{u}_m$ converges to a solution $\mathbf{u}$ to the original problem. For several applications of this technique in the context of nonlinear PDE we suggest the excellent monograph by J.-L. Lions [221].

In the case of the NSE the application of this method is not trivial, since the equations are *nonlinear* and some delicate compactness results are needed in order to pass to the limit as $m \to \infty$. Instead of a general theory of Faedo–Galerkin methods, we prefer to show how this method works in our particular case. Without going into detail, we will show the proof of the existence of weak solutions. We suggest the reader follows at least the main steps, to see (a) an explicit application of the Faedo–Galerkin method and (b) the energy estimates, that are common to many other problems of mathematical physics.

Proof (of Theorem 2.14). In the case of the NSE, we approximate the natural space L^2_σ, with $V_m \subset L^2_\sigma$ such that dim $V_m = m$, defined in the following way:

$$V_m := \mathrm{Span}\langle \mathcal{W}_1, \ldots, \mathcal{W}_m \rangle.$$

The functions $\mathcal{W}_i(\mathbf{x})$ are eigenfunctions of the stationary *Stokes* equations, *i.e.* they satisfy for $k \in \mathbb{N}$, the linear system:

$$\begin{cases} -\Delta \mathcal{W}_k + \nabla \mathcal{P}_k = \lambda_k \mathcal{W}_k & \text{in } \Omega, \\ \nabla \cdot \mathcal{W}_k = 0 & \text{in } \Omega, \\ \mathcal{W}_k = 0 & \text{on } \partial\Omega. \end{cases}$$

It is possible to prove (see Constantin and Foiaş [74] and Temam [295]) that the eigenvalues λ_k satisfy $0 < \lambda_1 \leq \cdots \leq \lambda_{n-1} \leq \lambda_n \leq \lambda_{n+1} \leq \ldots$, while $\mathcal{W}_i$ can be chosen to form an orthonormal "basis" of L_σ^2. The latter means that

$$\int_\Omega \mathcal{W}_i \mathcal{W}_j \, d\mathbf{x} = \delta_{ij} := \begin{cases} 1 \text{ if } i = j \\ \\ 0 \text{ if } i \neq j \end{cases}$$

and that finite linear combinations of $\mathcal{W}_i$ are dense in L_σ^2.

We consider the Faedo–Galerkin approximate function:

$$\mathbf{u}_m(\mathbf{x}, t) = \sum_{k=1}^m g_m^i(t) \mathcal{W}_i(\mathbf{x}),$$

and we look for a function $\mathbf{u}_m$ that satisfies the following initial value problem:

$$\begin{cases} \dfrac{d}{dt}\mathbf{u}_m - \dfrac{1}{Re}\Delta \mathbf{u}_m + P_m(\mathbf{u}_m \cdot \nabla \mathbf{u}_m) = P_m \mathbf{f} & \text{for } t \in (0, T) \\ \\ \mathbf{u}_m(0) = P_m \mathbf{u}_0, \end{cases} \tag{2.16}$$

where the operator P_m denotes the orthogonal projection onto V_m.

The weak form of (2.16) is similar to the weak formulation of the NSE, and it is a weak formulation in which the test functions belong to V_m. We have to solve, for $k = 1, \ldots, m$ the following Cauchy problem for a system of ODE (recall that we denote by $(.\,,.)$ the scalar product in $L^2(\Omega)$):

$$\begin{cases} \dfrac{d}{dt}(\mathbf{u}_m, \mathcal{W}_k) + \dfrac{1}{Re}(\nabla \mathbf{u}_m, \nabla \mathcal{W}_k) + (\mathbf{u}_m \cdot \nabla \mathbf{u}_m, \mathcal{W}_k) = (\mathbf{f}, \mathcal{W}_k) & \text{for } t \in (0, T) \\ \\ \mathbf{u}_m(\mathbf{x}, 0) = P_m(\mathbf{u}_0(\mathbf{x})), \end{cases} \tag{2.17}$$

with the $g_m^i(t) : [0, T] \to \mathbb{R}$ that are functions of class C^1. It is easily seen that the above system of ODE for the unknown $g_m^i(t)$, satisfy the hypotheses of the Cauchy–Lipschitz theorem. Consequently, the local existence and uniqueness of the solution can easily be proved with standard tools. This solution exists in some time interval $[0, T_m]$, and to prove that $T_m = T$ we will use an *a priori* estimate.

By multiplying (2.17)$_1$ by $g_m^k(t)$ and summing over k, we get the following identity:

$$\frac{1}{2}\frac{d}{dt}\|\mathbf{u}_m(t)\|^2 + \frac{1}{Re}\|\nabla \mathbf{u}_m(t)\|^2 = (\mathbf{f}(t), \mathbf{u}_m(t)), \quad t \in [0, T_m). \tag{2.18}$$

This procedure corresponds to multiplying the equation in (2.16)$_1$ by $\mathbf{u}_m$ and to integrating over Ω. Note that the nonlinear term disappears as in Sect. 2.4.3!

Then, by using the Hölder and Poincaré inequalities we get

$$\frac{1}{2}\frac{d}{dt}\|\mathbf{u}_m(t)\|^2 + \frac{1}{Re}\|\nabla\mathbf{u}_m(t)\|^2 \leq \|\mathbf{f}(t)\|\,\|\mathbf{u}_m(t)\| \leq C_P\|\mathbf{f}(t)\|\|\nabla\mathbf{u}_m(t)\|.$$

With the Young inequality

$$ab \leq \frac{a^p}{p} + \frac{b^{p'}}{p'}, \qquad \text{for} \quad \frac{1}{p} + \frac{1}{p'} = 1, \quad 1 < p < \infty, \tag{2.19}$$

we finally get

$$\frac{1}{2}\frac{d}{dt}\|\mathbf{u}_m(t)\|^2 + \frac{1}{Re}\|\nabla\mathbf{u}_m(t)\|^2 \leq \frac{C_P^2 Re}{2}\|\mathbf{f}(t)\|^2 + \frac{1}{2Re}\|\nabla\mathbf{u}_m(t)\|^2.$$

We can "absorb" the last term on the right-hand side into the second on the left-hand side, to deduce

$$\frac{d}{dt}\|\mathbf{u}_m(t)\|^2 + \frac{1}{Re}\|\nabla\mathbf{u}_m(t)\|^2 \leq C_P^2 Re\|\mathbf{f}(t)\|^2. \tag{2.20}$$

A first integration of the above inequality shows that, for each $t \in [0, T_m)$,

$$\|\mathbf{u}_m(t)\|^2 \leq \|P_m\mathbf{u}_0\|^2 + C_P^2 Re \int_0^T \|\mathbf{f}(t)\|^2\,dt \leq \|\mathbf{u}_0\|^2 + C_P^2 Re\|\mathbf{f}(t)\|_{L^2(0,T;L^2)}^2.$$

Recall also that $\|P_m\mathbf{u}_0\| \leq \|\mathbf{u}_0\|$, due to the fact that P_m is a projector.

Since the above bound is independent of m, a standard continuation argument for ODE implies that the maximal time of existence of solution, T_m, equals T. In fact, the above inequality proves that $\|\mathbf{u}_m(t)\|$ is bounded uniformly in $(0, T)$ and this contradicts the necessary condition for a blow-up of $g_m^i(t)$ as $t \to T_m$; see for instance Hartman [148].

Integrating (2.20) with respect to t on $(0, T)$ we also obtain

$$\|\mathbf{u}_m(T)\|^2 + \frac{1}{Re} \int_0^T \|\nabla\mathbf{u}_m(\tau)\|^2\,d\tau \leq \|\mathbf{u}_0\|^2 + C_P^2 Re\|\mathbf{f}\|_{L^2(0,T;L^2)}^2.$$

Remark 2.15. We derived in detail these estimates since they represent the core of the proof. We stress their importance since similar estimates can be derived, with the same techniques, for a wide range of different PDEs. The reader may also note that a similar estimate can be derived if $\mathbf{f}$ belongs just to $L^2(0, T; (H_{0,\sigma}^1)')$, where $(H_{0,\sigma}^1)'$ is the topological dual of $H_{0,\sigma}^1$.

We have now proved (recall Lemma 2.9) that the sequence $\{\mathbf{u}_m\}_{m\geq 1}$ is uniformly (in m) bounded in

$$L^\infty(0, T; L_\sigma^2) \cap L^2(0, T; H_{0,\sigma}^1).$$

Then, we can use classical weak compactness[5] results (see Brezis [45]) to show that from the sequence $\{\mathbf{u}_m\}_{m\geq 1}$ we can extract a subsequence (relabeled again as $\{\mathbf{u}_m\}_{m\geq 1}$) such that

$$\begin{cases} \mathbf{u}_m \overset{*}{\rightharpoonup} \mathbf{u} \text{ in } L^\infty(0,T;L_\sigma^2) \\[2mm] \mathbf{u}_m \rightharpoonup \mathbf{u} \text{ in } L^2(0,T;H_{0,\sigma}^1). \end{cases}$$

In the above expression $\rightharpoonup$ denotes the weak convergence, while $\overset{*}{\rightharpoonup}$ denotes the weak-$*$ convergence. These properties can be expressed, respectively, as

$$\begin{cases} \displaystyle\int_0^T \int_\Omega \mathbf{u}_m \mathbf{v}\,d\mathbf{x}\,dt \rightarrow \int_0^T \int_\Omega \mathbf{u}\mathbf{v}\,d\mathbf{x}\,dt \qquad \forall\, \mathbf{v} \in L^1(0,T;L_\sigma^2) \\[4mm] \displaystyle\int_0^T \int_\Omega \nabla\mathbf{u}_m \nabla\mathbf{v}\,d\mathbf{x}\,dt \rightarrow \int_0^T \int_\Omega \nabla\mathbf{u}\,\nabla\mathbf{v}\,d\mathbf{x}\,dt \qquad \forall\, \mathbf{v} \in L^2(0,T;H_{0,\sigma}^1). \end{cases}$$

The limit function $\mathbf{u}$ has the required regularity for a weak solution, but the most difficult point is to show now that such $\mathbf{u}$ is indeed a weak solution to the NSE, and in particular that (2.9) is satisfied.

The difficult technical point (this is one of the challenges in the study of nonlinear PDE) is now to analyze the following:

$$\int_0^T \int_\Omega \mathbf{u}_m \cdot \nabla\mathbf{u}_m\,\phi\,d\mathbf{x}\,dt \overset{?}{\longrightarrow} \int_0^T \int_\Omega \mathbf{u}\cdot\nabla\mathbf{u}\,\phi\,d\mathbf{x}\,dt. \qquad (2.21)$$

---

[5] We have no space here to review the basic results needed to extract weakly (or weakly-$*$) converging subsequences. Essentially they derive from the Banach–Alaoglu–Bourbaki theorem and other classical results on Banach spaces, see Brezis [45]. The reader should be acquainted at least with the following theorem: let $(X, \|\,\|, \|\|_X)$ be a Banach space. If $\{\phi_n\}_n \subset X'$ is a bounded sequence, then it is possible to extract a subsequence ϕ_{n_k} weakly-$*$ converging to some $\phi \in X'$, that is

$$\lim_{k\to+\infty} \langle \phi_{n_k}, x\rangle = \langle \phi, x\rangle \qquad \forall\, x \in X \subset X'' = (X')'.$$

Furthermore if X is also reflexive, i.e., $X'' = X$ then the convergence is weak and not weak-$*$.

A typical case of a reflexive space is a Hilbert space. If H, endowed with the scalar product $(\,.\,,\,.\,)$, is a Hilbert space and $\{x_n\}_n \subset H$ is a sequence such that $\|x_n\|_H \leq C$, then there exists $x \in H$ and a subsequence x_{n_k} such that

$$\lim_{k\to+\infty} (x_{n_k}, y) = (x, y), \qquad \forall\, y \in H.$$

In our case we are using that $L^2(0,T;H_{0,\sigma}^1)$ is a Hilbert space, while the nonreflexive Banach space $L^\infty(0,T;L_\sigma^2)$ is the topological dual (see [221]) of $L^1(0,T;L_\sigma^2)$.

To pass to the limit in the nonlinear expression it is necessary to have some kind of strong convergence. In fact, the product of a couple of sequences both weakly converging may not converge. Hopf succeeded in proving that the sequence $\{\mathbf{u}_m\}_{m \geq 1}$ satisfies also the following property of strong convergence:

$$\mathbf{u}_m \to \mathbf{u} \quad \text{in } L^2(0, T; L^2(C)), \qquad \text{for each cube } C \subset \Omega \subseteq \mathbb{R}^d. \qquad (2.22)$$

This property is obtained by Hopf as an elegant consequence of the Friederichs inequality (see page 114). With (2.22) it is rather easy to pass to the limit in the nonlinear term (see for instance Galdi [121] page 20).

The energy inequality is finally proved by using the fact that the smooth $\mathbf{u}_m$ satisfies the energy equality. By passing to the limit and using the lower semicontinuity of the norm we finally arrive at the inequality (2.10). Note that it is in this technical limit procedure that we pass from energy equality to energy inequality and *spurious* energy dissipation may take place. The diligent reader may also observe that in the extraction of the subsequence we used the *axiom of choice* and this Galerkin procedure is not at all constructive, since we do not know if the entire sequence $\{\mathbf{u}_m\}_{m \geq 1}$ converges to $\mathbf{u}$! (In particular this may happen provided we have a uniqueness result.)

We observe that *a posteriori* it is possible to show that the time derivative of $\mathbf{u}$ satisfies certain properties. In particular

$$\mathbf{u}_t \in \begin{cases} L^{4/3}(0, T; (H_{0,\sigma}^1)') & \text{if } \Omega \subset \mathbb{R}^3 \\[2mm] L^2(0, T; (H_{0,\sigma}^1)') & \text{if } \Omega \subset \mathbb{R}^2. \end{cases}$$

We also consider the initial datum: in which sense does $\mathbf{u}(\mathbf{x}, 0) = \mathbf{u}_0(\mathbf{x})$ since the function $\mathbf{u}$ is not continuous in t? It is possible to prove (possibly after redefining the velocity on a set of zero Lebesgue measure) that $\mathbf{u}$ is weakly continuous in L_σ^2, *i.e.*

$$\lim_{t \to t_0} \int_\Omega \mathbf{u}(t, \mathbf{x})\, \mathbf{v}(\mathbf{x})\, d\mathbf{x} = \int_\Omega \mathbf{u}(t_0, \mathbf{x})\, \mathbf{v}(\mathbf{x})\, d\mathbf{x}, \quad \forall t_0 \in [0, T], \ \forall \mathbf{v} \in L_\sigma^2.$$

This implies, together with the energy inequality and the semicontinuity of the norm (see for instance Galdi [121] page 21), that

$$\lim_{t \to 0} \|\mathbf{u}(t) - \mathbf{u}_0\| = 0. \qquad \qquad \square$$

On uniqueness of weak solutions. Let us see in a heuristic way why, at present, uniqueness of weak solutions is still an open problem. Let us consider two weak solutions $\mathbf{u}_1$ and $\mathbf{u}_2$, corresponding to the same initial datum $\mathbf{u}_0$ and to the same external force $\mathbf{f}$. Let us take the difference of the equation satisfied by $\mathbf{u}_2$ from that satisfied by $\mathbf{u}_1$ to get

$$(\mathbf{u}_1 - \mathbf{u}_2)_t - \frac{1}{Re}\Delta(\mathbf{u}_1 - \mathbf{u}_2) + \mathbf{u}_1 \cdot \nabla\mathbf{u}_1 - \mathbf{u}_2 \cdot \nabla\mathbf{u}_2 + \nabla(p_1 - p_2) = 0.$$

By subtracting and adding the term $\mathbf{u}_1 \cdot \nabla \mathbf{u}_2$ we can rewrite the latter identity in terms of $\mathbf{w} = \mathbf{u}_1 - \mathbf{u}_2$ and $q = p_1 - p_2$ as follows:

$$\mathbf{w}_t - \frac{1}{Re}\Delta\mathbf{w} + \nabla q + \mathbf{u}_1 \cdot \nabla\mathbf{w} + \mathbf{w} \cdot \nabla\mathbf{u}_2 = \mathbf{0}.$$

By multiplying by $\mathbf{w}$ and with integration by parts (note that $\int_\Omega \mathbf{u}_1 \cdot \nabla\mathbf{w}\,\mathbf{w}\,dx = 0$) we get

$$\frac{1}{2}\frac{d}{dt}\|\mathbf{w}\|^2 + \frac{1}{Re}\|\nabla\mathbf{w}\|^2 = -\int_\Omega \mathbf{w} \cdot \nabla\mathbf{u}_2\,\mathbf{w}\,dx. \tag{2.23}$$

This calculation is purely formal, but it gives a feeling for the difficulties in dealing with weak solutions. To estimate the integral on the right-hand side we need the following interpolation results. The proof of the following proposition (that is a particular case of the Gagliardo–Nirenberg inequalities) can be given with elementary tools, namely a clever application of Hölder inequality; see Ladyžhenskaya [197]. Due to its importance it is stated as Lemma 1 of Chap. 1 in [197].

Proposition 2.16. *Let Ω be any open subset of $\mathbb{R}^d$. Then, for any function belonging to $H_0^1(\Omega)$*

$$\|u\|_{L^4} \leq \begin{cases} 2^{1/4}\|u\|^{1/2}\|\nabla u\|^{1/2} & \text{if } \Omega \subset \mathbb{R}^2, \\[2mm] 4^{1/4}\|u\|^{1/4}\|\nabla u\|^{3/4} & \text{if } \Omega \subset \mathbb{R}^3. \end{cases} \tag{2.24}$$

By using the Hölder inequality with exponents 4, 2, and 4, the above proposition (in $\Omega \subset \mathbb{R}^3$) and Young inequality (2.19) with exponents 4 and 4/3 we get

$$\left|\int_\Omega \mathbf{w} \cdot \nabla\mathbf{u}_2\,\mathbf{w}\,dx\right| \leq \|\mathbf{w}\|_{L^4}^2\|\nabla\mathbf{u}_2\|^2 \leq c\,\|\mathbf{w}\|^{1/2}\|\nabla\mathbf{w}\|^{3/2}\|\nabla\mathbf{u}_2\|$$

$$\leq \frac{1}{Re}\|\nabla\mathbf{w}\|^2 + c_1\|\mathbf{w}\|^2\|\nabla\mathbf{u}_2\|^4.$$

Now we recall another fundamental tool in the analysis of time-dependent PDE.

Lemma 2.17 (Gronwall lemma). *Let $f, g : [\alpha, \beta] \to \mathbb{R}^+$ be two nonnegative, continuous functions and let $C \geq 0$ a given real constant. Let us suppose that*

$$f(t) \leq C + \int_\alpha^t f(\tau)\,g(\tau)\,d\tau, \qquad \forall\,t \in [\alpha, \beta].$$

Then,

$$f(t) \leq C\,e^{\int_\alpha^t g(\tau)\,d\tau}.$$

The hypotheses of the above lemma can be weakened, by requiring, for instance, g to be simply an $L^1(0,T)$ function, instead that of a continuous function.

From integration of (2.23), together with the "Ladyženskaya inequality" (2.24) we obtain

$$\|\mathbf{w}(t)\|^2 + \frac{1}{Re}\int_0^t \|\nabla\mathbf{w}(\tau)\|^2\,d\tau \leq \|\mathbf{w}(0)\|^2 + 2c\int_0^t \|\nabla\mathbf{u}_2(\tau)\|^4\|\mathbf{w}(\tau)\|^2\,d\tau,$$

and then the Gronwall lemma will imply that

$$\|\mathbf{w}(t)\|^2 \leq \|\mathbf{w}(0)\|^2\, \mathrm{e}^{2c\int_0^t \|\nabla\mathbf{u}_2(\tau)\|^4\,d\tau}.$$

Since $\mathbf{w}(0) = \mathbf{0}$, this will prove that $\mathbf{w} = \mathbf{u}_1 - \mathbf{u}_2 \equiv \mathbf{0}$ on $[0,T]$, provided that

$$\int_0^T \|\nabla\mathbf{u}_i(\tau)\|^4\,d\tau < \infty \quad \Longleftrightarrow \quad \nabla\mathbf{u}_i \in L^4(0,T;L^2(\Omega)), \text{ for } i = 1,2.$$

However, we do not know whether it is true, since $\mathbf{u}_2$ is a weak solution and $\nabla\mathbf{u}_2$ belongs just to $L^2(0,T;L^2(\Omega))$. We recall that $L^p(0,T;X) \subset L^q(0,T;X)$, provided that $p \geq q$, and $L^4(0,T;L^2)$ is a subspace of $L^2(0,T;L^2)$.

Remark 2.18. From the above calculation we can see that it is sufficient to require that only $\mathbf{u}_2$ is smoother than a weak solution, to prove uniqueness, even if a proof cannot be carried on in this way. Some smoothing to justify the calculations is necessary, but the following result is true: if at least one of the two weak solutions $\mathbf{u}_i$ belong to $L^4(0,T;H^1_{0,\sigma})$, then $\mathbf{u}_1 = \mathbf{u}_2$. The proof is due to Sather and Serrin, see Serrin [275] and also Temam [295].

Remark 2.19. By using the same procedure and by using (2.24) for a two-dimensional domain, it can be shown that the following estimate holds:

$$\frac{d}{dt}\|\mathbf{w}(t)\|^2 + \frac{1}{Re}\|\nabla\mathbf{w}(t)\|^2 \leq 2c\,\|\nabla\mathbf{u}_2(t)\|^2\|\mathbf{w}(t)\|^2.$$

In this case it is possible to apply the Gronwall lemma to deduce that, if $\Omega \subseteq \mathbb{R}^2$, then weak solutions are unique. In the 2D case the above calculations are not formal, due to the fact that, for instance, $\mathbf{u}\cdot\nabla\mathbf{u} \in L^2(0,T;L^2)$ and since $\mathbf{u} \in L^2(0,T;H^1_{0,\sigma})$ and $\mathbf{u}_t \in L^2(0,T;(H^1_{0,\sigma})')$, then

$$\int_0^t \langle\mathbf{u}_t(\tau),\mathbf{u}(\tau)\rangle\,d\tau = \frac{1}{2}\|\mathbf{u}(t)\|^2 - \frac{1}{2}\|\mathbf{u}_0\|^2 \quad \forall t \in [0,T],$$

where $\langle\,.\,,\,.\,\rangle$ denotes the duality pairing between $H^1_{0,\sigma}$ and its dual $(H^1_{0,\sigma})'$.

This is the remarkable difference between the 2D and the 3D case and this uniqueness result was proved for the first time in Kiselev and Ladyženskaya [190]. In the same paper the reader can find the interesting estimate that proves how the 3D problem for the *vector Burgers* equations (the NSE without pressure and the divergence-free constraint) is well-posed in the 3D case, see also Galdi [121].

Final Considerations

In this section we presented the basic tools needed to prove existence of a weak solution. These tools will be used extensively in the following chapters. As mentioned in the introduction, we did not give mathematical details, since it should also be considered as a "playground" for nonmathematicians. People coming from other areas should focus on the basic strategy of the Faedo–Galerkin method, since it can be applied to many other problems: the approximation of the problem, the *a priori* estimate (obtained by multiplying the solution by itself), and the use of the Gronwall lemma (an elementary but extremely powerful result).

2.4.5 More Regular Solutions

Since at present it is not possible to prove the uniqueness of weak solutions, we investigate the existence of more regular solutions.

If the initial data are more regular, say $\mathbf{u}_0 \in H_{0,\sigma}^1$, then it is possible to prove the local-in-time existence of more regular solutions, the so called *strong solutions*.

Definition 2.20. *We say that a weak solution* $\mathbf{u}$ *is a strong solution if*

$$\begin{cases} \mathbf{u} \in L^\infty(0,T;H_{0,\sigma}^1) \cap L^2(0,T;H_{0,\sigma}^1 \cap [H^2(\Omega)]^d), \\ \mathbf{u}_t \in L^2(0,T;L_\sigma^2), \end{cases}$$

where $H^2(\Omega) \subset L^2(\Omega)$ *is the space of (classes of equivalence of) functions in* $L^2(\Omega)$ *with derivatives up to the second order in* $L^2(\Omega)$.

The concept of strong solution is very important, for the following reasons:

(a) strong solutions are unique, also in the wider class of weak solutions;
(b) strong solutions satisfy the energy equality;
(c) a strong solution becomes smooth (for each positive time) in space-time variables if $\partial\Omega$, $\mathbf{u}_0$ and $\mathbf{f}$ are smooth.

Unfortunately, we are able to prove the existence of strong solutions only for small times, or small data.

Theorem 2.21. *Let* $\mathbf{u}_0 \in H_{0,\sigma}^1$ *and* $\mathbf{f} \in L^2(0,T;L_\sigma^2)$. *Then there exists* $0 < T_0 \leq T$ *such that there exists a unique strong solution in* $[0,T_0)$. *The time* T_0 *depends on* $\mathbf{f}$, $\|\nabla\mathbf{u}_0\|$, *and Re; see* (2.29).

Proof. We do not give the complete proof, but we show the basic *a priori* estimates involved in the proof of Theorem 2.21. If we multiply (2.17) by $\lambda_k g_m^k(t)$, sum over k (note that since $\mathcal{W}_i$ is an eigenfunction of the Stokes operator, this corresponds to multiplying the equations by $-P\Delta\mathbf{u}_m$, see [261]), and integrate by parts over Ω we obtain

$$\frac{1}{2}\frac{d}{dt}\|\nabla\mathbf{u}_m\|^2 + \frac{1}{Re}\|P\Delta\mathbf{u}_m\|^2 = (\mathbf{f}, -P\Delta\mathbf{u}_m) + (\mathbf{u}_m \cdot \nabla\mathbf{u}_m, P\Delta\mathbf{u}_m). \quad (2.25)$$

We estimate the first term on the right-hand side by the Schwartz inequality

$$|(\mathbf{f}, -P\Delta\mathbf{u}_m)| \le \|\mathbf{f}\|\,\|P\Delta\mathbf{u}_m\| \le \frac{1}{4Re}\|P\Delta\mathbf{u}_m\|^2 + Re\|\mathbf{f}\|^2.$$

The second term requires two inequalities that will be very useful in the sequel. For a proof see, for instance, Adams [4].

Proposition 2.22 (Convex-interpolation inequality). *Let $f \in L^r(\Omega) \cap L^s(\Omega)$ with $1 \le r < s \le \infty$. Then, $f \in L^p(\Omega)$ for each $r \le p \le s$ and the following inequality holds*

$$\|f\|_{L^p} \le \|f\|_{L^r}^{\theta}\|f\|_{L^s}^{1-\theta}, \quad \text{with } \theta \text{ satisfying } \quad \frac{1}{p} = \frac{\theta}{r} + \frac{1-\theta}{s}. \quad (2.26)$$

Proposition 2.23 (A special case of the Sobolev embedding). *Let $f \in H^1(\Omega)$, with $\Omega \subset \mathbb{R}^3$. Then, there exists a positive constant $C = C(\Omega)$ (independent of f) such that*

$$\|f\|_{L^6} \le C(\Omega)\|\nabla f\| \qquad \forall f \in H^1(\Omega). \quad (2.27)$$

In particular, in Sect. 1 of [197], the reader can find an elementary proof for the fact that if in addition $f \in H_0^1(\Omega)$, then the estimate holds with $C = 48^{1/6}$, for any open set $\Omega \subset \mathbb{R}^3$. We can now estimate the last term in (2.25) as follows: apply the Hölder inequality (with exponents 6, 3, and 2) to get

$$|(\mathbf{u}_m \cdot \nabla\mathbf{u}_m, P\Delta\mathbf{u}_m)| \le \|\mathbf{u}_m\|_{L^6}\|\nabla\mathbf{u}_m\|_{L^3}\|P\Delta\mathbf{u}_m\|.$$

Then, apply the interpolation inequality (2.26) to the second term to get

$$|(\mathbf{u}_m \cdot \nabla\mathbf{u}_m, P\Delta\mathbf{u}_m)| \le \|\mathbf{u}_m\|_{L^6}\|\nabla\mathbf{u}_m\|^{1/2}\|\nabla\mathbf{u}_m\|_{L^6}^{1/2}\|P\Delta\mathbf{u}_m\|.$$

Finally an application of the Sobolev embedding (2.27) and the Young inequality (with exponents 4 and 4/3) shows that

$$|(\mathbf{u}_m \cdot \nabla\mathbf{u}_m, P\Delta\mathbf{u}_m)| \le C\|\nabla\mathbf{u}_m\|^{3/2}\|P\Delta\mathbf{u}_m\|^{3/2}$$

$$\le \frac{1}{4Re}\|P\Delta\mathbf{u}_m\|^2 + C_1 Re^3\|\nabla\mathbf{u}_m\|^6.$$

The final differential inequality is then

$$\frac{d}{dt}\|\nabla\mathbf{u}_m\|^2 + \frac{1}{Re}\|P\Delta\mathbf{u}_m\|^2 \le 2C_1 Re^3\|\nabla\mathbf{u}_m\|^6 + 2Re\|\mathbf{f}\|^2. \quad (2.28)$$

To avoid inessential calculations, we consider from now on only the case $\mathbf{f} \equiv 0$. The results that can be obtained are essentially the same as those that can be proved if $\mathbf{f} \neq \mathbf{0}$. In Chap. 7, we shall analyze a very similar situation and we refer the reader to that chapter for further details.

If we set $y(t) = \|\nabla \mathbf{u}(t)\|^2$ we are left with the differential inequality

$$\begin{cases} \dfrac{d}{dt} y(t) \leq CRe^3 [y(t)]^3 \\ \\ y(0) = y_0 = \|\nabla \mathbf{u}_0\|^2, \end{cases}$$

which implies

$$y(t) \leq \frac{y_0}{\sqrt{1 - 2y_0^2 CRe^3 t}}, \quad \text{for} \quad 0 \leq t < T_0 := \frac{1}{2y_0^2 CRe^3}. \tag{2.29}$$

In fact, $Y(t) := y_0/\sqrt{1 - 2y_0^2 CRe^3 t}$ is the solution of the Cauchy problem

$$\begin{cases} \dfrac{d}{dt} Y(t) = CRe^3 [Y(t)]^3 \\ \\ Y(0) = y_0 \end{cases}$$

and since $y(0) = Y(0)$ and the slope of y is smaller than that of Y ($y' \leq Y'$) we get that $y(t) \leq Y(t)$. A comparison argument like this one is at the basis of many results on nonlinear evolution PDE.

This argument gives an estimate on the *life-span* of the function $\mathbf{u}_m$ and shows that $\{\mathbf{u}_m\}_{m \geq 1}$ is bounded uniformly (in m) in

$$L^\infty(0, \overline{T}; H^1_{0,\sigma}), \qquad \forall \overline{T} < T_0.$$

Integration in time of (2.28) shows that $\{P\Delta \mathbf{u}_m\}_{m \geq 1}$ is bounded uniformly (in m) in

$$L^2(0, \overline{T}; L^2(\Omega)), \qquad \forall \overline{T} < T_0.$$

Then, by using a result of elliptic regularity for the Stokes equations (see for instance Beirão da Veiga [21] for an elementary proof) we obtain that in $H^1_{0,\sigma} \cap [H^2(\Omega)]^d$ the norm $\|P\Delta g\|$ is equivalent to $\|g\|_{H^2(\Omega)}$. This implies that $\{\mathbf{u}_m\}_{m \geq 1}$ is bounded uniformly in

$$L^\infty(0, \overline{T}; H^1_{0,\sigma}) \cap L^2(0, \overline{T}; H^1_{0,\sigma} \cap [H^2(\Omega)]^d), \qquad \forall \overline{T} < T_0.$$

By a limit procedure we can show that $\mathbf{u}_m$ converges to some $\mathbf{u}$ satisfying the same properties and then we construct a strong solution as claimed. Uniqueness can be proved by using exactly the argument that fails on weak solutions. The fact that the energy equality is satisfied is rather technical. □

The additional regularity of the strong solutions is summarized in the following theorem:

Theorem 2.24. *Let **u** be a strong solution in* $[0,T]$. *If* Ω *is of class* C^∞ *and if* $\mathbf{f} \in C^\infty((0,T] \times \overline{\Omega})$ *then*

$$\mathbf{u} \in C^\infty([\varepsilon,T] \times \overline{\Omega}), \quad \forall \varepsilon > 0.$$

We do not give the proof of this result; we just sketch the idea of the technique of *bootstrapping* that can be used in the proof. The main idea is to consider the *linear* evolution problem

$$\mathbf{u}_t - \frac{1}{Re}\Delta\mathbf{u} + \mathbf{v} \cdot \nabla\mathbf{u} + \nabla p = \mathbf{f} \quad \text{in } \mathbb{R}^3 \times (0,T)$$
$$\nabla \cdot \mathbf{u} = 0 \quad \text{in } \mathbb{R}^3 \times (0,T), \tag{2.30}$$

where $\mathbf{v}$ is a given function that has the same regularity of the strong solution $\mathbf{u}$, and $\nabla \cdot \mathbf{v} = 0$. Then, due to the fact that it is a simpler problem (being linear), it is possible to prove for the solution to (2.30) more regularity than the original one known on $\mathbf{v}$. This holds for each $\mathbf{v}$ with a given regularity and in particular also for $\mathbf{v} = \mathbf{u}$. Due to the uniqueness of strong solutions, this shows how the strong solution $\mathbf{u}$ has more regularity, namely the regularity of the solution to (2.30). Using again the same argument, with a now smoother $\mathbf{v}$ we can go further. Unfortunately this arguments fails if $\mathbf{v}$ belongs just to $L^2(0,T;L^2_\sigma) \cap L^2(0,T;H^1_{0,\sigma})$, since in this case we cannot prove for the solution of (2.30) more regularity than that of weak solutions.

In this respect we note that to start the bootstrap argument it will also be sufficient to know that

$$\mathbf{u} \in L^r(0,T;L^s(\Omega)) \quad \text{for} \quad \frac{2}{r} + \frac{d}{s} = 1. \tag{2.31}$$

Weak solutions satisfying the above property are unique and smooth. Note that it is not known whether weak solutions do satisfy condition (2.31) if $d = 3$, while it is proved that they satisfy it for $d = 2$. The justification of the above condition can be understood in the light of the *scaling invariance*. In fact, (forget the boundaries and imagine functions in $\mathbb{R}^d$) if $(\mathbf{u}(\mathbf{x},t),p(\mathbf{x},t))$ is a solution to the NSE then also the family

$$(\mathbf{u}_\lambda,p_\lambda) = (\lambda\mathbf{u}(\lambda x, \lambda^2 t), \lambda^2 p(\lambda x, \lambda^2 t)) \quad \text{for each } \lambda > 0$$

is a solution. They are the so-called *self-similar* solutions. In particular the $L^r(0,T;L^s(\Omega))$-norms that are independent of λ are those and only those satisfying (2.31).

Remark 2.25. In the above theorem the regularity up to $t = 0$ cannot be obtained even if $\mathbf{u}_0 \in [C^\infty(\Omega)]^d \cap H^1_{0,\sigma}$. To have smoothness at the initial time, some additional *compatibility* conditions must be satisfied, see Temam [296].

Remark 2.26. In the 2D case we can follow the same path, to obtain the following estimate (note that the results of Proposition 2.16 hold also for

$f \in H^1$ (instead of belonging to H_0^1), but on the right-hand side there is now a positive number $C = C(\Omega)$ depending on Ω:

$$|(\mathbf{u}_m \cdot \nabla \mathbf{u}_m, -P\Delta \mathbf{u}_m)| \leq \|\mathbf{u}_m\|_{L^4} \|\nabla \mathbf{u}_m\|_{L^4} \|P\Delta \mathbf{u}_m\|$$

$$\leq C\|\mathbf{u}_m\|^{1/2} \|\nabla \mathbf{u}_m\| \|P\Delta \mathbf{u}_m\|^{3/2}$$

$$\leq \frac{1}{4Re} \|P\Delta \mathbf{u}_m\|^2 + c_1 Re^3 \|\mathbf{u}_m\|^2 \|\nabla \mathbf{u}_m\|^4.$$

Now, from the energy equality we have

$$\|\mathbf{u}_m(t)\| < +\infty, \ t \in [0, T] \qquad \text{and} \qquad \int_0^T \|\nabla \mathbf{u}_m(\tau)\|^2 \, d\tau < +\infty$$

and consequently we derive the following inequality for $y(t) = \|\nabla \mathbf{u}_m(t)\|^2$:

$$\frac{d}{dt} y(t) \leq c[y(t)]^2 = c\, y(t) \cdot y(t), \qquad \text{with} \quad \int_0^T y(\tau)\, d\tau < +\infty.$$

This implies (with the Gronwall lemma)

$$y(t) \leq y(0) \ e^{\displaystyle c \int_0^t y(\tau)\, d\tau} < +\infty \qquad \forall\, t \in [0, T],$$

showing that the life-span of strong solutions is all the positive half-line. In two dimensions, if we start from a smooth datum, we have a smooth solution for each positive time (provided the external force is smooth). In the end, these results on strong solutions show the main difference between the 2D and the 3D cases!

On the Possible Loss of Regularity

We have shown that for a smooth enough initial datum we can construct a unique strong solution in a time interval $[0, T_0)$ and we have given an explicit estimate on T_0 in terms of the H_0^1-norm of the initial datum and of the Reynolds number (recall (2.29)). We want to analyze what should happen at a time T^* at which a solution loses its regularity, if such a T^* exists! The first result, that is a clever application of the information hidden in the energy inequality and in differential inequality (2.28) is the so called *Théorème de Structure* of Leray, that furnishes preliminary, but deep insight into the structure of weak solutions.

Theorem 2.27 (Leray [213]). *Let $\mathbf{u}$ be a weak solution. Then, there exists a set $\mathcal{U} \subset (0, \infty)$, that is a union of disjoint intervals, such that*

1. *the Lebesgue measure of* $(0, \infty) \backslash \mathcal{U}$ *vanishes;*
2. $\mathbf{u} \in C^\infty(\overline{\Omega} \times \mathcal{U})$;
3. *there exists* $T_R \in (0, \infty)$ *such that* $\mathcal{U} \supset (T_R, \infty)$;
4. *there exists* $\lambda = \lambda(\Omega, Re) > 0$ *such that, if* $\|\nabla \mathbf{u}_0\| < \lambda$, *then* $\mathcal{U} = (0, \infty)$.

We do not give the proof here, even though it uses only elementary tools coupled with deep observations, since it is outside the scope of this book. The reader can find Leray's proof [213], with a modern explanation, in Galdi [121]. Essentially in the proof it is enough to show that the above properties are satisfied by strong solutions, since they become smooth, whenever they exist.

Theorem 2.27 states that the irregularity set is very small, and in particular that any weak solution, after a possible (transient) period of irregularity and nonuniqueness, becomes regular. In fact it is smooth for $t > T_R$. Furthermore, provided $\mathbf{u}_0 \in H^1_{0,\sigma}$ then $\mathbf{u}$ is smooth on a set that contains (but it is much bigger than) $(0, T_0) \cup (T_R, \infty)$, for a strictly positive T_0. The time T_0 can be estimated; on the contrary the proof of the existence of T_R is done by contradiction and so it does not give estimates for T_R.

We know now that the set of possible singularities is very small. We shall now give further results that can be obtained, squeezing out all the information from the energy inequality and (2.28).

Definition 2.28. *We say that a solution* $\mathbf{u}$ *becomes irregular at the time* T^* *if and only if*

(a) $T^* < \infty$;
(b) $\mathbf{u} \in C^\infty((s, T^*) \times \overline{\Omega})$, *for some* $s < T^*$;
(c) it is not possible to extend $\mathbf{u}$ *to a regular solution in any interval* (s, T^{**}), *with* $T^{**} > T^*$.

The number T^* is called the *epoch of irregularity* ("époque de irrégularité" in Leray [213]).

Theorem 2.29 (Leray [213], Scheffer [270]). *Let* $\mathbf{u}$ *be a weak solution and let* T^* *be an epoch of irregularity. Then the following properties hold:*

1. $\|\nabla \mathbf{u}(t)\| \to \infty$ *as* $t \to T^*$ *in such a way that,*

$$\exists C = C(\Omega) > 0 : \qquad \|\nabla \mathbf{u}(t)\| \leq \frac{C}{Re^{3/4}(T^* - t)}, \qquad \forall t < T^*;$$

2. *the 1/2-dimensional Hausdorff dimension of the set of (possible) epochs of irregularity is equal to zero.*

The above theorem gives an explicit lower bound on the growth of the $H^1_{0,\sigma}$ norm of the solution, near a singularity. Furthermore, it shows that the set of possible singularities lives in a small *fractal set*. We refer the reader to the cited references for the definition of Hausdorff measure, the proof of the theorem, and further comments.

The Leray Approach

We noted that the Galerkin procedure that we sketched is not the one used by Leray in the 1934 paper. In particular, Leray followed a completely different approach, that, to some extent, can be considered the first LES model. See the recent papers by Guermond, Oden, and Prudhomme [145] and by Cheskidov *et al.* [59].

First, we recommend any reader interested in the mathematics of NSE to read [213], since it can be considered one of the milestones in the history of mathematics, but at the same time is fully understandable.

The idea of Leray is to approximate the NSE with a family of linear transport problems and then to pass to the limit. Let us see with some details at least how the procedure starts.

To approximate nonsmooth functions with smooth ones there is a filtering technique that is very often used: convolution with smooth functions. Since in Chap. 3 we will use it extensively to derive LES models, we start with a definition and some preliminary results.

Proposition 2.30 (Basic property of convolution). *Let $f \in L^1(\mathbb{R}^d)$ and $g \in L^p(\mathbb{R}^d)$, for $1 \leq p \leq \infty$. Then, the convolution $f * g$*

$$(f * g)(\mathbf{x}) := \int_{\mathbb{R}^d} f(\mathbf{x} - \mathbf{y}) \, g(\mathbf{y}) \, d\mathbf{y},$$

is well-defined since almost everywhere (with respect to the Lebesgue measure in $\mathbb{R}^d$) the function $\mathbf{x} \mapsto f(\mathbf{x} - \mathbf{y}) \, g(\mathbf{y})$ belongs to $L^1(\mathbb{R}^d)$. Furthermore, the following estimates holds:

$$\|f * g\|_{L^p(\mathbb{R}^d)} \leq \|f\|_{L^1(\mathbb{R}^d)} \|g\|_{L^p(\mathbb{R}^d)}.$$

Definition 2.31. *[Friederichs mollifiers] A sequence of mollifiers $\{\rho_n\}_{n \geq 1}$ is any sequence of real functions defined on $\mathbb{R}^d$ such that:*

$$\rho_n \in C_0^\infty(\mathbb{R}^d), \qquad \mathrm{supp}\, \rho_n \subset \overline{B(0, 1/n)} := \overline{\{x \in \mathbb{R}^d : \ |\mathbf{x}| < 1/n\}},$$

$$\int_{\mathbb{R}^d} \rho_n(\mathbf{x}) \, d\mathbf{x} = 1, \quad and \quad \rho_n(\mathbf{x}) \geq 0 \quad \forall \, \mathbf{x} \in \mathbb{R}^d.$$

The classical example is obtained by starting with function

$$\rho(\mathbf{x}) = \begin{cases} \mathrm{e}^{1/(|\mathbf{x}|^2 - 1)} & \text{if } |\mathbf{x}| < 1 \\ 0 & \text{if } |\mathbf{x}| \geq 1, \end{cases}$$

and by defining

$$\rho_n(\mathbf{x}) := \frac{1}{\int_{\mathbb{R}^d} \rho(\mathbf{x}) \, d\mathbf{x}} \, n^d \rho(n\mathbf{x}) \qquad \forall \, \mathbb{N}.$$

We recall without proof the following results; see for instance Brezis [45], Chap. 4.

Proposition 2.32. *(i) Let $f \in C(\mathbb{R}^d)$. Then, $(\rho_n * f) \overset{n \to \infty}{\longrightarrow} f$, uniformly on compact subsets of $\mathbb{R}^d$;*
(ii) Let $f \in L^1_{loc}(\mathbb{R}^d)$. Then, $\rho_n * f \in C^\infty(\mathbb{R}^d)$ and

$$D^\alpha(\rho_n * f) = (D^\alpha \rho_n) * f,$$

where $\alpha = (\alpha_1, \ldots, \alpha_d)$ is a multi-index and

$$D^\alpha \phi = \frac{\partial^{\alpha_1} \ldots \partial^{\alpha_d} \phi}{\partial x_1^{\alpha_1} \ldots \partial x_1^{\alpha_d}} \qquad \forall \phi \in C^\infty(\mathbb{R}^d);$$

(iii) Let $f \in L^p(\mathbb{R}^d)$ for $1 \le p < \infty$. Then $(\rho_n * f) \overset{n \to \infty}{\longrightarrow} f$ in $L^p(\mathbb{R}^d)$.

The system studied by Leray to approximate the NSE is the following:

$$\mathbf{v}_t - \frac{1}{Re}\Delta\mathbf{v} + \mathbf{v}_n \cdot \nabla\mathbf{v} + \nabla p = \mathbf{f} \quad \text{in } \mathbb{R}^3 \times (0, T) \tag{2.32}$$

$$\nabla \cdot \mathbf{v} = 0 \quad \text{in } \mathbb{R}^3 \times (0, T) \tag{2.33}$$

$$\mathbf{v}_n = \rho_n * \mathbf{v} \quad \text{in } \mathbb{R}^3 \times (0, T). \tag{2.34}$$

The regularization consists in the fact that the transport is not realized by the velocity itself (as in the Euler equations and NSE) but by a spatial mean of the velocity on a region of diameter $2/n$ cfr. with the Leray α-model studied in [59, 145].

The existence theory for (2.32)–(2.34) is based on the fact that $\mathbf{v}_n$ is still a divergence-free vector (check it) and consequently

$$\int_{\mathbb{R}^d} \mathbf{v}_n \cdot \nabla\mathbf{v} \, \mathbf{v} \, d\mathbf{x} = 0.$$

In this way it is possible to obtain again, energy equality (for smooth solutions). The idea is then to prove the existence of smooth solutions, for arbitrary positive times, by keeping n fixed. Then, to use the energy equality to pass to the limit as $n \to \infty$ to show convergence (on some sequence) of the solutions $\mathbf{v}$ of the approximate problem, toward a solution of the NSE. The convergence

$$\mathbf{v} \to \mathbf{u}, \quad \text{as} \quad n \to \infty$$

takes place in weak spaces and the result is that the very smooth family $\mathbf{v}$ converges just to a weak solution $\mathbf{u}$, that satisfies the properties stated in Definition 2.11.

In [213] Leray proves existence, smoothness, and the *a priori* estimates for the solution of (2.32)–(2.34) (with fixed n) by using the technique of Green functions, more precisely, the fundamental solution of the heat equation in $\mathbb{R}^3$. The basic property of the smoothed transport theorem is that

$$\|\rho_n * \mathbf{v}\|_{L^\infty(\mathbb{R}^d)} \le \|\mathbf{v}\|_{L^\infty(\mathbb{R}^d)}, \tag{2.35}$$

and this allows one to obtain the desired smoothness of the solution. We refer the reader to Leray and also to Gallavotti [124] for other explanations.

Remark 2.33. The core of the proof of the result of Leray is the energy inequality, since it is the fundamental tool to pass to the limit as $n \to \infty$. Again, even with a completely different proof, we can see how this estimate is the "fundamental ingredient" in the mathematical theory of NSE.

Remark 2.34. Since uniqueness of weak solutions is still an open problem, the Hopf procedure and the Leray procedure may lead to different weak solutions!

2.5 Some Remarks on the Euler Equations

Together with the existence problems, there are several outstanding, open questions related to the mathematical theory of fluid mechanics. Among others, we may cite the problem of the long-time behavior of solutions, the stability questions (that are also connected with the numerical approximation), and the vanishing viscosity limits.

Regarding the latter point, the fundamental question is: "do the solutions to the NSE converge to those of the Euler equations as $Re \to \infty$?"

The Euler equations for incompressible ideal fluids can be written as

$$\mathbf{u}_t + \mathbf{u} \cdot \nabla \mathbf{u} + \nabla p = \mathbf{f} \quad \text{in } \Omega \times (0, T) \tag{2.36}$$
$$\nabla \cdot \mathbf{u} = 0 \quad \text{in } \Omega \times (0, T) \tag{2.37}$$
$$\mathbf{u}(\mathbf{x}, 0) = \mathbf{u}_0 \quad \text{in } \Omega. \tag{2.38}$$

Now the boundary conditions are not the same as for the NSE, since the problem involves only space derivatives of the first order. The natural condition is then

$$\mathbf{u} \cdot \mathbf{n} = 0 \quad \text{on } \partial \Omega \times [0, T]. \tag{2.39}$$

This fact is very important since this difference of boundary conditions, contributes to make the limit

$$\text{NSE} \quad \to \quad \text{Euler} \quad \text{as} \quad Re \to \infty$$

a *strongly singular limit.*

Concerning the mathematical theory of the Euler equations, the situation is very similar to that of the NSE. In fact, in the 2D case we know global existence and uniqueness of smooth solutions. In the 3D case, the one that is really interesting from the physical point of view, it is possible to prove just local existence and uniqueness of smooth solutions.

The mathematical theory of the Euler equations is more difficult (with respect to the NSE) because there is no smoothing term (the Laplacian) and the nature of the equation is *hyperbolic* instead of *parabolic*. We briefly sketch some results for the Euler equations and we refer to the bibliography for proofs and for more details.

Theorem 2.35. *Let Ω be a domain such that $\Omega = \mathbb{R}^3$ or Ω is smooth and bounded. Let $\mathbf{u}_0 \in [H^3(\Omega)]^3$ with $\nabla \cdot \mathbf{u}_0 = 0$ and $\mathbf{u}_0 \cdot \mathbf{n} = 0$ on $\partial\Omega$. Let also $\mathbf{f} \in L^1(0, T; [H^3(\Omega)]^3)$. Then, there exists a strictly positive $T_0 = T_0(\|\mathbf{u}_0\|_{H^3}, \|\mathbf{f}\|_{L^1(0,T;H^3)}) \leq T$ such that there exists a unique solution to the Euler equations (2.36)–(2.39) in the time interval $[0, T_0)$. This solution satisfies*

$$\mathbf{u}(\mathbf{x}, t) \in C(0, T_0; H^3) \cap C^1(0, T_0; H^2) \cap C^2(0, T_0; H^1) \cap C^3(0, T_0; L^2).$$

In the above theorem $H^k(\Omega)$ denotes the space of functions with distributional derivatives up to the k-order in L^2 (see page 10), while the symbol $C^k(0, T; X)$ denotes the space of C^k functions on $(0, T)$ with values in X.

Remark 2.36. Functions belonging to the space H^3 may be identified with smooth functions, say $C^{0,1/2}$-Hölder continuous functions, see Adams [4]. This shows that the above solutions of the Euler equations are indeed classical solutions.

To give the flavor of the proof, and to understand where the limitation of small times comes from, we write the *a priori* estimate that can be established for the Faedo–Galerkin approximate functions. In this case we need a different basis, since we have to deal with function $\mathcal{W}_k$ that are eigenfunctions of the Stokes operator, subject to the boundary condition $\mathcal{W}_k \cdot \mathbf{n} = 0$. By applying the differential operator D^α, for $|\alpha| \leq 3$, to (2.36) and by multiplying by $D^\alpha \mathbf{u}_m$ with suitable integration by parts it is possible to show that (see Temam [294], but proofs using other methods are known, see the references at the end of the section)

$$\frac{1}{2}\frac{d}{dt}\|\mathbf{u}_m\|_{H^3}^2 \leq C\left(\|\mathbf{u}_m\|_{H^3}^3 + \|\mathbf{f}\|_{H^3}\|\mathbf{u}_m\|_{H^3}\right). \tag{2.40}$$

Consequently, $\|\mathbf{u}_m\|_{H^3} \leq Y$, where $Y(t)$ satisfies the differential inequality

$$\begin{cases} Y'(t) = C\left[Y(t)^2 + \|\mathbf{f}\|_{H^3}\right] \\[2mm] Y(0) = Y_0 = \|\mathbf{u}_0\|_{H^3}, \end{cases}$$

whose life-span may be bounded from below by an expression depending on C, $\|\mathbf{u}_0\|_{H^3}$, and $\|\mathbf{f}\|_{L^1(0,T;H^3)}$.

In the case $\Omega = \mathbb{R}^3$ it is also possible to prove that in the time interval[6] $(0, T_0)$ the unique smooth solution of the NSE, $\mathbf{u}^{Re}$, converges to those of the Euler equation $\mathbf{u}^\infty$ (if the initial data and the external force are the same) in such a way that

---

[6] Note that in this time interval unique smooth solutions for both the NSE and the Euler equations do exist. Furthermore, the time T_0 is independent of Re. In this way we have a common time-interval in which we can study both problems, see Kato [180].

$$\left.\begin{array}{ll} \mathbf{u}^{Re} \to \mathbf{u}^\infty & \text{in} \quad H^2 \\ \mathbf{u}^{Re} \rightharpoonup \mathbf{u}^\infty & \text{in} \quad H^3 \end{array}\right\} \quad \text{uniformly in } t, \text{ as} \quad Re \to \infty.$$

In the presence of boundaries the situation is much more complicated, due also to the fact that we do not know the existence of reasonably weak solutions to the Euler equations. In the case of weak solutions $\mathbf{u}^{Re}$ to the NSE Kato [181] proved that

$$\mathbf{u}^{Re} \to \mathbf{u}^\infty \quad \text{in} \quad L^2(\Omega), \text{ uniformly in } t \in [0, T'], \quad \text{as} \quad Re \to \infty,$$

if and only if

$$\frac{1}{Re} \int_0^{T'} \|\nabla \mathbf{u}^{Re}(\tau)\|_{L^2(\Omega^{Re})} \, d\tau \to 0, \quad \text{as} \quad Re \to \infty,$$

where Ω^{Re} is a boundary strip of width $1/Re$.

In the case of 2D fluids Theorem 2.35 may be improved to show that there is no restriction on the life span of smooth solutions. The fact that given $\mathbf{u}_0 \in [H^3(\Omega)]^2$ then there exists a unique smooth solution for all positive times comes from an accurate study of the equation of the vorticity. In fact in the 2D case, if we take the curl of the equation (2.36) we may derive the equation satisfied by the scalar $\omega = \nabla \times \mathbf{u} := \partial_1 u^2 - \partial_2 u^1$:

$$\omega_t + \mathbf{u} \cdot \nabla\omega = \nabla \times \mathbf{f} \qquad \text{in } \mathbb{R}^2 \times (0, T).$$

In this case (suppose that $\mathbf{f}$ vanishes for simplicity) the vorticity is simply transported by the flow. So if the vorticity is bounded at time $t = 0$, then it follows the following estimate

$$\|\omega(t)\|_{L^\infty} \leq \|\omega_0\|_{L^\infty} \qquad \forall \, t \geq 0.$$

This fact, together with the Biot–Savart law that allows one to write the velocity in terms of the vorticity, is the main tool used to construct global in time smooth solutions for the 2D Euler equations.

This argument fails in the three-dimensional case, since the dynamical equation for the vector $\nabla \times \mathbf{u} = \boldsymbol{\omega}$ is now

$$\boldsymbol{\omega}_t + \mathbf{u} \cdot \nabla\boldsymbol{\omega} = \boldsymbol{\omega} \cdot \nabla\mathbf{u} + \nabla \times \mathbf{f} \qquad \text{in } \mathbb{R}^3 \times (0, T).$$

The term $\boldsymbol{\omega} \cdot \nabla\mathbf{u}$ on the right-hand side is responsible for an increase of vorticity and also for changes of its direction: vorticity is no longer simply transported by the velocity field. This causes the lack of global estimates needed to prove existence of smooth solutions for all positive times! The results cited in this section have been proved by, among others, Lichtenstein [218], Wolibner [317], Yudovich [318], Kato [179], Ebin and Marsden [103], and Bourguignon and Brezis [42]. See also the review in Marchioro and Pulvirenti [230].

2.6 The Stochastic Navier–Stokes Equations

Among other mathematical methods used to describe the chaotic behavior of turbulent fluids there is also the stochastic approach. In this section we briefly introduce the stochastic Navier–Stokes equations (SNSE):

$$\mathbf{u}_t + (\mathbf{u} \cdot \nabla)\,\mathbf{u} - \frac{1}{Re}\Delta\mathbf{u} + \nabla p = \mathbf{f} + \mathbf{G}_t \quad \text{in } \Omega \times (0,T) \qquad (2.41)$$

$$\nabla \cdot \mathbf{u} = 0 \quad \text{in } \Omega \times [0,T] \qquad (2.42)$$

$$\mathbf{u} = 0 \quad \text{on } \partial\Omega \times (0,T) \qquad (2.43)$$

$$\mathbf{u}(\mathbf{x},0) = \mathbf{u}_0(\mathbf{x}) \quad \text{in } \Omega. \qquad (2.44)$$

The body forces are split into two terms: $\mathbf{f}$ is a classical term, and may represent a slowly (differentiable) varying force, while $\mathbf{G}_t$ correspond to fast fluctuations of the force. It is possible to make different assumptions to describe rapid fluctuations. We assume that $\mathbf{G}$ is continuous, but not differentiable. Another possible choice is to take generalized stochastic processes, but we shall not enter into details; overview on stochastic partial differential equations can be found in Da Prato and Zabczyk [81, 82].

The introduction of the SNSE is reasonable since the nonlinear nature of the equation leads naturally to the study of chaotic dynamical systems (see Wiggins [314]). A heuristic justification of the study of SNSE can be the following, see Chorin [63]:

> ... we shall now consider random fields $\mathbf{u}(\mathbf{x},\omega)$ which, for each ω (i.e., for each experiment that produces them), satisfy the NSE. $\mathbf{u}$ depends also on the time t; we shall usually not exhibit this dependence explicitly.
>
> There is an interesting question of principle that must be briefly discussed: why does it make sense to view solutions of the deterministic NSE as being random? It is an experimental fact that the flow one obtains in the laboratory at a given time is a function of the experiment. The reason must be that the flow described by the NSE for large Re is chaotic; microscopic perturbations, even at a molecular scale, are amplified to macroscopic scales; no two experiments are truly identical and what one gets is a function of the experiment. The applicability of our constructions is plausible even if we do not know how to formalize the underlying probability space.

Another justification is given by Barenblatt [15] by considering the solution of the NSE at high Reynolds number as a realization of a *turbulent* flow:

> ...the flow properties for supercritical values of the Reynolds number undergo sharp and disorderly variations in space and in time, and the fields of flow properties, – pressure, velocity etc. – can to a good approximation be considered random. Such a regime of flow is called turbulent...

and extensive overview on the statistical study of NSE can be found in Monin and Yaglom [241] and in Višik and Fursikov [305].

We show how it is possible to define the concept of weak solutions for the SNSE, together with an existence proof.

We assume that $\mathbf{u}_0 \in L^2_\sigma$ and that $\mathbf{f} \in L^2(0, T; (H^1_{0,\sigma})')$. Furthermore, we assume that

$$\mathbf{G} \in C([0, T]; H^1_{0,\sigma}) \quad \text{and} \quad \mathbf{G}(0) = \mathbf{0}.$$

The equation (2.41) can have meaning only in an integral sense. To construct a weak solution, we project the SNSE onto the space spanned by the first m eigenvectors of the Stokes operator and we consider the following integral system in $V_m := P_m(L^2_\sigma)$:

$$\mathbf{u}_m(t) - \frac{1}{Re} \int_0^t P_m \Delta \mathbf{u}_m(s)\, ds + \int_0^t P_m(\mathbf{u}_m(s) \cdot \nabla \mathbf{u}_m(s))\, ds = P_m\, \mathbf{u}_0$$

$$+ \int_0^t P_m\, \mathbf{f}(s)\, ds + P_m\, \mathbf{G}(t), \quad t \geq 0,$$

which has a unique maximal solution $\mathbf{u}_m \in C(0, T; V_m)$. Next, we define $\mathbf{v}_m := \mathbf{u}_m - P_m\, \mathbf{G} \in C(0, T; V_m)$, that satisfies

$$\mathbf{v}_m(t) - \int_0^t P_m \Delta \mathbf{v}_m(s)\, ds$$

$$+ \frac{1}{Re} \int_0^t P_m\left[(\mathbf{v}_m(s) + P_m\, \mathbf{G}(s)) \cdot \nabla(\mathbf{v}_m(s) + P_m\, \mathbf{G}(s)) \right] ds$$

$$= P_m(\mathbf{u}_0 - \mathbf{G}(0)) + \int_0^t P_m\, \mathbf{f}(s)\, ds - \frac{1}{Re} \int_0^t P_m \Delta \mathbf{G}(s)\, ds.$$

We can use the "energy method" (multiply by $\mathbf{u}_m$ and perform suitable integration by parts) to obtain

$$\frac{1}{2}\frac{d}{dt}\|\mathbf{v}_m\|^2 + \frac{1}{Re}\|\nabla \mathbf{v}_m\|^2 \leq \left| \int_\Omega (\mathbf{v}_m + P_m\, \mathbf{G}) \cdot \nabla P_m\mathbf{G}\ \ \mathbf{v}_m\, dx \right|$$

$$+ \|\nabla \mathbf{v}_m\|\|\mathbf{f}\|_{(H^1_{0,\sigma})'} + \frac{1}{Re}\|\nabla \mathbf{v}_m\|\,\|\nabla \mathbf{G}\|.$$

By using the usual Hölder inequality (2.26) and Proposition 2.16 one can show that

$$\frac{1}{2}\frac{d}{dt}\|\mathbf{v}_m\|^2 + \frac{1}{2Re}\|\nabla \mathbf{v}_m\|^2 \leq C \Bigg(\|\mathbf{v}_m\|^2 \|P_m\, \mathbf{G}\|^8_{L^4(D)}$$

$$+ \|P_m\, \mathbf{G}\|^4_{L^4(D)} + \|\nabla \mathbf{G}\|^2 + 2\|\mathbf{f}\|_{(H^1_{0,\sigma})'} \Bigg).$$

From the last equation, if some estimate on $P_m \mathbf{G}$ is given, we can extract (as in the deterministic case) subsequences $\mathbf{v}_{m_k}$ that converge to some $\mathbf{v}$, which satisfies $\forall\, t \geq t_0 \geq 0$ and $\forall\, \phi \in H_{0,\sigma}^1$:

$$\langle \mathbf{v}(t) - \mathbf{v}(t_0), \phi \rangle + \int_{t_0}^t \langle \nabla \mathbf{v}(s), \nabla \phi \rangle \, ds$$

$$+ \int_{t_0}^t \langle (\mathbf{v}(s) + \mathbf{G}(s)) \cdot \nabla (\mathbf{v}(s) + \mathbf{G}(s)), \phi \rangle \, ds$$

$$= \int_{t_0}^t \langle \mathbf{f}(s), \phi \rangle \, ds + \int_{t_0}^t \langle \nabla \mathbf{G}(s), \nabla \phi \rangle \, ds,$$

where $\langle \,.\,,\,.\, \rangle$ denotes the duality paring between $H_{0,\sigma}^1$ and its dual space. Now by recalling that, for $m \in \mathbb{N}$, we defined $\mathbf{v}_m := \mathbf{u}_m - P_m \mathbf{G}$, we can state the following theorem, with $\mathbf{u} := \mathbf{v} + \mathbf{G}$.

Theorem 2.37. *Let* $\mathbf{f} \in L^2(0,T;(H_{0,\sigma}^1)')$, $\mathbf{G} \in C([0,T];H_{0,\sigma}^1)$, *and* $\mathbf{u}_0 \in L_\sigma^2$. *Then, there exists a weak solution to the SNSE* (2.41), *i.e. a function* $\mathbf{u}$ *belonging to* $L^\infty(0,T;L_\sigma^2) \cap L^2(0,T;H_{0,\sigma}^1)$, *which satisfies the regularity property*

$$\text{if } d = 3 \text{ then} \qquad \frac{d}{dt}(\mathbf{u} - \mathbf{G}) \in L^{4/3}(0,T;(H_{0,\sigma}^1)')$$

$$\text{if } d = 2 \text{ then} \qquad \frac{d}{dt}(\mathbf{u} - \mathbf{G}) \in L^2(0,T;(H_{0,\sigma}^1)')$$

and such that
(1): $\forall\, t \geq t_0 \geq 0$ *and* $\forall\, \phi \in H_{0,\sigma}^1$

$$\langle \mathbf{u}(t) - \mathbf{u}(t_0), \phi \rangle + \int_{t_0}^t \langle \nabla \mathbf{u}(s), \nabla \phi \rangle \, ds + \int_{t_0}^t \langle \mathbf{u}(s) \cdot \nabla \mathbf{u}(s)), \phi \rangle \, ds$$

$$= \langle \mathbf{G}(t) - \mathbf{G}(t_0), \phi \rangle + \int_{t_0}^t \langle \mathbf{f}(s), \phi \rangle \, ds;$$

(2) for almost all t and t_0, with $t \geq t_0 \geq 0$ it holds

$$\|\mathbf{u}(t) - \mathbf{G}(t)\|^2 \leq e^{\int_{t_0}^t (-\lambda_1 + C \|\mathbf{G}(s)\|_{L^4}^8) \, ds} \|\mathbf{u}(t_0) - \mathbf{G}(t_0)\|^2$$

$$+ \int_{t_0}^t e^{\int_\sigma^t (-\lambda_1 + C \|\mathbf{G}(s)\|_{L^4}^8) \, ds}$$

$$\times C \left[\|\mathbf{G}(\sigma)\|_{L^4}^4 + \|\Delta \mathbf{G}(\sigma)\|_{(H_{0,\sigma}^1)'}^2 + \|\mathbf{f}(\sigma)\|_{(H_{0,\sigma}^1)'}^2 \right] d\sigma;$$

(3) for almost all t and t_0, with $t \geq t_0 \geq 0$ it holds

$$\|\mathbf{u}(t) - \mathbf{G}(t)\|^2 + \int_{t_0}^{t} \|\mathbf{u}(s) - \mathbf{G}(s)\|^2 \, ds \leq \|\mathbf{u}(t_0) - \mathbf{G}(t_0)\|^2$$

$$+ C \int_{t_0}^{t} \left[\|\mathbf{u}(\sigma) - \mathbf{G}(\sigma)\|^2 \|\mathbf{G}(\sigma)\|_{L^4}^8 + 4\|\mathbf{G}(\sigma)\|_{L^4}^4 \right.$$

$$\left. + 4\|\Delta\mathbf{G}(\sigma)\|_{(H_{0,\sigma}^1)'}^2 + 4\|\mathbf{f}(\sigma)\|_{(H_{0,\sigma}^1)'}^2 \right] d\sigma.$$

This is the first result in the study of the SNSE, see Bensoussan and Temam [24]. This shows how it is possible to make sense of the NSE with a non-smooth forcing term. The further, very technical step is to study the probabilistic properties of the solution that may considered as

$$\mathbf{u} = \mathbf{u}(\mathbf{x}, t, \mathbf{G})$$

with the last argument being a random variable. This can be considered as a starting point in the program explained in [63]. Further results can be found in the references cited in this section. Furthermore, in the 2D case it is possible to prove uniqueness of this class of solutions, the existence of suitable *random attractors*, and that an ergodic theorem holds.

2.7 Conclusions

We started this chapter by presenting a (nonrigorous) derivation of the equations for fluid flows and by showing some connections with the K41 theory. Then, we summarized the main available mathematical results regarding existence, uniqueness, and regularity for solutions of the equations for viscous and ideal fluids. We also briefly recalled some results on the SNSE that the reader will find helpful in connection with turbulence modeling.

The results in this chapter should be useful in understanding what could be reasonable to try proving, what can be done rigorously, and where the mathematical theory reaches its limits. We also hope that this chapter will give the LES practitioners at least a flavor of the mathematical analysis of fluid flows.

Part II

Eddy Viscosity Models

3

Introduction to Eddy Viscosity Models

3.1 Introduction

Since the basic problem in LES is to predict $\overline{\mathbf{u}}$, in cases where predicting $\mathbf{u}$ accurately is not possible, it is natural to begin by deriving equations for $\overline{\mathbf{u}}$. The problem of filtering on a bounded domain is very important, but we will postpone it until Chap. 9. Thus, we begin with the NSE without boundaries, *i.e.* either the Cauchy problem or (our choice) with periodic boundary conditions defined by (2.3). For $d = 2$ or 3 and $\Omega = (0, L)^d$, we seek a velocity $\mathbf{u} : \Omega \times [0, T] \to \mathbb{R}^d$ and a pressure $p : \Omega \times (0, T] \to \mathbb{R}$ satisfying

$$\mathbf{u}_t + \nabla \cdot (\mathbf{u}\, \mathbf{u}^T) - \frac{1}{Re} \Delta \mathbf{u} + \nabla p = \mathbf{f} \quad \text{in } \Omega \times (0, T), \tag{3.1}$$

$$\nabla \cdot \mathbf{u} = 0 \quad \text{in } \Omega \times (0, T), \tag{3.2}$$

subject to the initial conditions $\mathbf{u}(\mathbf{x}, 0) = \mathbf{u}_0(\mathbf{x})$.

Note that we have written the nonlinear term in a way that is different from the standard one we used in Chap. 2. We use the above expression for the NSE since it will turn that this formulation is more useful in the study of LES models. The two formulations are clearly equivalent in the case of divergence-free functions, since we have the following equality:

$$[\nabla \cdot (\mathbf{u}\, \mathbf{u}^T)]_i := \sum_{j=1}^{d} \frac{\partial u_i u_j}{\partial x_j} = \sum_{j=1}^{d} \frac{\partial u_i}{\partial x_j} u_j = [\mathbf{u} \cdot \nabla \mathbf{u}]_i \quad \text{for } i = 1, \ldots, d.$$

To derive the space-filtered NSE, we convolve the NSE with the chosen filter function $g_\delta(\mathbf{x})$ (operationally, consider the NSE as a function of $\mathbf{x}' \in \Omega$, multiply (3.1) by $g_\delta(\mathbf{x} - \mathbf{x}')$ and then integrate over Ω with respect to $\mathbf{x}'$, recall Sect. 1.2). Using the fact that (for constant $\delta > 0$ and in the absence of boundaries) filtering commutes with differentiation, gives the system, often called the space-filtered Navier–Stokes equations (SFNSE):

$$\overline{\mathbf{u}}_t + \nabla \cdot (\overline{\mathbf{u}\,\mathbf{u}^T}) - \frac{1}{Re}\Delta\overline{\mathbf{u}} + \nabla\overline{p} = \overline{\mathbf{f}} \quad \text{in } \Omega \times (0,T), \qquad (3.3)$$

$$\nabla \cdot \overline{\mathbf{u}} = 0 \quad \text{in } \Omega \times (0,T). \qquad (3.4)$$

This system is not closed, since it involves both $\mathbf{u}$ and $\overline{\mathbf{u}}$; it is usual to rewrite it in a way that focuses attention on the closure problem. Define the tensor $\boldsymbol{\tau} = \boldsymbol{\tau}(\mathbf{u}, \mathbf{u})$ by

$$\boldsymbol{\tau}(\mathbf{u}, \mathbf{u}) = \overline{\mathbf{u}\,\mathbf{u}^T} - \overline{\mathbf{u}}\,\overline{\mathbf{u}}^T \quad \text{or} \quad \tau_{ij}(\mathbf{u}, \mathbf{u}) = \overline{\mathbf{u}_i\,\mathbf{u}_j} - \overline{\mathbf{u}}_i\,\overline{\mathbf{u}}_j. \qquad (3.5)$$

This tensor $\boldsymbol{\tau}(\mathbf{u}, \mathbf{u})$ is often called the subgrid-scale stress tensor, subfilter-scale stress tensor, or the Reynolds stress tensor. There is disagreement about the latter so it is safest to call it the subgrid-scale or subfilter-scale stress tensor. In the sequel, we will use the latter. Following Leonard [212], terms in the subfilter-scale tensor are generally grouped in the so called *triple decomposition* (see Chap. 3 in [267]) in which there is the cross-stress tensor $\mathbf{C}$, the Leonard stress tensor $\mathbf{L}$, and the proper Reynolds stress tensor $\mathbf{R}$ defined respectively by

$$\mathbf{C} := \overline{\overline{\mathbf{u}}(\mathbf{u} - \overline{\mathbf{u}})^T} + \overline{(\mathbf{u} - \overline{\mathbf{u}})\overline{\mathbf{u}}^T}$$

$$\mathbf{L} := \overline{\overline{\mathbf{u}}\,\overline{\mathbf{u}}^T} - \overline{\mathbf{u}}\,\overline{\mathbf{u}}^T$$

$$\mathbf{R} := \overline{(\mathbf{u} - \overline{\mathbf{u}})(\mathbf{u} - \overline{\mathbf{u}})^T}.$$

Then, (3.3) and (3.4) can be rewritten as

$$\overline{\mathbf{u}}_t + \nabla \cdot (\overline{\mathbf{u}}\,\overline{\mathbf{u}}^T) - \frac{1}{Re}\Delta\overline{\mathbf{u}} + \nabla \cdot \boldsymbol{\tau}(\mathbf{u}, \mathbf{u}) + \nabla\overline{p} = \overline{\mathbf{f}} \quad \text{in } \Omega \times (0,T), \quad (3.6)$$

$$\nabla \cdot \overline{\mathbf{u}} = 0 \quad \text{in } \Omega \times (0,T), \quad (3.7)$$

and the problem is now to write the subfilter-scale stress tensor $\boldsymbol{\tau}$ in terms of filtered variables.

Definition 3.1. *The <u>interior closure</u> problem in LES is to specify a tensor $\mathcal{S} = \mathcal{S}(\overline{\mathbf{u}}, \overline{\mathbf{u}})$ to replace $\boldsymbol{\tau}(\mathbf{u}, \mathbf{u})$ in equation (3.6).*

There are many proposals for "solving" the closure problem. The workhorse of LES is still, however, the eddy viscosity (EV) model. EV models are motivated by the idea that the global effect of the subfilter-scale stress tensor $\boldsymbol{\tau}(\mathbf{u}, \mathbf{u})$, in the mean, is to transfer energy from resolved to unresolved scales through inertial interactions. With this phenomenology in mind, we now consider eddy viscosity models in LES.

3.2 Eddy Viscosity Models

The first closure problem of LES is thus to find a tensor $\mathcal{S}(\overline{\mathbf{u}}, \overline{\mathbf{u}})$ approximating $\boldsymbol{\tau}(\mathbf{u}, \mathbf{u})$ or at least approximating its effects in the SFNSE. To do this, it is useful to have some understanding of the effects of those turbulent fluctuations.

EV models are motivated by the following observed experimental behavior (paraphrased from Frisch [117] who cites it as one of the two experimental laws of turbulence):

> *Suppose, in an experiment, all control parameters are kept fixed except the viscosity is reduced as far as possible and the energy dissipation is measured (typically by measuring drag). While the flow is laminar, then energy dissipation is reduced proportional to the reduction in ν. When the flow is turbulent, the energy dissipation does not vanish as $\nu \to 0$ but approaches a finite, positive limit.*

This experimental law is part of Kolmogorov's (K-41) theory, whose essential aspects were presented in Sect. 2.4.3. We will not present here the details of Kolmogorov's theory (the interested reader is referred to the exquisite presentations in Frisch [117] – Chaps. 6 and 7, Pope [258] – Chap. 6, and Sagaut [267]). Instead, we will briefly sketch the idea of *energy cascade*, introduced by Richardson in 1922 [263].

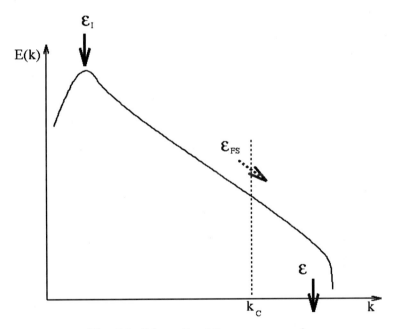

Fig. 3.1. Schematic of the energy cascade

The essence of the energy cascade is that kinetic energy enters the turbulent flow at the largest scales of motion, and is then transferred (by inviscid processes) to smaller and smaller scales, until is eventually dissipated through viscous effects. A schematic of the energy cascade is presented in Fig. 3.1. As explained in Sect. 2.4.3, the energy cascade has a suggestive illustration in the wavenumber space. The quantity plotted in Fig. 3.1 is the *energy spectrum* $E(k)$, from which one can obtain the energy contained in the wavenumber range (k_1, k_2) through

$$k_{(k_1, k_2)} = \int_{k_1}^{k_2} E(k)dk. \tag{3.8}$$

For more details on the energy spectrum $E(k)$ and its rigorous mathematical definition, the reader is referred to the thorough presentations in Pope [258] (Sects 3.7, 6.1, and 6.5) and Frisch [117] (Sect. 4.5).

Figure 3.1 contains the log–log plot of the energy spectrum $E(k)$ against the wavenumber k, and illustrates the energy cascade: ε_I represents the energy that enters the flow at the largest scales (smallest wavenumbers), ε_{FS} represents the energy transferred to smaller and smaller scales (larger and larger wavenumbers), and ε is the energy eventually dissipated through viscous effects at the smallest scales (largest wavenumbers).

Thus, the action of the subfilter-scale stress τ is thought of as having a dissipative effect on the mean flow: the action of the scales uncaptured on the numerical mesh (above the cut-off wavenumber k_c) on the large scales (below the cut-off wavenumber k_c) should replicate the effect of ε_{FS}, the so-called *forward-scatter*.

Boussinesq Hypothesis

In 1877, Boussinesq [43] first formulated the EV/Boussinesq hypothesis based upon an analogy between the interaction of small eddies and the perfectly elastic collision of molecules (*e.g.* molecular viscosity or heat), stating:

"Turbulent fluctuations are dissipative in the mean."

The mathematical realization is the model

$$\nabla \cdot \boldsymbol{\tau}(\overline{\mathbf{u}}, \overline{\mathbf{u}}) \approx -\nabla \cdot (\nu_T \nabla^s \overline{\mathbf{u}}) + \text{ terms incorporated into } \overline{p},$$

where $\nu_T \geq 0$ is the "turbulent viscosity coefficient."
This yields the simple model for the divergence-free $\mathbf{w} \cong \overline{\mathbf{u}}$:

$$\mathbf{w}_t + \nabla \cdot (\mathbf{w}\,\mathbf{w}^T) - \nabla \cdot \left(\left[\frac{2}{Re} + \nu_T \right] \nabla^s \mathbf{w} \right) + \nabla \overline{q} = \overline{\mathbf{f}} \text{ in } \Omega \times (0, T), \tag{3.9}$$

$$\nabla \cdot \mathbf{w} = 0 \text{ in } \Omega \times (0, T). \tag{3.10}$$

The modeling problem then reduces to determining one parameter: the turbulent viscosity coefficient ν_T:

Closure Problem: Find $\nu_T = \nu_T(\overline{\mathbf{u}}, \delta)$.

EV models are very appealing, since the global energy balance is very simple and clear.

Proposition 3.2. *Let* $(\mathbf{w}, q)$ *be a classical solution to* (3.9) *and* (3.10) *subject to either periodic or no-slip boundary conditions. Let* $\nu_T = \nu_T(\mathbf{w}, \delta) > 0$. *Then,*

$$k(t) + |\Omega| \int_0^t \epsilon_{\text{model}}(\tau) \, d\tau = k(0) + \int_0^t P(\tau) \, d\tau,$$

where $k(t) = \dfrac{1}{2} \displaystyle\int_\Omega |\mathbf{w}|^2(t) \, d\mathbf{x}, \quad P(t) := \displaystyle\int_\Omega \overline{\mathbf{f}} \cdot \mathbf{w} \, d\mathbf{x}$ *and*

$$\epsilon_{\text{model}} = \frac{1}{|\Omega|} \int_\Omega \left[\frac{2}{Re} + \nu_T(\mathbf{w}, \delta) \right] \nabla^s \mathbf{w} : \nabla^s \mathbf{w} \, d\mathbf{x}.$$

Proof. This property is obtained simply by multiplying the equations by $\mathbf{w}$ and integrating by parts over Ω. $\square$

The most common EV model is known in LES as the Smagorinsky model in which

$$\nu_T = \nu_{\text{Smag}}(\mathbf{w}, \delta) := (C_S \delta)^2 |\nabla^s \mathbf{w}|.$$

The term $-\nabla \cdot ((C_S \delta)^2 |\nabla^s \mathbf{w}| \nabla^s \mathbf{w})$ was studied in 1950 by von Neumann and Richtmyer [306] as a nonlinear artificial viscosity in gas dynamics and by Smagorinsky [277] in 1963 for geophysical flow calculations. A complete mathematical theory for PDEs involving this term was constructed around 1964 by Ladyžhenskaya (see [195, 196]), who considered that term as a correction term for the linear stress–strain relation, for flows with larger stresses. For further mathematical and numerical development of the model we refer to the work of Du and Gunzburger [94, 95], Parés [249, 250], Layton [201], and John and Layton [177].

The modeling difficulty now shifts to determining the non negative constant C_S. The first major result in LES is due to Lilly [219], who showed (under a number of optimistic assumptions) that C_S has a simple, universal value 0.17 and is not a "tuning" constant.

Lilly's Estimation of C_S

The idea of Lilly is to equate $\langle \epsilon \rangle = \langle \epsilon_{\text{model}} \rangle$ and from this to determine a value for C_S. This approach is very natural: if the model is to give the correct statistics, according to the K-41 theory, it must exactly replicate $\langle \epsilon \rangle$. To explain this idea, we follow closely the presentation of Hughes, Mazzei, and Jansen [160].

Ignoring the viscous dissipation in ϵ_{model} (and suppressing the time averaging of each term in each step) we can approximate

$$\epsilon_{\text{model}} \cong \int_\Omega \nu_{\text{Smag}}(\mathbf{w}, \delta) |\nabla^s \mathbf{w}|^2 \, d\mathbf{x}$$

$$= \int_\Omega (C_S \delta)^2 |\nabla^s \mathbf{w}|^3 \, d\mathbf{x} = (C_S \delta)^2 \|\nabla^s \mathbf{w}\|_{L^3}^3.$$

If we assume that the time average of $\mathbf{w}$ is exactly the same as that of $\mathbf{u}$, restricted to the frequencies $0 \leq |\mathbf{k}| \leq k_c := \pi/\delta$, then Plancherel's theorem and the Kolmogorov relation $E(k) \cong \alpha\,\epsilon^{2/3}k^{-5/3}$ give (in the time averaging sense)

$$\|\nabla^s \mathbf{w}\|^2 \cong 2 \int_0^{k_c} k^2 E(k)\, dk \cong 2 \int_0^{k_c} k^2 (\alpha\epsilon^{2/3}k^{-5/3})\, dk = \frac{3}{2}\,\alpha\epsilon^{2/3}k_c^{4/3}.$$

If we assume that for homogeneous, isotropic turbulence, after time averaging, $\|\nabla^s \mathbf{w}\|_{L^3}^3 \cong \|\nabla^s \mathbf{w}\|^3$, we can write

$$\|\nabla^s \mathbf{w}\|_{L^3}^3 \cong \left(\frac{3\alpha}{2}\right)^{3/2} \epsilon\, k_c^2,$$

where α is the Kolmogorov constant. Then, we have the following expression for ϵ_{model}:

$$\epsilon_{\text{model}} \cong (C_S\delta) \left(\frac{3\alpha}{2}\right)^{3/2} \epsilon\, k_c^2.$$

Since $k_c = \pi/\delta$, we finally[1] have

$$\epsilon_{\text{model}} \cong C_S^2 \pi^2 \left(\frac{3}{2}\right)^{3/2} \alpha^{3/2}\epsilon.$$

Equating $\epsilon = \epsilon_{\text{model}}$, the dependence on ϵ in the equation cancels out giving

$$C_S = \frac{1}{\pi}\left(\frac{4}{3}\right)^{3/4} \alpha^{-3/4} \cong 0.17, \quad \text{for} \quad \alpha \cong 1.6.$$

The universal value 0.17, independent of the particular flow, is obtained. This is often expressed as

"Smagorinsky is consonant with Kolmogorov."

Interestingly, this universal value $C_S = 0.17$ has almost universally (in numerical experiments) been found to be too large. There have been many other criticisms of the Smagorinsky model associated with it being too dissipative.

---

[1] The classical estimate of α is $\alpha = 1.4$. More recent studies suggest α should be a bit larger, around 1.6. Part of this variation might be due to normal experimental errors and part might be because the K41 theory is an asymptotic theory at very high Reynolds numbers, while experiments and calculations occur at high but finite Re. In addition, the value $C_S = 0.17$ is too large for almost all shear flows. The reason is that the mean shear, which is not taken into account in the local isotropy hypothesis that leads to the 0.17 value will be accounted for in the evaluation of $\nabla^s \overline{\mathbf{u}}$. Since the subgrid dissipation associated with the Smagorinsky model is $(C_S\delta^2)\|\nabla^s \overline{\mathbf{u}}\|_{L^3}^3$, the overestimation in the resolved gradient must be balanced by a decrease in the constant.

Rather than summarizing them here, we present in Fig. 3.2 two simulations from Sahin [269] of a $2D$ flow over an obstacle: one a DNS and the other with the Smagorinsky model. In both simulations, slip-with-friction boundary conditions were used, see p. 259. It is clear from these pictures that the dissipation in this model is too powerful.

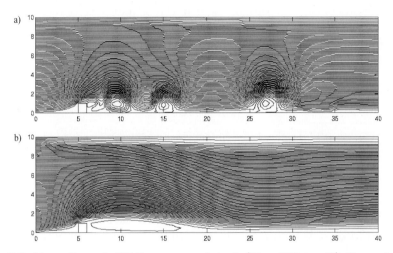

Fig. 3.2. Streamlines of a $2D$ flow over an obstacle ($Re = 700, t = 40$). True solution (*top*) and Smagorinsky model (3.11) (*bottom*)

3.3 Variations on the Smagorinsky Model

The Smagorinsky Model

$$\mathbf{w}_t + \nabla \cdot (\mathbf{w}\,\mathbf{w}^T) + \nabla q - \nabla \cdot \left(\frac{2}{Re} \nabla^s \mathbf{w} + (C_S \delta)^2 |\nabla^s \mathbf{w}| \nabla^s \mathbf{w} \right) = \overline{\mathbf{f}} \quad (3.11)$$

$$\nabla \cdot \mathbf{w} = 0, \quad (3.12)$$

where $C_S \approx 0.17$ seems to be a universal answer in LES. It is very easy to implement, very stable, and (under "optimistic" assumptions) it well replicates energy dissipation rates, according to the analysis of Lilly [219] reviewed in the previous section. Unfortunately, it is also quite inaccurate for many problems. Thus, there has been a lot of work testing modifications of (3.11) which are easy to implement and more accurate. Usually "more accurate" means the modifications are made to try to limit excessive amounts of extra dissipation in (3.11). Thus, variations of the Smagorinsky model have been derived not with the idea of increasing the accuracy of the approximation of the subfilter-scale stress tensor (3.5), but rather of ameliorating the overly diffused predictions of the model.

3.3.1 Van Driest Damping

Near walls, boundary layers in $\mathbf{w}$ introduce large amounts of dissipation in (3.11). This extra dissipation prevents the formation of eddies and can eliminate any turbulence from beginning. The idea of van Driest [302] was to reduce the Smagorinsky constant C_S to 0 as the boundary is approached such that averages of the flow variables satisfy the boundary layer theory (a logarithmic law-of-the-wall). For more details on the physical insight behind the van Driest damping, the reader is referred to the presentations in Pope [258] and Sagaut [267]. The van Driest scaling reads

$$C_S = C_S(\mathbf{y}) = \left[C_S \, \delta \left(1 - e^{-y^+/A} \right) \right]^2, \tag{3.13}$$

where $C_S = 0.17$ is the Lilly–Smagorinsky constant and y^+ is the non-dimensional distance from the wall

$$y^+ = \frac{u_\tau (H - |y|)}{\nu}, \tag{3.14}$$

which determines the relative importance of viscous and turbulent phenomena. In (3.14), H is the channel half-width, u_τ is the wall shear velocity (for more details on the derivation and interpretation of u_τ, the reader is referred to Sect. 12.2.2), and $A = 25$ is the van Driest constant.

The van Driest scaling (3.13) improves the performance of the model in predicting statistics of turbulent flow in simple geometries, where Boundary Layer theory holds (*e.g.* flow past a flat plate and pipe flow). Numerical simulations for the Smagorinsky model (3.11) equipped with the van Driest scaling (3.13) in turbulent channel flow simulations are presented in Sect. 12.2. We will just mention here that without a scaling of the form $C_S(\mathbf{x}) \to 0$ (such as the van Driest damping (3.13)), numerical simulations with the Smagorinsky model (3.11) are generally reported to be very unstable with commonly used time-stepping schemes.

With $C_S(\mathbf{x}) \to 0$ as $\mathbf{x} \to \partial\Omega$, most of the standard mathematical results, such as Körn's inequality, the Poincaré–Friederichs inequality, and Sobolev's inequality, no longer hold. Thus, the mathematical development of the Smagorinsky model under no-slip boundary conditions with van Driest damping (3.13) is an important open problem. For recent mathematical results in this direction, see Swierczewska [293].

3.3.2 Alternate Scalings

To start this section, think of a flow as composed of eddies of different sizes in different places. If we are in a region of large eddies then the velocity changes over an $O(1)$ distance and the velocity deformation is $O(1)$ as well. In a region of smaller eddies the velocity changes over a distance of $O(\text{eddy length scale})$

so the local deformation is $O(1/\text{eddy length scale})$. Hence, the Smagorinsky model (3.11) introduces a turbulent viscosity coefficient $\nu_T = (C_S\delta)^2|\nabla^s\mathbf{w}|$ with the relative magnitude:

$$\nu_T = \begin{cases} O(\delta^2) & \text{in regions where } |\nabla^s\mathbf{w}| = O(1), \\ O(\delta) & \text{in the smallest resolved eddies wherein } |\nabla^s\mathbf{w}| = O(\delta^{-1}). \end{cases}$$

Thus, it is most successful when used with second–order finite difference methods for which it gives a perturbation of $O(\text{discretization error})$ in the smooth/laminar flow regions. For higher order methods (say, order r) the natural generalization is thus, see Layton [201],

$$\nu_T = (C_S\delta)^r\, |\nabla^s\mathbf{w}|^{r-1} = \begin{cases} O(\delta^r) & \text{in smooth regions,} \\ O(\delta) & \text{in the smallest resolved eddies.} \end{cases}$$

This scaling is motivated by experiments with central difference approximations to linear convection diffusion problems. Another scaling, again motivated by the interface between models and higher order numerics, was proposed in [201]. In three dimensions and using a numerical method of order $O(h^r)$ accuracy, the natural choice is

$$\nu_T = C_{r,p}\, |\log(\delta)|^{-\frac{2}{3}(p-1)}\, \delta^{\frac{3}{2}r-\frac{3}{2}}\, |\nabla^s\mathbf{w}|^{p-2}, \qquad \text{with } p \geq \frac{2}{3}r + 1.$$

As p increases, these formulas concentrate the eddy viscosity more and more in the regions in which the gradient is large. These regions include the smallest resolved eddies and also regions with large shears.

A third rescaling follows directly from interpreting the turbulent viscosity coefficient ν_T micro-locally in K-41 theory. In this theory, setting the estimate of the smallest persistent eddy to equal δ, yields a scaling formula for the turbulent viscosity coefficient (see [201]) as follows: consider,

$$\nu_T = (C_r\,\delta)^r|\nabla^s\mathbf{w}|^{p-2}.$$

In the smallest resolved eddy, $\mathbf{w}$ undergoes an $O(1)$ change over a distance $O(\delta)$. Thus, $|\nabla^s\mathbf{w}| = O(\delta^{-1})$ therein and $\nu_T = O(\delta^{r-p+2})$ in the smallest resolved eddies.

Considering this to be the (effective local Reynolds number)$^{-1}$, then the K-41 theory predicts that the smallest persistent eddy is $O(\nu_T^{3/4})$, in three dimensions. Equating $\delta = \nu_T^{3/4}$, we get an equation

$$\delta \approx (\delta^{r-p+2})^{3/4}.$$

Since r is fixed to be the order of the underlying numerical method, this determines p via

$$1 = \frac{3}{4}(r - p + 2) \quad \text{or} \quad p = r - \frac{2}{3}.$$

Thus, the K-41 theory suggests the higher order Smagorinsky-type model

$$\nu_T = (C_r\delta)^r \; |\nabla^s \mathbf{w}|^{r-4/3} \approx \begin{cases} O(\delta^r) & \text{in smooth regions,} \\ O(\delta^{4/3}) & \text{in the smallest resolved eddies.} \end{cases}$$

In two dimensions, the appropriate modification, applying the theory of Kraichnan [193], suggests that this should be modified to read

$$\nu_T = (C_r\delta)^r |\nabla^s \mathbf{w}|^{r-2}.$$

Other improved EV methods will be discussed in Chap. 4.

3.3.3 Models Acting Only on the Smallest Resolved Scales

In Variational Multiscale Methods introduced by Hughes and his collaborators [160, 161, 162], a model for the fluctuations $\mathbf{u}'$ is derived and discretized. In this work (see Chap. 11 for details), a Smagorinsky model acting only on the fluctuations $\mathbf{u}'$

$$-\nabla' \cdot ((C_S \; \delta)^2 \; |\nabla^s \mathbf{u}'| \; \nabla^s \mathbf{u}')$$

has been used successfully in the numerical simulation of decay of homogeneous isotropic turbulence [161] and turbulent channel flows [162]. Further developments were presented in Collis [68]. In Layton [203], an analogous idea was discussed. In the model for $\mathbf{w}$ (which approximates $\overline{\mathbf{u}}$) a Smagorinsky model acting only on the smallest scales of $\mathbf{w}$ was proposed. This model can be written in a natural variational way (for smooth enough $\mathbf{v}$ and $\mathbf{w}$) as:

$$(C_S \; \delta)^2 \; (|\nabla^s(\mathbf{w} - \overline{\mathbf{w}})| \; \nabla^s(\mathbf{w} - \overline{\mathbf{w}}), \nabla^s(\mathbf{v} - \overline{\mathbf{v}})).$$

These refinements are very promising in that they aim to improve the over-damping of the large structures seen in Fig. 3.2. This will likely yield improvements in turbulent flow simulations and great improvements in transitional flow simulations.

3.3.4 Germano's Dynamic Model

The "dynamic model" introduced by Germano *et al.* [129], is currently one of the best performing models in LES. In this model, the Smagorinsky model's "constant" C_S is chosen locally in space and time, so $C_S = C_S(\mathbf{x}, t)$, to make the Smagorinsky model agree in a least squares sense as closely as possible with the Bardina scale similarity model, which we will analyze in Chap. 8.

The mathematical development of the dynamic model is an open problem: since it can produce turbulent viscosities that can change sign, it seems beyond present mathematical tools.

Germano's idea of dynamic parameter selection, see Germano *et al.* [129], gives a big improvement in the performance of the Smagorinsky model. We will not delve into dynamic models here for two reasons. First, dynamic parameter selection is really a way to *improve* the performance of almost any model (and is not specific to the Smagorinsky model). Second, its mathematical foundation seems to be beyond the tools presently available.

3.4 Mathematical Properties of the Smagorinsky Model

In the 1966 International Congress of Mathematicians (ICM), the Russian mathematician O.A. Ladyžhenskaya described her work on three new models for fluids that are undergoing large stresses. One motivation for her work was the famous gap in the theory of the NSE in three dimensions: strong solutions are unique but the best efforts of mathematicians have not been able to prove their global existence, while weak solutions were proved to exist in 1934 by Leray and yet their uniqueness has been similarly elusive, see Chap. 2. Since this was the case for long time, large data, and small viscosity coefficients, and since all three are connected with the physical phenomenon of turbulence, it seemed (and still seems) possible that there might be a breakdown in the physical model of the NSE. The breakdown point, if any, seems to be the assumption of a linear stress–strain relation for larger stresses.

The following assumptions were made by Ladyžhenskaya:

1. the viscous stress tensor $\boldsymbol{\sigma}$ depends only on the deformation tensor $\nabla^s \mathbf{w}$;
2. the viscous stress tensor $\boldsymbol{\sigma}$ is invariant under rotation;
3. the viscous stress tensor $\boldsymbol{\sigma}$ is a smooth function of $\nabla^s \mathbf{w}$ and viscous forces are dissipative, that is, its Taylor expansion is dominated by odd powers of $\nabla^s \mathbf{w}$;
4. the material is incompressible.

The simplest such case is when $\boldsymbol{\sigma}(\nabla^s \mathbf{w})$ is an odd cubic polynomial in $\nabla^s \mathbf{w}$ with

$$\boldsymbol{\sigma}(\nabla^s \mathbf{w}) : \nabla^s \mathbf{w} \geq 0.$$

These assumptions lead immediately to the model (3.11), (3.12) also studied by Smagorinsky.

If the conditions are slightly generalized, for example by dropping the analyticity and still seeking the simplest interesting example, the model becomes

$$\mathbf{w}_t + \nabla \cdot (\mathbf{w}\,\mathbf{w}^T) + \nabla q - \nabla \cdot \left(\frac{2}{Re} \nabla^s \mathbf{w} + (C_S \delta)^2 |\nabla^s \mathbf{w}|^{r-2} \nabla^s \mathbf{w} \right) = \bar{\mathbf{f}}, \quad (3.15)$$

$$\nabla \cdot \mathbf{w} = 0, \quad (3.16)$$

with $r \geq 0$, and from now on we will call it the Smagorinsky–Ladyžhenskaya Model (SLM). For $C_S \delta > 0$, Ladyžhenskaya proved in [195, 196] existence of weak solutions of (3.15) and (3.16) for any $r \geq 2$, and uniqueness of weak solutions in three dimensions for any $r \geq 5/2$, including the case (3.11) and (3.12). For fixed parameters, $i.e.$ $C_S^2 \delta^2$ is $O(1)$, the theory of (3.11) and (3.15) is quite complete. For example, uniqueness of weak solutions has been extended to $r \geq 12/5$ [94, 95, 249, 250, 228]. The case when Re is fixed and $C_S \delta \to 0$, the interesting case for our purposes, is much murkier.

Many of the basic functional analytic tools used in this mathematical work are fundamental also in the further development of eddy viscosity models. We will therefore present the tools and show that they can also be used to give a clearer understanding of the SLM (3.15) and (3.16) itself.

Function Spaces

For this problem we need some spaces that are a little bit more sophisticated than those required in the variational formulation of the NSE.

Definition 3.3. *The Sobolev space $W^{1,p}(\Omega)$ is defined by*

$$W^{1,p}(\Omega) := \left\{ \begin{array}{l} u \in L^p(\Omega) : \ \exists\, g_i \in L^p(\Omega), \ i = 1, \ldots, d \ such\ that \\[2mm] \displaystyle\int_\Omega u\, \frac{\partial \phi}{\partial x_i}\, d\mathbf{x} = -\int_\Omega g_i\, \phi\, d\mathbf{x}, \quad \forall\, \phi \in C_0^\infty(\Omega) \end{array} \right\}.$$

In other words, $W^{1,p}(\Omega)$ denotes the subspace of functions belonging to $L^p(\Omega)$, together with their first-order distributional derivatives. The space $W^{1,p}(\Omega)$, for $1 \le p \le \infty$, is a Banach space, endowed with the norm

$$\|u\|_{W^{1,p}(\Omega)} = \begin{cases} \left[\|u\|_{L^p}^p + \|\nabla u\|_{L^p}^p \right]^{1/p} & if\ 1 \le p < \infty, \\[4mm] \max\left\{ \sup_{\mathbf{x}\in\Omega} |u|,\ \sup_{\mathbf{x}\in\Omega} |\nabla u| \right\} & if\ p = \infty, \end{cases}$$

and it is also reflexive and separable, provided $1 < p < \infty$. As usual we can define the space of functions vanishing on the boundary.

Definition 3.4. *We say that $W_0^{1,p}(\Omega)$ is the closure of $C_0^\infty(\Omega)$ with respect to the norm $\| . \|_{W^{1,p}}$.*

The space $W_0^{1,p}(\Omega)$ is the subspace of $W^{1,p}(\Omega)$ of functions vanishing on the boundary. These functions vanish in the <u>traces sense</u>, *i.e.* in the sense of $W^{1-1/p,p}(\partial\Omega)$. We refer again to [4] for the introduction and properties of fractional Sobolev spaces.

First, we note that the <u>Poincaré inequality</u> also holds in the L^p-setting: Let Ω be a bounded subset of $\mathbb{R}^d$. Then, there exists a positive constant C (depending now on Ω and p) such that

$$\|u\|_{L^p} \le C\|\nabla u\|_{L^p}, \qquad \forall\, u \in W_0^{1,p}(\Omega),\ with\ 1 \le p < \infty.$$

Consequently, $\|\nabla u\|_{L^p}$ is a norm on $W_0^{1,p}(\Omega)$ equivalent to $\| . \|_{W^{1,p}}$.

We will use a generalized version of the following lemma:

Lemma 3.5. *The semi-norm*

$$|\mathbf{u}|_{W_0^{1,p}} := \|\nabla^s \mathbf{u}\|_{L^p}, \qquad 1 < p < \infty$$

is equivalent to the norm of $[W_0^{1,p}(\Omega)]^d$.

Proof. The proof of this lemma, is based on a generalization of a classical tool in *continuum mechanics*: the Körn inequality. This inequality states that if $\mathbf{u} \in [H^1(\Omega)]^d$ and vanishes on a measurable (with nonvanishing measure) portion of $\partial\Omega$, then

$$\exists\, C_K > 0: \quad \int_\Omega |\nabla^s \mathbf{u}|^2 \, d\mathbf{x} \geq C_K \|\nabla \mathbf{u}\|^2.$$

The generalization of this inequality to L^p spaces can be found in Nečas [245]. Finally, Lemma 3.5 follows by using the standard Poincaré inequality, see (for instance) Parés [249]. $\square$

We now define the basic function space of divergence-free vector fields that we will need in the sequel

$$W_{0,\sigma}^{1,p} := \left\{ \mathbf{u} \in [W_0^{1,p}(\Omega)]^d : \nabla \cdot \mathbf{u} = 0 \right\},$$

endowed with the norm $\|\cdot\|_{W_0^{1,p}}$.

Remark 3.6. We develop the theory for the SLM model mainly for (3.11), that is to say (3.15) with $r = 3$. We study this case since its analysis is the starting point for deeper results. Next, we will give other results related with different values of the parameter r and the reader can find complete details in the references cited throughout this chapter.

We start by giving the definition of weak solutions for the Smagorinsky model.

Definition 3.7. *A measurable function* $\mathbf{w} : \Omega \times [0, T] \to \mathbb{R}^d$ *is a weak solution to the SLM* (3.11) *if*

1. $\mathbf{w} \in H^1(0, T; L_\sigma^2) \cap L^3(0, T; W_{0,\sigma}^{1,3})$ *with* $\mathbf{w}(0) = \mathbf{w}_0$, *the latter space being endowed with the norm*

$$\|\mathbf{w}\|_{H^1(0,T;L_\sigma^2) \cap L^3(0,T;W_{0,\sigma}^{1,3})} = \|\nabla^s \mathbf{w}\|_{L^3(0,T;L^3(\Omega))} + \|\mathbf{w}_t\|_{H^1(0,T;L_\sigma^2)};$$

2. $\mathbf{w}$ *satisfies* (3.11) *in the weak sense, i.e. for each* $\phi \in C_0^\infty(\Omega \times [0, T))$ *with* $\nabla \cdot \phi = 0$, *the following identity holds:*

$$\int_0^\infty \int_\Omega \left[\mathbf{w}_t\, \phi + \frac{1}{Re} \nabla^s \mathbf{w}\, \nabla^s \phi + (C_S \delta)^2 |\nabla^s \mathbf{w}| \nabla^s \mathbf{w}\, \nabla^s \phi \right.$$

$$\tag{3.17}$$

$$\left. + \mathbf{w} \cdot \nabla \mathbf{w}\, \phi \right] d\mathbf{x}\, dt = \int_0^\infty \int_\Omega \bar{\mathbf{f}}\, \phi \, d\mathbf{x}\, dt.$$

Remark 3.8. In the above definition

$$H^1(0, T; L^2_\sigma) := \left\{ \mathbf{f} \in L^2(0, T; L^2_\sigma), \text{ with } \mathbf{f}_t \in L^2(0, T; L^2_\sigma) \right\},$$

where $\mathbf{f}_t$ is the derivative in the sense of distributions, *i.e.* it is a function in $L^2(0, T; L^2_\sigma)$ such that

$$\int_0^T \int_\Omega \mathbf{f} \, \mathbf{v} \, \phi_t(t) \, d\mathbf{x} \, dt = - \int_0^T \int_\Omega \mathbf{f}_t \, \mathbf{v} \, \phi(t) \, d\mathbf{x} \, dt, \quad \forall \mathbf{v} \in L^2_\sigma, \, \forall \phi \in C_0^\infty(0, T).$$

This definition of weak solution is slightly different from the one regarding the weak solutions for the NSE equations (Definition 2.11). In fact, since $\mathbf{w}_t$ belongs to $L^2(0, T; L^2_\sigma)$ it is not necessary to integrate by parts the term involving the time derivative. Furthermore, it is well-known (see [4]) that $H^1(0, T; L^2_\sigma) \subset C(0, T; L^2_\sigma)$ and the initial condition $\mathbf{w}(\mathbf{x}, 0) = \mathbf{w}_0(\mathbf{x})$ is satisfied in the usual sense. This model, due to the presence of a stronger dissipative term, has weak solutions that are much more regular than in the NSE case.

We now give the proof of the existence of weak solution with almost all the needed details.

Theorem 3.9 (Ladyžhenskaya [195, 196]). *Let $\Omega \subset \mathbb{R}^3$ be a bounded open set and let be given $\mathbf{w}_0 \in W_{0,\sigma}^{1,3}$ and $\overline{\mathbf{f}} \in L^2(0, T; L^2(\Omega))$. Then, the SLM (3.11), (3.12) possesses at least a weak solution $\mathbf{w}$.*

Proof. The proof of existence is based on the Faedo–Galerkin procedure. We start as in the proof of existence of weak solutions for the NSE (Sect. 2.4). In fact, in this case, we obtain different *a priori* estimates for the approximate functions

$$\mathbf{w}_m(\mathbf{x}, t) = \sum_{k=1}^m g_m^i(t) \mathcal{W}_i(\mathbf{x}),$$

where the functions $\mathcal{W}_i(\mathbf{x})$ form an orthonormal (with respect to $(\cdot, \cdot)$, the usual L^2-scalar product) "basis" of $W_{0,\sigma}^{1,3}$. In this case, we will not need to use special functions (recall that in Theorem 2.14 we used a basis of eigenfunctions) since, roughly speaking, we do not need to multiply the equations by $-P\Delta\mathbf{w}_m$.

The function $\mathbf{w}_m$ should satisfy, for each $1 \leq k \leq m$, the following system of ODE:

$$\frac{d}{dt}(\mathbf{w}_m, \mathcal{W}_k) + \frac{2}{Re}(\nabla^s \mathbf{w}_m, \nabla^s \mathcal{W}_k) + (C_S \delta)^2 (|\nabla^s \mathbf{w}_m| \nabla^s \mathbf{w}_m, \nabla^s \mathcal{W}_k)$$

(3.18)

$$+ (\mathbf{w}_m \cdot \nabla \mathbf{w}_m, \mathcal{W}_k) = (\overline{\mathbf{f}}, \mathcal{W}_k).$$

The energy estimate. The first *a priori* estimate is obtained, as usual, by using $\mathbf{w}_m$ itself as test function, to get

$$\frac{1}{2}\frac{d}{dt}\|\mathbf{w}_m(t)\|^2 + \frac{2}{Re}\|\nabla^s\mathbf{w}_m(t)\|^2 + (C_S\delta)^2\|\nabla^s\mathbf{w}_m(t)\|^3_{L^3} = (\overline{\mathbf{f}}, \mathbf{w}_m). \quad (3.19)$$

(For more details, the reader is referred to the derivation of (2.18).) With the Gronwall lemma it is easy to show that (3.19) implies

$$\sup_{0\leq t\leq T_m} \|\mathbf{w}_m(t)\| \leq \|\mathbf{w}_m(0)\| + \int_0^T \|\overline{\mathbf{f}}(\tau)\|\, d\tau. \quad (3.20)$$

Again, the functions $\mathbf{w}_m$ do exist in some time interval $[0, T_m)$, since they satisfy a system of ODE with a Lipschitz nonlinear term. Estimate (3.20) together with a standard argument implies that $T_m = T$; see also p. 49.

Now, integration with respect to time of (3.19) gives

$$\frac{1}{2}\|\mathbf{w}_m(T)\|^2 - \frac{1}{2}\|\mathbf{w}_m(0)\|^2 + \frac{2}{Re}\int_0^T \|\nabla^s\mathbf{w}_m(\tau)\|^2\, d\tau$$

$$+ (C_S\delta)^2\int_0^T \|\nabla^s\mathbf{w}_m(\tau)\|^3_{L^3}d\tau \leq \int_0^T \|\overline{\mathbf{f}}(\tau)\|\,\|\mathbf{w}_m(\tau)\|\, d\tau.$$

By using the bound on $\|\mathbf{w}_m\|$ from (3.20) and inserting it to increase the right-hand side of the latter estimate, it is easy to show that

$$\frac{2}{Re}\int_0^T \|\nabla^s\mathbf{w}_m(\tau)\|^2\, d\tau + (C_S\delta)^2\int_0^T \|\nabla^s\mathbf{w}_m(\tau)\|^3_{L^3}\, d\tau$$

$$\leq \|\mathbf{w}(0)\|^2 + \frac{3}{2}\left(\int_0^T \|\overline{\mathbf{f}}(\tau)\|\, d\tau\right)^2. \quad (3.21)$$

Another _a priori_ estimate. The second estimate is obtained by using as test function $\partial_t\mathbf{w}_m$. In fact, in this case the first estimate is not enough to prove existence of weak solutions. Multiplication by $\partial_t\mathbf{w}_m$ has to be understood as (a) multiplying (3.18) by $dg_m^k(t)/dt$ and (b) summing over k. This shows that

$$\|\partial_t\mathbf{w}_m\|^2 + \frac{d}{dt}\left(\frac{1}{Re}\|\nabla^s\mathbf{w}_m\|^2 + \frac{(C_S\delta)^2}{3}\|\nabla^s\mathbf{w}_m\|^3_{L^3}\right)$$

$$(3.22)$$

$$+ \int_\Omega \mathbf{w}_m\cdot\nabla\mathbf{w}_m\,\partial_t\mathbf{w}_m\, d\mathbf{x} = \int_\Omega \overline{\mathbf{f}}\,\partial_t\mathbf{w}_m\, d\mathbf{x}.$$

In particular, note that this holds since

(i) $\frac{1}{2}\frac{d}{dt}\int_\Omega |\nabla^s\mathbf{u}_m|^2 d\mathbf{x} = \int_\Omega \nabla^s\mathbf{u}_m\partial_t\nabla^s\mathbf{u}_m\, d\mathbf{x} = \int_\Omega \nabla^s\mathbf{u}_m\partial_t\nabla\mathbf{u}_m\, d\mathbf{x},$

(ii) $\frac{1}{3}\frac{d}{dt}\int_\Omega |\nabla^s\mathbf{u}_m|^3 d\mathbf{x} = \int_\Omega |\nabla^s\mathbf{u}_m|\nabla^s\mathbf{u}_m\partial_t\nabla\mathbf{u}_m\, d\mathbf{x},$

and the identities can easily be proved by recalling the following result of linear algebra: let A and B be $n \times n$ matrices. If A is symmetric and if $A\,B$ denotes the inner product of the real matrices A and B, then

$$A\,B = A\,\frac{(B + B^T)}{2} + A\,\frac{(B - B^T)}{2} = A\,\frac{(B + B^T)}{2},$$

since the inner product of a symmetric matrix and an anti-symmetric matrix vanishes (recall that $A\,B = \sum_{i,j=1}^{d} a_{ij} b_{ij}$).

The convective term can be estimated (by using the Hölder inequality) as follows

$$\left| \int_{\Omega} \mathbf{w}_m \cdot \nabla \mathbf{w}_m \, \partial_t \mathbf{w}_m \, dx \right| \leq \|\partial_t \mathbf{w}_m\| \, \|\nabla \mathbf{w}_m\|_{L^3} \|\mathbf{w}_m\|_{L^6}.$$

By using the Sobolev embedding $H^1(\Omega) \subset L^6(\Omega)$, and the continuous embedding of $L^3(\Omega)$ into $L^2(\Omega)$ (recall that Ω is bounded[2]), we get

$$\|\mathbf{g}\|_{L^6} \leq c_1 \|\nabla \mathbf{g}\| \leq c_2 \|\nabla \mathbf{g}\|_{L^3} \qquad \forall \mathbf{g} \in [W_0^{1,3}(\Omega)]^d.$$

Then, Lemma 3.5 and the Young inequality imply

$$\exists\, c = c(\Omega): \qquad \left| \int_{\Omega} \mathbf{w}_m \cdot \nabla \mathbf{w}_m \partial_t \mathbf{w}_m \, dx \right| \leq \frac{1}{4}\|\partial_t \mathbf{w}_m\|^2 + c\|\nabla^s \mathbf{w}_m\|_{L^3}^4.$$

The term involving the external force is estimated as follows:

$$\left| \int_{\Omega} \overline{\mathbf{f}} \, \partial_t \mathbf{w}_m \, dx \right| \leq \|\overline{\mathbf{f}}\|^2 + \frac{1}{4}\|\partial_t \mathbf{w}_m\|^2.$$

By using the above estimates and by integrating (3.22) with respect to the time variable over $[0, t]$ (for $t \leq T$), we get

$$\int_0^t \|\partial_t \mathbf{w}_m(\tau)\|^2 \, d\tau + \frac{2}{Re}\|\nabla^s \mathbf{w}_m(t)\|^2 + \frac{2(C_S\delta)^2}{3}\|\nabla^s \mathbf{w}_m(t)\|_{L^3}^3$$

$$\leq \frac{2}{Re}\|\nabla^s \mathbf{w}_0\|^2 + \frac{2(C_S\delta)^2}{3}\|\nabla^s \mathbf{w}_0\|_{L^3}^3 + 2\int_0^T \|\overline{\mathbf{f}}(\tau)\|^2 d\tau$$

$$+ 2c\int_0^T \|\nabla^s \mathbf{w}_m(\tau)\|_{L^3}^4 d\tau.$$

---

[2] This fact follows from an application of the Hölder inequality

$$\int |f|^2 \, dx \leq \left[\left(\int |f|^2\right)^{3/2} dx\right]^{2/3} \left[\int 1 \, dx\right]^{1/3} = \|f\|_3^2 |\Omega|^{1/3},$$

where $|\Omega|$ denotes the measure of Ω. By taking the square root of both sides, we finally get the desired inequality.

An application of the Gronwall Lemma 2.17 with

$$f(t) = \frac{2(C_S\delta)^2}{3}\|\nabla^s\mathbf{w}_m(t)\|_{L^3}^3, \qquad g(t) = c\frac{3}{(C_S\delta)^2}\|\nabla^s\mathbf{w}_m(t)\|_{L^3}$$

$$C = \frac{2}{Re}\|\nabla^s\mathbf{w}_0\|^2 + \frac{2(C_S\delta)^2}{3}\|\nabla^s\mathbf{w}_0\|_{L^3}^3 + 2\int_0^T\|\bar{\mathbf{f}}(\tau)\|^2 d\tau$$

shows that

$$\sup_{0\leq t\leq T}\frac{(C_S\delta)^2}{3}\|\nabla^s\mathbf{w}_m(t)\|_{L^3}^3 \leq Ce^{\frac{3c}{2(C_S\delta)^2}\int_0^T\|\nabla^s\mathbf{w}_m(\tau)\|_{L^3}d\tau}.$$

By Hölder inequality, we get

$$\int_0^T\|\nabla^s\mathbf{w}_m(\tau)\|_{L^3}d\tau \leq \|\nabla^s\mathbf{w}_m\|_{L^3(0,T;L^3(\Omega))}^3 T^{2/3}.$$

By using the *a priori* estimate (3.21) to bound the latter integral, we finally get that there exists a positive constant C, depending on T, but independent of m, such that

$$\int_0^T\|\partial_t\mathbf{w}_m(\tau)\|^2 d\tau + \sup_{0\leq t\leq T}\frac{2(C_S\delta)^2}{3}\|\nabla^s\mathbf{w}(t)\|_{L^3}^3 \leq C. \qquad (3.23)$$

Remark 3.10. The core of the proof relies again on some *a priori* estimates. In this case we obtained the second estimate with a different tool, namely multiplication by $\partial_t\mathbf{w}_m$. In the sequel we will see other, more sophisticated, techniques that are required by LES models. We want to stress again the importance of *a priori* estimates in variational problems of mathematical physics and especially in fluid mechanics.

With the previous estimates, we have then proved that the sequence $\{\mathbf{w}_m\}_{m\geq 1}$ is uniformly bounded in

$$H^1\left(0,T;L_\sigma^2\right) \cap L^\infty\left(0,T;W_{0,\sigma}^{1,3}\right).$$

By using results of weak compactness (see footnote on p. 50) it is possible to prove that there exists a subsequence, relabeled again as $\{\mathbf{w}_m\}_{m\geq 1}$, and a function $\mathbf{w} \in H^1(0,T;L_\sigma^2) \cap L^\infty(0,T;W_{0,\sigma}^{1,3})$ such that:

$$\begin{cases} \mathbf{w}_m \rightharpoonup \mathbf{w} & \text{in } H^1(0,T;L_\sigma^2); \\ \mathbf{w}_m \overset{*}{\rightharpoonup} \mathbf{w} & \text{in } L^\infty(0,T;W_{0,\sigma}^{1,3}); \\ \nabla\mathbf{w}_m \rightharpoonup \nabla\mathbf{w} & \text{in } L^3(0,T;L^3(\Omega)). \end{cases}$$

Since we know a bound on the time derivative, we can use a standard and useful compactness tool; see, for instance, Lions [221], Chap. 1.

Lemma 3.11 (Aubin–Lions). *Let, for some $p > 1$, the set $\mathcal{Y}$ be bounded in*

$$\mathcal{X} := \left\{ u \in L^p(0, T; X_1) : \quad \frac{du}{dt} \in L^p(0, T; X_3) \right\}.$$

If $X_1 \subset X_2 \subset X_3$ are reflexive Banach spaces and the first inclusion is compact, while the second one is continuous, then $\mathcal{Y}$ is compactly included in $L^p(0, T; X_2)$.

By recalling the bounds proven for $\mathbf{w}_m$ and $\partial_t \mathbf{w}_m$, we get that the hypotheses of the above lemma are satisfied with $p = 2$ and $X_1 = X_3 = L_\sigma^2$. Then, the sequence $\{\mathbf{w}_m\}_{m \geq 1}$ belongs to a set that is compactly included in $L^2(0, T; L_\sigma^2)$ and consequently we can extract a subsequence, relabeled again as $\{\mathbf{w}_m\}_{m \geq 0}$, such that

$$\mathbf{w}_m \to \mathbf{w} \quad \text{in } L^2(0, T; L_\sigma^2). \tag{3.24}$$

Together with these properties, we also have another strong convergence property, that is more powerful than (2.22) that we used in Chap. 2.

Lemma 3.12. *The sequence $\{\mathbf{w}_m\}_{m \geq 1}$ satisfies*

$$\mathbf{w}_m \to \mathbf{w} \quad \text{in } L^q(0, T; L^q(\Omega)), \quad 1 \leq q < 4.$$

Proof. The proof is based on the classical Ladyžhenskaya inequality (2.24) (note that we can replace ∇ with ∇^s in the second term, thanks to Lemma 3.5)

$$\|\mathbf{w}\|_{L^4} \leq c \|\mathbf{w}\|^{1/4} \|\nabla^s \mathbf{w}\|^{3/4}.$$

By using the previously proved uniform bounds, $\mathbf{w}_m$ is uniformly bounded in $L^4(0, T; L^4(\Omega))$.

An application of the Hölder inequality shows, for $q < 4$, that

$$\|\mathbf{w} - \mathbf{w}_m\|^q_{L^q(0,T;L^q(\Omega))} = \int_0^T \int_\Omega |\mathbf{w} - \mathbf{w}_m|^{2-\varepsilon} |\mathbf{w} - \mathbf{w}_m| |\mathbf{w} - \mathbf{w}_m| \, d\mathbf{x} \, dt$$

$$\leq \|(\mathbf{w} - \mathbf{w}_m)^{2-\varepsilon}\|_{L^2(0,T;L^2)} \|\mathbf{w} - \mathbf{w}_m\|^2_{L^4(0,T;L^4)}.$$

The last term on the right-hand side is bounded, while the first one can be written as

$$\|(\mathbf{w} - \mathbf{w}_m)^{2-\varepsilon}\|^2_{L^2(0,T;L^2)} = \int_0^T \int_\Omega |\mathbf{w} - \mathbf{w}_m|^{2-2\varepsilon} |\mathbf{w} - \mathbf{w}_m| |\mathbf{w} - \mathbf{w}_m| \, d\mathbf{x} \, dt,$$

and the final terms may be bounded as in the previous case.

After a finite number ($k \in \mathbb{N}$) of steps, since the number $\varepsilon > 0$ is fixed, we get an expression involving $0 < 2 - k\varepsilon \leq 1$. Then, by using the result of (3.24) we finally get

$$\|(\mathbf{w} - \mathbf{w}_m)^{2-k\varepsilon}\|^2_{L^2(0,T;L^2)} \leq c \|\mathbf{w} - \mathbf{w}_m\|^2_{L^2(0,T;L^2)} \overset{m \to \infty}{\longrightarrow} 0.$$

This proves the lemma. $\square$

While all the other terms can be treated as in the study of the NSE (see p. 50) it is necessary to have an auxiliary tool to analyze the convergence

$$\int_0^T \int_\Omega |\nabla^s \mathbf{w}_m| \nabla^s \mathbf{w}_m \nabla^s \phi \, d\mathbf{x} \, dt \stackrel{?}{\longrightarrow} \int_0^T \int_\Omega |\nabla^s \mathbf{w}| \nabla^s \mathbf{w} \nabla^s \phi \, d\mathbf{x} \, dt.$$

In this case, we introduce the so called "Minty-trick" that is a very powerful tool to study monotone operators, see Minty [237] and Browder [46].

Remark 3.13. Historical remarks on the theory of monotone operators may be found in the introduction to Chap. 26 in Zeidler [320], where many authors are named as fundamental contributors to this field. Among them, we may cite Golomb (1935), Zarantonello (1960), Vainberg (1956), Kačurovskii (1960), and Leray and J.-L. Lions (1965). The *folklore comment* in [320] is: "The truly new ideas are extremely rare in mathematics"!

First we note that, uniformly in m,

$$\frac{2}{Re} \nabla^s \mathbf{w}_m + (C_S \delta)^2 |\nabla^s \mathbf{w}_m| \nabla^s \mathbf{w}_m \quad \text{is bounded in } L^{3/2}(0, T; L^{3/2}(\Omega))$$

and consequently there exists $\mathbf{B} \in L^{3/2}(0, T; L^{3/2}(\Omega))$ such that

$$\frac{2}{Re} \nabla^s \mathbf{w}_m + (C_S \delta)^2 |\nabla^s \mathbf{w}_m| \nabla^s \mathbf{w}_m \rightharpoonup \mathbf{B} \quad \text{in } L^{3/2}(0, T; L^{3/2}(\Omega)).$$

This finally shows that

$$\int_0^\infty \int_\Omega [\mathbf{w}_t \, \phi + \mathbf{B} \, \nabla^s \phi + \mathbf{w} \cdot \nabla \mathbf{w} \, \phi] \, d\mathbf{x} \, dt = \int_0^\infty \int_\Omega \overline{\mathbf{f}} \, \phi \, d\mathbf{x} \, dt. \tag{3.25}$$

We have proved this equality for smooth functions ϕ written as $\phi = \sum_{k=1}^m \beta^k(t) \mathcal{W}_k(\mathbf{x})$, with β^k absolutely continuous functions. By a density argument (see Ladyžhenskaya [197], p. 159) it is possible to show that:

identity (3.25) holds also for $\phi \in H^1(0, T; L^2_\sigma) \cap L^3(0, T; W^{1,3}_{0,\sigma})$.

Remark 3.14. This property is both nontrivial and crucial since we really need it to use $\mathbf{w}$ itself as a test function. Identity (3.25) involves $\mathbf{w}$ and not simply the approximate functions $\mathbf{w}_m$. As we have seen before, multiplying the equation satisfied by $\mathbf{w}$ by the solution itself may involve calculations that are formal and not justified.

To better explain the monotonicity argument, we collect it into a lemma, from which it is possible to deduce that $\mathbf{w}_m$ converges to a solution to SLM (3.11), (3.12).

Lemma 3.15. *Let $\psi \in H^1\left(0, T; L^2_\sigma\right) \cap L^3\left(0, T; W^{1,3}_{0,\sigma}\right)$. Then*

$$
-\int_0^T \int_\Omega (\mathbf{w}_t + \mathbf{w} \cdot \nabla \mathbf{w} - \bar{\mathbf{f}})(\mathbf{w} - \psi) \, d\mathbf{x} \, dt
$$

$$
-\int_0^T \int_\Omega \left(\frac{2}{Re}\nabla^s\psi + (C_S\delta)^2|\nabla^s\psi|\nabla^s\psi\right)(\nabla^s\mathbf{w} - \nabla^s\psi) \, d\mathbf{x} \, dt \geq 0.
$$
$$\tag{3.26}$$

Proof. In this lemma we will use the operator

$$
T(\mathbf{v}) = \frac{2}{Re}\nabla^s\mathbf{v} + (C_S\delta)^2|\nabla^s\mathbf{v}|\nabla^s\mathbf{v},
\tag{3.27}
$$

defined for a smooth enough vector field $\mathbf{v}$. We will use this notation to focus on the properties of the subfilter-scale term and also to stress other abstract properties.

First, note that by (3.18) $\mathbf{w} = \mathbf{w}_m$ satisfies (3.25) with $\mathbf{B} = T(\mathbf{w}_m)$. We then subtract the identity (3.25) for $\mathbf{w}_m$, with test function $\phi = \mathbf{w}_m$, from the same identity with test function $\phi = \psi$, to obtain

$$
\int_0^T \int_\Omega (\partial_t\mathbf{w}_m + \mathbf{w}_m \cdot \nabla\mathbf{w}_m - \mathbf{f}, \mathbf{w}_m - \psi) \, d\mathbf{x} \, dt
$$

$$
+\int_0^T \int_\Omega T(\mathbf{w}_m)(\nabla^s\mathbf{w}_m - \nabla^s\psi) \, d\mathbf{x} \, dt = 0.
$$

We subtract and add on the left-hand side of the previous equality the term

$$
\int_0^T \int_\Omega T(\psi)(\nabla^s\mathbf{w}_m - \nabla^s\psi) \, d\mathbf{x} \, dt,
$$

to get

$$
\int_0^T \int_\Omega (\partial_t\mathbf{w}_m + \mathbf{w}_m \cdot \nabla\mathbf{w}_m - \mathbf{f}, \mathbf{w}_m - \psi) \, d\mathbf{x} \, dt
$$

$$
+\int_0^T \int_\Omega (T(\mathbf{w}_m) - T(\psi))(\nabla^s\mathbf{w}_m - \nabla^s\psi) \, d\mathbf{x} \, dt \tag{3.28}
$$

$$
+\int_0^T \int_\Omega T(\psi)(\nabla^s\mathbf{w}_m - \nabla^s\psi) \, d\mathbf{x} \, dt = 0.
$$

Now, we shall use the following fundamental fact (for stronger, more refined, properties see also Sect. 3.4.1)

$$
\int_\Omega [T(\mathbf{w}_m) - T(\psi)]\,(\nabla^s\mathbf{w}_m - \nabla^s\psi) \, d\mathbf{x} \geq \frac{2}{Re}\|\nabla^s(\mathbf{w} - \psi)\|^2 \geq 0. \tag{3.29}
$$

For the moment we claim this result and we will prove it in a more general form after the proof of the present theorem. The main property, that is essentially the definition of a monotone operator, is

$$\int_{\Omega} \left[|\nabla^s \mathbf{w}_m| \nabla^s \mathbf{w} - |\nabla^s \boldsymbol{\psi}| \nabla^s \boldsymbol{\psi}) \right] (\nabla^s \mathbf{w}_m - \nabla^s \boldsymbol{\psi}) \, d\mathbf{x} \geq 0,$$

see Proposition 3.22.

Since the middle term in (3.28) is nonnegative, we finally get

$$\int_0^T (\partial_t \mathbf{w}_m + \mathbf{w}_m \cdot \nabla \mathbf{w}_m - \mathbf{f}, \mathbf{w}_m - \boldsymbol{\psi}) + (T(\boldsymbol{\psi}), \nabla^s \mathbf{w}_m - \nabla^s \boldsymbol{\psi}) \, dt \leq 0. \quad (3.30)$$

It is now rather standard to pass to the limit as $m \to +\infty$ in almost all the terms of (3.30), to prove (3.26). The challenge is now

$$\int_0^T (\mathbf{w}_m \cdot \nabla \mathbf{w}_m, \mathbf{w}_m) \, dt \quad \overset{?}{\longrightarrow} \quad \int_0^T (\mathbf{w} \cdot \nabla \mathbf{w}, \mathbf{w}) \, dt,$$

since it involves three times the function $\mathbf{w}_m$ (note the difference from (2.21), where the terms involved were $\mathbf{u}_m$ twice and the test function once).

In this case, to prove that such convergence takes place, it is necessary to use an additional result proved in lemma 3.12. In particular, since we can use that lemma with $q = 3$ we have the strong convergence

$$\mathbf{w}_m \mathbf{w}_m \to \mathbf{w}\,\mathbf{w} \quad \text{in } L^{3/2}(0, T; L^{3/2}(\Omega)).$$

We write component-wise

$$\int_0^T \int_{\Omega} \mathbf{w}_m \cdot \nabla \mathbf{w}_m \mathbf{w}_m \, d\mathbf{x} \, dt = \sum_{k,l=1}^{d} \int_0^T \int_{\Omega} w_m^k \frac{\partial w_m^l}{\partial x_k} w_m^l \, d\mathbf{x} \, dt$$

$$= \sum_{k,l=1}^{d} \int_0^T \int_{\Omega} w^k \frac{\partial w_m^l}{\partial x_k} w^l \, d\mathbf{x} \, dt + \sum_{k,l=1}^{d} \int_0^T \int_{\Omega} \frac{\partial w_m^l}{\partial x_k} \left(w_m^k w_m^l - w^k w^l \right) \, d\mathbf{x} \, dt.$$

The first integral converges to $\int_0^T \int_{\Omega} \mathbf{w} \cdot \nabla \mathbf{w} \, d\mathbf{x} \, dt$ as $m \to \infty$, since $\nabla \mathbf{w}_m \rightharpoonup \nabla \mathbf{w}$ in $L^3(0, T; L^3(\Omega))$. The second one vanishes, as $m \to \infty$, due to the strong convergence of $\mathbf{w}_m \mathbf{w}_m$ in $L^{3/2}(0, T; L^{3/2}(\Omega))$ and to the uniform bound of $\nabla \mathbf{w}_m$ in $L^3(0, T; L^3(\Omega))$. Passing to the limit as m goes to ∞ we finally get inequality (3.26). $\square$

The Monotonicity Trick

At this point it is possible to conclude the proof of Theorem 3.9, namely we have to show that $\mathbf{B} = T(\mathbf{w})$. This will be done by using the following argument: add together (3.26) and (3.25) and set $\boldsymbol{\phi} = \mathbf{w} - \boldsymbol{\psi}$ (at this point it is really necessary to use $\mathbf{w}$ as a test function) to obtain

$$\int_0^T \int_{\Omega} (\mathbf{B} - T(\boldsymbol{\psi})) \, (\nabla^s \mathbf{w} - \nabla^s \boldsymbol{\psi}) \, d\mathbf{x} \, dt \geq 0.$$

Choose now an arbitrary $\boldsymbol{\eta} \in H^1(0,T;L_\sigma^2) \cap L^3(0,T;W_{0,\sigma}^{1,3})$ and set, for $\epsilon > 0$,

$$\psi = \mathbf{w} - \epsilon\boldsymbol{\eta}$$

to get

$$\epsilon \int_0^T \int_\Omega (\mathbf{B} - T(\mathbf{w} - \epsilon\boldsymbol{\eta})) \nabla^s \boldsymbol{\eta} \, dx \, dt \geq 0.$$

By dividing by $\epsilon > 0$ and by taking the limit[3] as $\epsilon \to 0^+$ we get

$$\int_0^T \int_\Omega (\mathbf{B} - T(\mathbf{w})) \nabla^s \boldsymbol{\eta} \, dx \, dt \geq 0. \tag{3.31}$$

Since (3.31) holds for an arbitrary $\boldsymbol{\eta} \in H^1(0,T;L_\sigma^2) \cap L^3(0,T;W_{0,\sigma}^{1,3})$, it holds also for $-\boldsymbol{\eta}$ (note that we are using the fact that $H^1(0,T;L_\sigma^2) \cap L^3(0,T;W_{0,\sigma}^{1,3})$ is a linear space). Thus the integral in (3.31) is both nonnegative and non-positive. This implies that it vanishes identically, for each $\boldsymbol{\eta} \in H^1(0,T;L_\sigma^2) \cap L^3(0,T;W_{0,\sigma}^{1,3})$ and proves the equality:

$$\mathbf{B} = \frac{2}{Re}\nabla^s \mathbf{w} + (C_S\delta)^2|\nabla^s \mathbf{w}|\nabla^s \mathbf{w}.$$

□

The SLM model (3.11), (3.12) shares also a uniqueness property and a stability estimate. We collect them in the following theorem; see [195].

Theorem 3.16. *Let $\mathbf{w}^1$ and $\mathbf{w}^2$ be weak solutions to the SLM model (3.11), (3.12), corresponding respectively to the data $(\mathbf{w}_0^1, \mathbf{f}^1) \in W_{\sigma,0}^{1,3} \times L^2(0,T;L_\sigma^2)$ $(\mathbf{w}_0^2, \mathbf{f}^2) \in W_{\sigma,0}^{1,3} \times L^2(0,T;L_\sigma^2)$. Then, the following estimate holds*

$$\|\mathbf{w}^1 - \mathbf{w}^2\|_{L^\infty(0,T;L^2)}$$

$$\leq \left[\|\mathbf{w}_0^1 - \mathbf{w}_0^2\|^2 + \frac{1}{2c_1}\int_0^T \|\mathbf{f}^1 - \mathbf{f}^2\|^2 \, dt\right] e^{\left(c_2\|\nabla\mathbf{w}^1\|_{L^2(0,T;L^3)}^2 + \frac{c_1 T}{2}\right)},$$

for some positive constants c_1, c_2.

The theorem implies that if $\mathbf{w}_0^1 = \mathbf{w}_0^2$ and $\mathbf{f}^1 = \mathbf{f}^2$, then there exists a unique solution to the problem.

---

[3] In this case we are using the property that

$$\mathbb{R} \ni \lambda \mapsto \int_\Omega T(\mathbf{u} + \lambda\mathbf{v})\nabla^s\mathbf{w} \, dx \, dt \in \mathbb{R} \text{ is a continuous function } \forall \, \mathbf{u},\mathbf{v},\mathbf{w} \in W^{1,3}.$$

This property means that the operator is <u>hemicontinuous</u>, see Sect. 3.4.1 for further details.

Proof. We give just the proof of uniqueness, the other result may be achieved with the same technique. Let us subtract the equation satisfied by $\mathbf{w}^2$ from that satisfied by $\mathbf{w}^1$ and multiply the resulting equation by $\mathbf{W} = \mathbf{w}^1 - \mathbf{w}^2$:

$$\frac{1}{2}\frac{d}{dt}\|\mathbf{W}\|^2 + \int_\Omega \left[T(\mathbf{w}^1) - T(\mathbf{w}^2)(\mathbf{w}^1 - \mathbf{w}^2)\right] d\mathbf{x} = \int_\Omega \mathbf{W} \cdot \nabla \mathbf{w}^1 \mathbf{W} \, d\mathbf{x}.$$

By using again (3.29) and the usual estimates for the term on the right-hand side

$$\left|\int_\Omega \mathbf{W} \cdot \nabla \mathbf{w}^1 \mathbf{W} \, d\mathbf{x}\right| \leq \|\mathbf{W}\|\,\|\mathbf{W}\|_{L^6}\|\nabla \mathbf{w}^1\|_{L^3} \leq C_S\|\mathbf{W}\|\,\|\nabla^s \mathbf{W}\|\,\|\nabla \mathbf{w}^1\|_{L^3},$$

we easily get

$$\frac{1}{2}\frac{d}{dt}\|\mathbf{W}\|^2 + \frac{2}{Re}\|\mathbf{W}\|^2 \leq C\|\mathbf{W}\|^2\|\nabla \mathbf{w}^1\|_{L^3}^2.$$

Due to the known regularity of $\mathbf{w}^1$, which is better than that usually known for the NSE, we can use the Gronwall lemma to deduce that

$$\|\mathbf{W}(t)\|^2 \leq \|\mathbf{W}_0\|^2 \, e^{2C\int_0^T \|\nabla \mathbf{w}^1\|_{L^3}^2 \, dt}.$$

This implies $\|\mathbf{W}(t)\| \equiv 0$. $\square$

3.4.1 Further Properties of Monotone Operators

In this section we introduce some concepts regarding monotone operators. This is motivated by the fact that the extra term

$$-\nabla \cdot |\nabla^s \mathbf{w}|^{r-2}\nabla^s \mathbf{w} \qquad (3.32)$$

in (3.15), (3.16) is one of the most relevant particular cases. Due to its importance, the operator in (3.32) is called the *r-Laplacian* and it has a number of important mathematical properties.

We start with some generalities on monotone operators, since they represent a well-known part in the theory of PDE.

Definition 3.17. *Let* $(X, \|\cdot\|_X)$ *be a Banach space with topological dual* X'. *We say that the operator* $A: X \to X'$ *is monotone if*

$$\langle Au - Av, u - v\rangle \geq 0, \qquad \forall\, u, v \in X, \qquad (3.33)$$

where $\langle\cdot,\cdot\rangle$ *denotes the duality pairing between* X *and* X'.

In the case of $X = X' = \mathbb{R}$ (*i.e.* A is a real function), it is easy to see that

$$A \text{ monotone} \iff A \text{ is a monotone increasing function,}$$

so the concept of monotone operator is a generalization of the concept of monoton increasing functions.

The following property connects monotone operators and monotone increasing functions

Proposition 3.18. *Let $A : X \to X'$ be a given operator and set*

$$f(t) = \langle A(u + tv), v \rangle, \qquad \forall t \in \mathbb{R}.$$

Then, the following statements are equivalent

(a) The operator A is monotone.
(b) The function $f : [0,1] \to \mathbb{R}$ is monotone increasing for any u and v belonging to X.

To detect if an operator is monotone there is a well-known result that connects monotone operators and convex functionals of the calculus of variations.

Proposition 3.19. *Let $f : X \to \mathbb{R}$ be a Gateaux-differentiable functional. Then, the following two conditions are equivalent*

(i) f is a convex functional, i.e.

$$f((1 - t) u + tv) \le (1 - t) f(u) + t f(v), \qquad \forall t \in [0, 1], \ \forall u, v \in X.$$

(ii) $f' : X \to X'$ is monotone, where $f'(u) \in X'$ denotes the Gateaux-derivative defined as

$$\langle f'(u), h \rangle = \lim_{\lambda \to 0} \frac{f(u + \lambda h) - f(u)}{\lambda}, \qquad \forall h \in X.$$

Proof. First we observe that if the functional f is convex, then the function $\phi(t) = f((1-t)u+tv)$, defined for $t \in [0, 1]$ and $u, v \in X$, is a real convex function. Furthermore, a function ϕ is a differentiable, real, and convex function if and only if ϕ' is monotone increasing.

Differentiation gives $\phi'(t) = f'((1 - t)u + tv)(v - u)$. Since $f'(w) \in X'$, for each $w \in X$, we can write

$$\phi'(t) = \langle f'((1 - t)u + tv), v - u \rangle.$$

We can now prove that $(i) \Rightarrow (ii)$. If f is a convex functional, then ϕ is convex and consequently $\phi'(t)$ is a monotone increasing function. This implies that $\phi'(0) \le \phi'(1)$, namely

$$\langle f'(u) - f'(v), u - v \rangle \ge 0 \qquad \forall u, v \in X,$$

i.e. f' is monotone.

Conversely, to prove that $(ii) \Rightarrow (i)$ let $f' : X \to X'$ be monotone. Then, if $t > s$

$$\phi'(t) - \phi'(s) = \langle f'(u + t(u - v)) - f'(u + s(u - v)), v - u \rangle \geq 0.$$

Thus the real function ϕ' is monotone increasing, which implies that ϕ is a convex function and hence f is a convex functional. □

The above result may be applied to the functional

$$J(\mathbf{v}) = \frac{1}{r} \int_\Omega |\nabla^s \mathbf{v}|^r \, d\mathbf{x} \qquad \forall \, \mathbf{v} \in [W_0^{1,p}(\Omega)]^d,$$

which is convex (check it!) and its Gateaux-derivative

$$\lim_{\lambda \to 0} \frac{J(\mathbf{v} + \lambda \mathbf{w}) - J(\mathbf{v})}{\lambda} = \int_\Omega |\nabla^s \mathbf{v}|^{r-2} \nabla^s \mathbf{v} \nabla^s \mathbf{w} \, d\mathbf{x}, \qquad \forall \, \mathbf{v}, \, \mathbf{w} \in [W_0^{1,3}(\Omega)]^d$$

turns out to be the variational definition of the r-Laplacian. In other words, by using the Riesz representation theorem we can define the nonlinear operator

$$T_r(\mathbf{v}) = -\nabla \cdot (|\nabla^s \mathbf{v}|^{r-2} \nabla^s \mathbf{v}) : \ [W_0^{1,r}(\Omega)]^d \to [W^{-1,r/(r-1)}(\Omega)]^d$$

as

$$\langle T_r(\mathbf{v}_1), \mathbf{v}_2 \rangle = \int_\Omega |\nabla^s \mathbf{v}_1|^{r-2} \nabla^s \mathbf{v}_1 \nabla^s \mathbf{v}_2 \, d\mathbf{x} \qquad \forall \, \mathbf{v}_1, \mathbf{v}_2 \in [W_0^{1,r}(\Omega)]^d.$$

$$(3.34)$$

Note that $W^{-1,r/(r-1)}(\Omega) = W^{-1,r'}(\Omega)$ is the dual space of $W_0^{1,r}(\Omega)$.

The nonlinear operator T_r shares other good properties. In fact, from the Definition (3.34) it is easy to see that

(a) T_r is bounded, *i.e.* $\|T_r(\mathbf{v})\|_{[W^{-1,r'}(\Omega)]^d} \leq c \|\mathbf{v}\|_{[W_0^{1,r}(\Omega)]^d}^{r-1}$,

(b) T_r is coercive, *i.e.* $\langle T_r(\mathbf{v}), \mathbf{v} \rangle \geq \|\nabla^s \mathbf{v}\|_{[W_0^{1,r}(\Omega)]^d}^3$.

Furthermore, it is possible to show that $T_r(\mathbf{v})$ is hemicontinuous, *i.e.* $\forall \, \mathbf{u}, \mathbf{v}, \mathbf{w} \in [W_0^{1,3}(\Omega)]^d$ the map

$$\lambda \mapsto \int_\Omega T_r(\mathbf{u} + \lambda \, \mathbf{v}) \nabla^s \mathbf{w} \, d\mathbf{x} \qquad \text{is a continuous real function,}$$

see Proposition 1.1, Chap. 2 in [221].

Remark 3.20. These properties are very important since, by following the same path as Theorem 3.9, it is possible to prove an abstract result.

The application of the "Monotonicity trick" argument can be found in all its generality for instance in Lions [221]. Just to give a flavor of the basic properties we state a typical theorem that can be proved for monotone operators.

Theorem 3.21. *Let $(V, \| \cdot \|_V)$ be a reflexive Banach space and let V' be its topological dual, with duality pairing denoted by $\langle \cdot, \cdot \rangle$. Let $A : V \to V'$ be an operator (possibly nonlinear) such that*

(i)   A is monotone;
(ii) A is bounded: $\exists\, c > 0$ such that $\|Aw\|_{V'} \leq c\|w\|_V^{p-1}$, for each $w \in V$;
(iii) A is hemicontinuous;
(iv) A is coercive: $\exists\, \alpha > 0 :$ such that $\langle Aw, w \rangle \geq \alpha\|w\|_V^p$, for $1 < p < \infty$.

Then, for each $f \in V'$, the equation

$$Au = f$$

has a solution $u \in V$. Furthermore, if A is strictly monotone, i.e.

$$\langle Au - Av, u - v \rangle > 0 \qquad \forall\, u, v \in V,$$

then such a solution is unique.

To study the time evolution problem, suppose that H is a Hilbert space such that [4]

$$V \hookrightarrow H \quad \text{with continuous and dense inclusion.}$$

Then, if A satisfies $(i), (ii), (iii), (iv)$ and if $f \in L^{p'}(0, T; V')$, then equation

$$u_t + Au = f \quad \text{with} \quad u(0) = u_0$$

has a unique solution $u \in L^p(0, T; V)$.

The proof can be found in [221] and [320]. Note that, since $u_t = -Au + f$, it follows that $u_t \in L^{p'}(0, T; V')$. By using an interpolation result (see for instance [84]) it follows that

$$u \in C(0, T; V')$$

and the initial condition makes sense (at least in this space).

Basic Properties of the r-Laplacian

To simplify the exposition of the following result we select the simplest boundary conditions: periodic boundary conditions with zero mean imposed upon all data and the solution. In this case we define the functions spaces $\mathbf{X}^1$ and $\mathbf{X}_r^1$ by

$$\mathbf{X}^1 := \text{ closure in } [H^1(\Omega)]^3 \text{ of } \mathbf{w} \in [C^1(\Omega)]^3 : \mathbf{w} \text{ satisfying } (2.3),$$
$$\mathbf{X}_r^1 := \text{ closure in } [W^{1,r}(\Omega)]^3 \text{ of } \mathbf{w} \in [C^1(\Omega)]^3 : \mathbf{w} \text{ satisfying } (2.3).$$

---

[4] A practical example is $V = W^{1,p}(\Omega)$, with $p \geq 2$, and $H = L^2(\Omega)$.

By using the Riesz representation theorem, we can define a nonlinear operator $T_r(\cdot) : \mathbf{X}_r^1 \to (\mathbf{X}_r^1)'$ by the correspondence

$$\langle T_r(\mathbf{v}_1), \mathbf{v}_2 \rangle = (|\nabla^s \mathbf{v}_1|^{r-2} \nabla^s \mathbf{v}_1, \nabla^s \mathbf{v}_2), \qquad \forall\, \mathbf{v}_1, \mathbf{v}_2 \in \mathbf{X}_r^1. \tag{3.35}$$

As we pointed out before T_r is an abstract representation of the operator

$$T_r(\mathbf{v}) \sim -\nabla \cdot (|\nabla^s \mathbf{v}|^{r-2} \nabla^s \mathbf{v}).$$

Proposition 3.22. *For $r \geq 2$, $\mathbf{v}_1, \mathbf{v}_2 \in \mathbf{X}_r^1$, the operator $T_r(\cdot)$ satisfies:*

$$\langle T_r(\mathbf{v}_1) - T_r(\mathbf{v}_2), \mathbf{v}_1 - \mathbf{v}_2 \rangle \geq \xi(\|\nabla^s(\mathbf{v}_1 - \mathbf{v}_s)\|_{L^r})\, \|\nabla^s(\mathbf{v}_1 - \mathbf{v}_s)\|_{L^r} \tag{3.36}$$

$$\|T_r(\mathbf{v}_1) - T_r(\mathbf{v}_2)\|_{(\mathbf{X}_r^1)'} \leq \Gamma(\rho)\, \|\nabla^s(\mathbf{v}_1 - \mathbf{v}_s)\|_{L^r}, \tag{3.37}$$

for $\|\nabla^s \mathbf{v}_j\|_{L^r} \leq \rho$, where $\Gamma(\rho) = C\,(2r-3)\rho^{r-1}$ and $\xi(s) = C\left(\frac{1}{2}\right)^{r-2} s^{r-1}$.

Remark 3.23. The property (3.37) is the local Lipschitz continuity. Property (3.36) is a monotonicity condition which is called "strong monotonicity" by Vainberg [298] and "uniform monotonicity" by Zeidler [320]. If $1 < r \leq 2$, the operator $T_r(\cdot)$ satisfies the weaker monotonicity condition (called "strict monotonicity" by Zeidler [320])

$$(T_r(\mathbf{v}_1) - T_r(\mathbf{v}_2), \mathbf{v}_1 - \mathbf{v}_2) > 0 \qquad \text{for all} \quad \mathbf{v}_1, \mathbf{v}_2 \in \mathbf{X}_r^1, \mathbf{v}_1 \neq \mathbf{v}_2.$$

The proposition shows that it is possible to get both upper and lower bounds on quantities involving the operator $T_r(\cdot)$ and with these a fairly complete analysis of its effects is possible. The proof of Proposition 3.22 is based upon the following algebraic inequality.

Lemma 3.24. *Consider the function $\psi : \mathbb{R} \to \mathbb{R}$ defined by*

$$\psi(u) = \begin{cases} |u|^{p-2}u & \text{if } u \neq 0 \\ 0 & \text{if } u = 0. \end{cases}$$

Then:

(a) if $p > 1$, then ψ is strictly monotone;
(b) if $p = 2$, then ψ is strongly monotone, i.e.

$$\exists\, c > 0: \quad < \psi(u) - \psi(v), u - v >\, \geq c|u - v|^2 \qquad \forall\, u, v \in \mathbb{R};$$

(c) if $p \geq 2$, then ψ is uniformly monotone, i.e.

$$(\psi(u) - \psi(v))(u - v) \geq a(|u - v|)|u - v|^2 \qquad \forall\, u, v \in \mathbb{R},$$

where the function $a : \mathbb{R}_+ \to \mathbb{R}_+$ is continuous and strictly monotone increasing, with $a(0) = 0$ and $\lim_{t \to +\infty} a(t) = +\infty$. A typical example is $a(t) = \alpha\, t^{p-1}$, with $\alpha > 0$ and $p > 1$.

Proof. The proofs of (a) and (b) are straightforward. Regarding part (c), it easily follows from the algebraic inequality

$$\exists c > 0: \quad (|u|^{p-2}u - |v|^{p-2}v)(u - v) \geq c|u - v|^p \quad \forall u, v \in \mathbb{R}$$

and fixed $p \geq 2$.

The proof of the above inequality is given by considering first the case $0 \leq v \leq u$. Then,

$$u^{p-1} - v^{p-1} = \int_0^{u-v} (p-1)(t+v)^{p-2}\, dt$$

$$\geq \int_0^{u-v} (p-1)\, t^{p-2}\, dt = (u-v)^{p-1}.$$

In the case $v \leq 0 \leq u$ we can use the inequality

$$\exists c > 0: \quad \left(\sum_{i=1}^N \zeta_i\right)^r \leq c \sum \zeta_i^r \quad \forall 0 < r < \infty, \ \forall \zeta_i \in \mathbb{R}_+$$

to obtain

$$u^{p-1} + |v|^{p-1} \geq c(u + |v|)^{p-1},$$

concluding the proof. □

The proof of Proposition 3.22 is then an easy consequence of Lemma 3.24.

To illustrate the role applicability of Proposition 3.22, we consider two questions. First, it is known from basic properties of averaging, see Hirschman and Widder [152] or Hörmander [158], that

$$\overline{\mathbf{u}} \to \mathbf{u} \quad \text{as} \quad \delta \to 0$$

in various spaces, including $L^\infty(0,T; L^2(\Omega))$. Since the solution $\mathbf{w}$ to SLM eddy viscosity model (3.15), (3.16), with periodic boundary conditions approximates $\overline{\mathbf{u}}$, it is reasonable to ask if this limit consistency condition (see Chap. 6) holds for $\mathbf{w}$ as well, see Kaya and Layton [186, 188].

Theorem 3.25. *Let $\mathbf{u}$ be the solution to the NSE under periodic boundary conditions and let $\mathbf{w}$ be a solution to the SLM (3.15), (3.16). Suppose the energy dissipation rate is regular:*

$$\varepsilon(t) := \frac{2}{|\Omega|Re} \int_\Omega |\nabla^s \mathbf{u}|^2\, dx \in L^2(0,T), \tag{3.38}$$

and that $\mathbf{u} \in L^3(0,T; \mathbf{X}_3^1)$. Then, provided $\mathbf{f} \in L^2(\Omega \times (0,T))$,

$$\mathbf{w} \to \mathbf{u} \quad \text{as} \quad \delta \to 0, \quad \text{in} \quad L^\infty(0,T; [L^2(\Omega)]^3) \cap L^2(0,T; \mathbf{X}^1).$$

Remark 3.26. Condition (3.38) is a natural one, but it is unnecessarily strong. Using an inequality of Serrin [275], it can easily be relaxed be to the condition (2.31) we encountered in the study of smooth solutions to the NSE.

Proof. The idea is to rewrite the NSE satisfied by $\mathbf{u}$ to resemble (3.15), *i.e.* we add to both sides the term $(C_S\delta)^2(T_3(\mathbf{u}), \mathbf{v})$ to get

$$(\mathbf{u}_t, \mathbf{v}) + \frac{2}{Re}(\nabla \mathbf{u}, \nabla \mathbf{v}) + (\mathbf{u} \cdot \nabla \mathbf{u}, \nabla \mathbf{v}) + (C_S\delta)^2(T_3(\mathbf{u}), \mathbf{v})$$
$$= (\mathbf{f}, \mathbf{v}) + (C_S\delta)^2(T_3(\mathbf{u}), \mathbf{v}).$$

Next, (3.15) is subtracted from this to obtain an equation for $\phi = \mathbf{u} - \mathbf{w}$. The estimates in Proposition 3.22 are then used to show that $\|\mathbf{u} - \mathbf{w}\| \to 0$ as $\delta \to 0$. To proceed, subtraction gives

$$(\phi_t, \mathbf{v}) + \frac{2}{Re}(\nabla^s\phi, \nabla^s\mathbf{v}) + (\mathbf{u} \cdot \nabla\mathbf{u} - \mathbf{w} \cdot \nabla\mathbf{w}, \mathbf{v})$$
$$+ (C_S\delta)^2(T_3(\mathbf{u}) - T_3(\mathbf{w}), \mathbf{v}) = (\mathbf{f} - \bar{\mathbf{f}}, \mathbf{v}) + (T_3(\mathbf{u}), \mathbf{v}).$$

By setting $\mathbf{v} = \phi$, we note that

$$|(\mathbf{u} \cdot \nabla\mathbf{u} - \mathbf{w} \cdot \nabla\mathbf{w}, \mathbf{v})| = |(\phi \cdot \nabla\mathbf{u}, \phi)| \leq C\|\phi\|^{1/2}\|\nabla\mathbf{u}\| \, \|\nabla\phi\|^{3/2}$$
$$\leq \frac{1}{Re}\|\nabla^s\phi\|^2 + C(Re)\|\nabla\mathbf{u}\|^4\|\phi\|^2.$$

Thus, substituting $\mathbf{v} = \phi$ in the equation satisfied by ϕ gives

$$\frac{1}{2}\frac{d}{dt}\|\phi\|^2 + \frac{1}{Re}\|\nabla^s\phi\|^2 + (C_S\delta)^2(T_3(\mathbf{u}) - T_3(\mathbf{w}), \mathbf{u} - \mathbf{w})$$
$$\leq C(Re)\|\phi\|^2\|\nabla\mathbf{u}\|^4 + \frac{1}{2}\|\mathbf{f} - \bar{\mathbf{f}}\|^2 + (C_S\delta)^2(T_3(\mathbf{u}), \phi).$$

By monotonicity (Proposition 3.22),

$$(T_3(\mathbf{u}) - T_3(\mathbf{w}), \mathbf{u} - \mathbf{w}) \geq (C_S\delta^2)\|\nabla^s\phi\|_{L^3}^3$$

and Hölder's inequality implies also

$$|(T_3(\mathbf{u}), \phi)| = |(C_S\delta)^2(|\nabla^s\mathbf{u}| \, \nabla^s\mathbf{u}, \nabla^s\phi)|$$

$$\leq (C_S\delta)^2\|\nabla^s\phi\|_{L^3}\| \, |\nabla^s\mathbf{u}|^2\|_{L^{3/2}} = (C_S\delta)^2\|\nabla^s\phi\|_{L^3}\|\nabla^s\mathbf{u}\|_{L^3}^2$$

$$\leq (C_S\delta)^2\|\nabla^s\phi\|_{L^3}^3 + \frac{(C_S\delta)^2}{4}\|\nabla^s\mathbf{u}\|_{L^3}^3.$$

Combining these estimates gives

$$\frac{1}{2}\frac{d}{dt}\|\phi\|^2 + \frac{1}{Re}\|\nabla^s\phi\|^2 \leq C(Re)\|\phi\|^2\|\nabla\mathbf{u}\|^4 + \|\mathbf{f} - \bar{\mathbf{f}}\|^2 + \frac{(C_S\delta)^2}{4}\|\nabla^s\mathbf{u}\|_{L^3}^3.$$

Now by the regularity assumption that $\varepsilon(t) \in L^2(0,T)$, it follows that $\|\nabla u\| \in L^4(0,T)$. Thus, Gronwall's inequality implies

$$\|\phi(t)\|^2 + \frac{2}{Re} \int_0^t \|\nabla^s \phi(s)\|^2 \, ds$$

$$\leq C \left[\|\phi(0)\|^2 + \|\mathbf{f} - \overline{\mathbf{f}}\|_{L^2(0,T;L^2(\Omega))}^2 + \frac{(C_S \delta)^2}{4} \int_0^t \|\nabla^s \mathbf{u}\|_{L^3}^3(s) \, ds \right] e^{\int_0^t \|\varepsilon(s)\|^2 \, ds}$$

By the hypotheses of the theorem and the properties of the averaging operator, all terms on the right-hand side vanish as $\delta \to 0$. $\square$

Remark 3.27. The assumption that $\mathbf{u} \in L^3(0,T;\mathbf{X}_3^1)$ is needed to ensure that the approximate subfilter-scale stress $\mathcal{S}(\mathbf{u},\mathbf{u}) := -(C_S\delta)^2 |\nabla^s \mathbf{u}| \nabla^s \mathbf{u}$ is regular enough that

$$(C_S\delta)^2 \int_0^T \|\mathcal{S}(\mathbf{u},\mathbf{u})\|_{L^{3/2}}^{3/2} \, dt = (C_S\delta)^2 \int_0^T \|\nabla^s \mathbf{u}\|_{L^3}^3 \, dt \to 0 \quad \text{as} \quad \delta \to 0.$$

Since $\text{Trace}[\mathcal{S}(\mathbf{u},\mathbf{u})] = 0$ in the SLM (3.15), (3.16) (because of the incompressibility condition $\nabla \cdot \mathbf{u} = 0$), it is important to calculate the modeling consistency error by comparing $\mathcal{S}(\mathbf{u},\mathbf{u})$ with

$$\boldsymbol{\tau}^*(\mathbf{u},\mathbf{u}) := \boldsymbol{\tau}(\mathbf{u},\mathbf{u}) - \frac{1}{3} \text{Trace}[\boldsymbol{\tau}(\mathbf{u},\mathbf{u})] \, \mathbb{I}.$$

A similar argument, using Proposition 3.22, can be used to bound the model error $\|\overline{\mathbf{u}} - \mathbf{w}\|$ in terms of the model's consistency $\|\boldsymbol{\tau}^*(\mathbf{u},\mathbf{u}) - \mathcal{S}(\overline{\mathbf{u}},\overline{\mathbf{u}})\|$. To develop this bound, recall that $\overline{\mathbf{u}}$ satisfies

$$\overline{\mathbf{u}}_t + \nabla \cdot (\overline{\mathbf{u}}\,\overline{\mathbf{u}}^T) - \frac{1}{Re}\triangle\overline{\mathbf{u}} + \nabla\left(\overline{p} + \frac{1}{3}\tau_{ii}(\mathbf{u},\mathbf{u})\right) + \nabla \cdot \boldsymbol{\tau}^*(\mathbf{u},\mathbf{u}) = \overline{\mathbf{f}}.$$

Let $\phi = \overline{\mathbf{u}} - \mathbf{w}$. Subtracting (3.15) from this equation gives, in a variational form,

$$(\phi_t, \mathbf{v}) + (\overline{\mathbf{u}} \cdot \nabla\overline{\mathbf{u}} - \mathbf{w} \cdot \nabla\mathbf{w}, \mathbf{v}) + \frac{2}{Re}(\nabla^s\phi, \nabla^s\mathbf{v})$$
$$- (\mathcal{S}(\overline{\mathbf{u}},\overline{\mathbf{u}}) - \mathcal{S}(\mathbf{w},\mathbf{w}), \nabla^s\mathbf{v}) = -(\mathcal{S}(\overline{\mathbf{u}},\overline{\mathbf{u}}) - \boldsymbol{\tau}^*(\mathbf{u}\,\mathbf{u}), \nabla^s\mathbf{v}),$$

for any $\mathbf{v} \in L^\infty(0,T;L_\sigma^2) \cap L^2(0,T;\mathbf{X}^1)$.

Setting $\mathbf{v} = \phi$, gives

$$\frac{1}{2}\frac{d}{dt}\|\phi\|^2 + \frac{2}{Re}\|\nabla^s\phi\|^2 + (\overline{\mathbf{u}} \cdot \nabla\overline{\mathbf{u}} - \mathbf{w} \cdot \nabla\mathbf{w}, \phi)$$
$$+ (C_S\delta)^2(T_3(\overline{\mathbf{u}}) - T_3(\mathbf{w}), \overline{\mathbf{u}} - \mathbf{w}) = (\boldsymbol{\tau}^*(\mathbf{u},\mathbf{u}) - \mathcal{S}(\overline{\mathbf{u}},\overline{\mathbf{u}}), \nabla^s\phi). \tag{3.39}$$

It is clear from this that the modeling error $\|\overline{\mathbf{u}} - \mathbf{w}\|$ satisfies an equation driven by the model's consistency

$$|||\boldsymbol{\tau}^*(\mathbf{u}\,\mathbf{u}) - \mathcal{S}(\overline{\mathbf{u}}\,\overline{\mathbf{u}})|||$$

in an appropriate norm $|||\,.\,|||$. This quantity is evaluated at the true solution of the NSE. Thus, it can be assessed by performing a DNS or taking experimental data to compute $\boldsymbol{\tau}^*(\mathbf{u}\,\mathbf{u}) - \mathcal{S}(\overline{\mathbf{u}}\,\overline{\mathbf{u}})$ directly. Furthermore, Eq. (3.39) suggests that if the model is stable to perturbations, then a small consistency error (which is observable) leads to a small modeling error. In other words, the model can be verified by experiment.

We have also the following result, see [186].

Theorem 3.28. *Let $\mathbf{u}$ be a solution to the NSE under periodic boundary conditions and suppose $\varepsilon(t) := \frac{2}{|\Omega|Re} \int_\Omega |\nabla^s \mathbf{u}|^2 \, d\mathbf{x} \in L^2(0,T)$. Suppose $\boldsymbol{\tau}^*(\overline{\mathbf{u}}, \overline{\mathbf{u}}) := (\overline{\mathbf{u}\,\mathbf{u}^T} - \overline{\mathbf{u}}\,\overline{\mathbf{u}}^T) - \frac{1}{3}\mathrm{Trace}[\overline{\mathbf{u}\,\mathbf{u}^T} - \overline{\mathbf{u}}\,\overline{\mathbf{u}}^T]\,\mathbb{I} \in L^2(\Omega \times (0,T))$ and that $\mathcal{S}(\overline{\mathbf{u}}, \overline{\mathbf{u}}) \in L^2(\Omega \times (0,T))$. Then,*

$$\begin{aligned}
&\|\overline{\mathbf{u}} - \mathbf{w}\|^2_{L^\infty(0,T;L^2)} + \frac{2}{Re}\|\nabla(\overline{\mathbf{u}} - \mathbf{w})\|^2_{L^2(0,T;L^2)} \\
&+ (C_S\delta)^2 \|\nabla^s(\overline{\mathbf{u}} - \mathbf{w})\|^3_{L^3(0,T;L^3)} \\
&\leq C(Re, \|\overline{\varepsilon}\|_{L^2(0,T)})\|\mathcal{R}^*(\mathbf{u},\mathbf{u}) - \mathcal{S}(\overline{\mathbf{u}}, \overline{\mathbf{u}})\|^2_{L^2(\Omega\times(0,T))}.
\end{aligned}$$

Proof. In (3.39) we use the lower monotonicity result for the r-Laplacian for $T_3(\,\cdot\,)$ and the hypotheses for the terms on the right-hand side. This gives, for $\phi = \overline{\mathbf{u}} - \mathbf{w}$:

$$\begin{aligned}
\frac{1}{2}\frac{d}{dt}\|\phi\|^2 &+ \frac{2}{Re}\|\nabla^s\phi\|^2 + (C_S\delta)^2\|\nabla^s\phi\|^3_{L^3} \\
&\leq |(\overline{\mathbf{u}}\cdot\nabla\overline{\mathbf{u}}, \phi) - (\mathbf{w}\cdot\nabla\mathbf{w}, \phi)| + \frac{1}{Re}\|\nabla^s\phi\|^2 \\
&+ C(Re)\,\|\boldsymbol{\tau}^*(\mathbf{u},\mathbf{u}) - \mathcal{S}(\overline{\mathbf{u}}, \overline{\mathbf{u}})\|^2.
\end{aligned}$$

As in the proof of Theorem 3.25,

$$|(\overline{\mathbf{u}}\cdot\nabla\overline{\mathbf{u}} - \mathbf{w}\cdot\nabla\mathbf{w}, \phi)| = |(\phi\cdot\nabla\overline{\mathbf{u}}, \phi)| \leq \frac{1}{2Re}\|\nabla\phi\|^2 + C(Re)\|\nabla\overline{\mathbf{u}}\|^4\|\phi\|^2.$$

Thus,

$$\begin{aligned}
\frac{d}{dt}\|\phi\|^2 &+ \frac{1}{Re}\|\nabla^s\phi\|^2 + (C_S\delta)^2\|\nabla^s\phi\|^3_{L^3} \\
&\leq C(Re)\Big[\|\nabla\overline{\mathbf{u}}\|^4\|\phi\|^2 + \|\boldsymbol{\tau}^*(\mathbf{u},\mathbf{u}) - \mathcal{S}(\overline{\mathbf{u}}, \overline{\mathbf{u}})\|^2\Big].
\end{aligned}$$

Most filtering processes are smoothing. Thus, for most filters

$$\|\nabla\overline{\mathbf{u}}\| \leq C(\delta)\|\mathbf{u}\| \leq C(\delta)\,C(\text{data}).$$

Thus, the assumption that $\|\nabla\overline{\mathbf{u}}\| \in L^4(0,T)$ can be unnecessary. However, some filters, such as the top-hat filter, are not smoothing and this assumption is necessary.

In all cases (due to the assumption or properties of filters) $\|\nabla\overline{\mathbf{u}}\|^4(t) \in L^1(0,T)$ and Gronwall's inequality can be applied to complete the proof. $\square$

The key ingredients of the proofs of Theorems 3.25 and 3.28 are:

Condition 1: Stability of the LES model to data perturbation.
Condition 2: Enough regularity of the true solution to apply Gronwall's inequality. To these two, should be added
Condition 3: A model with small modeling error.

The second condition holds at least over small time intervals. It is unknown whether it holds more generally (and it is connected with the famous uniqueness question for weak solutions in three dimensions). Thus, to obtain a result over $O(1)$ time intervals, we need an extra regularity assumption on the energy dissipation rate of the true solution of the NSE.

The first condition is generally satisfied by very stable models, such as eddy viscosity models. However, these models fail the third condition typically. Models which satisfy the third condition, typically fail the first one. The quest for "universality" in LES models can simply be stated to be a search for a model satisfying conditions 1 and 3!

Remark 3.29. We conclude this section by observing that there is intense activity in the study of existence and regularity of solutions for problems involving the r-Laplacian, also in the case in which $r-2$ is negative (this corresponds to $1 < p < 2$ in Lemma 3.24). The interest in such cases comes from the modeling of power-law fluids more than from the study of turbulence. Examples of fluids with governing equations involving $1 < p < 2$ are, for instance, electrorheological fluids, fluids with pressure-depending viscosities or, in general, fluids with the property of shear thickening, *i.e.* with viscosity increasing as $|\nabla^s \mathbf{u}|$ increases. The reader can find several results, together with an extensive bibliography on recent advances on this topic in Málek, Nečas, and Ružička [229] and in Frehse and Málek [116].

3.5 Backscatter and the Eddy Viscosity Models

We close this chapter with a few remarks on an interesting and important phenomenon – the backscatter.

While, *on average*, energy is transferred from large scales to smaller scales ("forward-scatter"), it has been proven that the inverse transfer of energy from small to large scales ("backscatter") may be quite significant and should be included in the LES model. The action of the backscatter in the energy cascade context is illustrated in Fig. 3.3. Backscatter does not contradict the energy cascade concept: the average energy transfer is from the large scales to the small ones (*i.e.* from the small wavenumbers to the large ones). This transfer is called forward-scatter and is denoted by ϵ_{FS} in Fig. 3.3. However, at certain instances in time and space, there is an inverse transfer of energy, denoted by ϵ_{BS} in Fig. 3.3.

Piomelli *et al.* [254] have performed DNS of transitional and turbulent channel flows and compressible isotropic turbulence. In all flows considered,

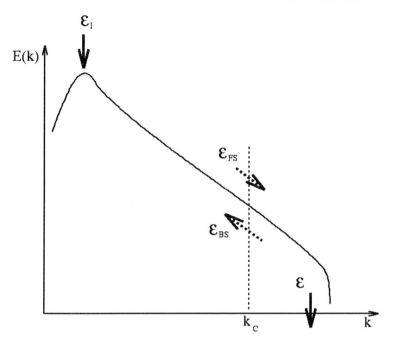

Fig. 3.3. Schematic of the energy cascade

approximately 50% of the grid points experienced backscatter when a Fourier cutoff filter was used, and somewhat less when a Gaussian filter or a box filter were used. It is now generally accepted that an LES model should include backscatter.

Eddy viscosity models, in their original form, cannot include backscatter, being purely dissipative. An example in this class is the Smagorinsky model (3.15), (3.16). Its inability to include backscatter is believed to be one of the sources of its relatively low accuracy in many practical flows. To include backscatter, the Smagorinsky model is usually used in its dynamic version, proposed by Germano *et al.* [129]. However, care needs to be taken, since the resulting model can be unstable in numerical simulations. Thus, the dynamic version of the Smagorinsky model is usually used with some limiters for C_S.

There are some LES models which introduce backscatter in a *natural* way. Some of these will be presented in Chaps. 7 and 8.

3.6 Conclusions

In this chapter we introduced eddy viscosity methods, probably the oldest methods for studying and describing turbulence. We mainly focused on the Smagorinsky model (3.11) and some of its simple variations.

The interest in this model is two-fold. First, due to its clear energy balance it works as a paradigm in the design of advanced and more sophisticated models, such as those we will introduce in Chap. 4. The link with the K41-theory is also very appealing and it will work again as a guideline.

Second, the mathematical analysis of this eddy viscosity method – initiated by Ladyžhenskaya – uses relevant tools (such as monotone operators) that can also be successfully employed in the analysis of different models. (The study of equations similar to (3.11) is currently a very active area of mathematical research.)

Even if some delicate mathematical points are not yet completely known, the stability estimates that can be derived for (3.11) are also useful to derive results of consistency. For the numerical analysis of this model, we refer to the exquisite presentation of John [175].

4

Improved Eddy Viscosity Models

4.1 Introduction

The connection between turbulent fluctuations and the choice

$$\nu_T = (C_S \delta)^2 \, |\nabla^s \overline{\mathbf{u}}|$$

of the Smagorinsky's model eddy viscosity seems tenuous. Thus, it is natural to seek other choices of ν_T with a more direct connection with turbulence modeling.

Boussinesq based his model upon the analogy between perfectly elastic collisions and interaction of small eddies. Within this reasoning (whose "optimism" he surely understood) it is clear that the amount of turbulent mixing should depend mainly on the local kinetic energy in the turbulent fluctuations, k':

$$\nu_T = \nu_T(\delta, \overline{k'}), \qquad k'(\mathbf{x}) = \frac{1}{2} |\mathbf{u}'|^2(\mathbf{x}).$$

The simplest functional form which is dimensionally consistent is the, so-called, Kolmogorov–Prandtl relation [258], given by

$$\nu_T \cong C \delta \sqrt{\overline{k'}}, \qquad k' = \frac{1}{2} |\mathbf{u}'|^2. \tag{4.1}$$

One method of estimating k' is to follow the approach taken in the $k - \varepsilon$ conventional turbulence model (see Sect. 4.3 for a brief description and Mohammadi and Pironneau [239] or Coletti [67] for further details) and solve an approximate energy equation. For conventional turbulence models, $O(1)$ structures are modeled and this extra work is justified. In LES, the idea is to use simple, economical models because only small structures are to be modeled. This has the advantage of avoiding the modeling steps needed in deriving the k and ε equation. Further, the recent important work of Duchon and Robert [97] on the energy equation of turbulence shows the correctness of solving a strong form of the energy equation for k to be unclear.

In keeping with the ideas of LES, a simple and direct estimate of k' is obtained by scale-similarity (Chap. 8): the best estimate for the kinetic energy in the unresolved scales is that of the smallest resolved scales. This approach was first taken (to our knowledge) by Horiuti [156, 157] and validated computationally by Sagaut and Lê [268]. Mathematical development and extension of these ideas was begun in Iliescu and Layton [170] and Layton and Lewandowski [208].

To illustrate this, recall that if we assume the turbulence is homogeneous and isotropic, then $E(k)$ is given by the K-41 theory in the inertial range by

$$E(k) \sim \alpha\, \varepsilon^{2/3}\, k^{-5/3} \qquad \text{for} \qquad 0 < k \le \eta \sim \left(\frac{\varepsilon}{\nu^3}\right)^{1/4}.$$

We have, by direct calculation

$$k' = \frac{1}{2} \int_{k_1}^{\eta} E(k)\, dk = (\text{inserting (4.1)}) = \frac{3}{4}\, \alpha\, \varepsilon^{2/3} \left[\pi^{-2/3} \delta^{2/3} - \nu^{-1/2}\, \varepsilon^{1/6}\right].$$

If we consider the asymptotic limit of very large Reynolds numbers (or very small ν), then we can estimate k' roughly by

$$k' \sim \frac{1}{2} \int_{k_1}^{\infty} E(k)\, dk = \frac{3}{4}\, \pi^{-2/3}\, \alpha\, \varepsilon^{2/3}\, \delta^{2/3}.$$

Consider now a scale-similarity (see Chap. 8 for further details) estimate of k':

$$k' = \text{energy in scales between } 0 \text{ and } O(\delta)$$
$$\approx \text{constant} \times [\text{energy in resolved scales between } O(\delta) \text{ and } O(2\delta)].$$

This gives, by direct calculation

$$
\begin{aligned}
k_{scale\text{-}similarity} &:= \frac{1}{2} \int_{\frac{1}{2}k_1}^{k_1} E(k)\, dk = \frac{1}{2} \int_{\frac{1}{2}k_1}^{k_1} \alpha\, \varepsilon^{2/3}\, k^{-5/3} \\
&= \left(\frac{1}{2}\, \alpha\, \varepsilon^{2/3}\right)\left(-\frac{3}{2}\right)\left[\left(\frac{\pi}{\delta}\right)^{-2/3} - \left(\frac{\pi}{\sqrt{2}\delta}\right)^{-2/3}\right] \\
&= \frac{3}{4}\, \alpha\, \varepsilon^{2/3}\, \pi^{-2/3}\, \delta^{2/3}\left(2^{1/3} - 1\right).
\end{aligned}
\tag{4.2}
$$

Thus, to within the accuracy of the K-41 theory [117], and the approximation (4.2),

$$k' = \left(2^{1/3} - 1\right) k_{scale\text{-}similarity}.$$

Now, $k_{scale\text{-}similarity} = \frac{1}{2} \int_{\Omega} |\overline{\mathbf{u}} - \overline{\overline{\mathbf{u}}}|^2\, d\mathbf{x}$ for a Gaussian filter. Thus, we have a computable estimation of k' in terms of resolved quantities given by

$$k'(\mathbf{x}, t) \sim \left(2^{1/3} - 1\right) \frac{1}{2} \left|\mathbf{u} - \overline{\mathbf{u}}\right|^2 (\mathbf{x}, t).$$

An alternate route to formulas of this type is to simply "model" $\frac{1}{2} \left|\mathbf{u} - \overline{\mathbf{u}}\right|^2$ by $\frac{1}{2} \left|\mathbf{u} - \overline{\overline{\mathbf{u}}}\right|^2$. This leads to the same approximation

$$\overline{k'} = \overline{\frac{1}{2} \left|\mathbf{u} - \overline{\mathbf{u}}\right|}^2 \sim (2^{1/3} - 1) \frac{1}{2} \mu_0 \left|\mathbf{u} - \overline{\mathbf{u}}\right|^2, \qquad \mu_0 = \text{constant}.$$

Inserting these approximations into (4.1) gives the LES eddy viscosity model:

$$\mathbf{w}_t + \nabla \cdot (\mathbf{w}\,\mathbf{w}^T) + \nabla q - \frac{1}{Re} \Delta \mathbf{w} - \nabla \cdot (\mu_0\, \delta \left|\mathbf{w} - \overline{\mathbf{w}}\right| \nabla^s \mathbf{w}) = \overline{\mathbf{f}}, \quad (4.3)$$

$$\nabla \cdot \mathbf{w} = 0. \quad (4.4)$$

The parameter μ_0 can either be determined dynamically or estimated by adapting the approach of Lilly [220].

Remark 4.1. It is possible to find improved estimates of k' by using more information from the resolved scales. For example, a more accurate approximation of $\mathbf{u}$ from the resolved scales is $\mathbf{u} \approx 3\,\overline{\mathbf{u}} - 3\,\overline{\overline{\mathbf{u}}} + \overline{\overline{\overline{\mathbf{u}}}}$ (see Chap. 8). This gives the approximation $\mathbf{u} - \overline{\mathbf{u}} \approx 2\,\overline{\mathbf{u}} - 3\,\overline{\overline{\mathbf{u}}} + \overline{\overline{\overline{\mathbf{u}}}}$. Thus, the first example more accurate than the above is

$$k' \approx \frac{1}{2} \left|2\,\overline{\mathbf{u}} - 3\,\overline{\overline{\mathbf{u}}} + \overline{\overline{\overline{\mathbf{u}}}}\right|^2.$$

The LES Eddy Viscosity Model (4.3)

The eddy viscosity model (4.3) is much less dissipative than the Smagorinsky model. Indeed, in smooth regions $|\nabla^s \mathbf{w}| = O(1)$, while where $\mathbf{w}$ undergoes an $O(1)$ change across the smallest length scale δ, $|\nabla^s \mathbf{w}| = O(\delta^{-1})$. Thus,

$$\nu_{\text{Smag}} = (C_s \delta)^2 \,|\nabla^s \mathbf{w}| = \begin{cases} O(\delta^2) & \text{in smooth regions,} \\ O(\delta) & \text{for fluctuations,} \end{cases}$$

while (recall that $\|\mathbf{w} - \overline{\mathbf{w}}\| = O(\delta^2)$ in smooth regions),

$$\nu_T = \mu_0 \,\delta \,|\mathbf{w} - \overline{\mathbf{w}}| = \begin{cases} O(\delta^3) & \text{in smooth regions,} \\ O(\delta) & \text{for fluctuations.} \end{cases}$$

Since (4.3) is an EV model, its energy budget is clear (Proposition 3.2). Nevertheless, the fact that $\nu_T(\mathbf{w})$ can be unbounded places the model (4.3) outside the usual Leray–Lions theory for verifying existence of a distributional solution to the model. The mathematical elucidation of model (4.3) was begun in Layton and Lewandowski [208]. It is again based upon the global energy equality of EV methods.

Definition 4.2. *Let $\mu_0 > 0$ be fixed and consider (4.3) and (4.4) subject to periodic boundary conditions. Then, $\mathbf{w}$ is a distributional solution of (4.3) and (4.4) if*

$$\mathbf{w} \in Y = \left\{ closure\ of\ C^\infty(0, T; C_{per}^\infty)\ in\ \mathbf{v} \in L^\infty(0, T; L_\sigma^2) \cap L^2(0, T; H_\sigma^1(\Omega)) \right\}$$

and for all $\phi \in C^\infty(0, T; C_{per}^\infty)$ such that $\nabla \cdot \phi = 0$ and $\phi(T, \cdot) = \mathbf{0}$,

$$\int_\Omega \mathbf{w}_0(\mathbf{x}) \cdot \phi(\mathbf{x}, 0)\, d\mathbf{x} - \int_0^T \int_\Omega \mathbf{w} \frac{\partial \phi}{\partial t}\, d\mathbf{x}\, dt$$

$$- \int_0^T \int_\Omega (\mathbf{w}\, \mathbf{w}^T) : \nabla \phi + \nu_T(\mathbf{w}) \nabla^s \mathbf{w} : \nabla^s \phi\, d\mathbf{x}\, dt = \int_0^T \int_\Omega \overline{\mathbf{f}} \cdot \phi\, d\mathbf{x}\, dt.$$

The symbol C_{per}^∞ denotes the space of smooth periodic functions.

In [208] existence of distributional solutions to the model (4.3), (4.4) was proven. Here we just state the main result and give some ideas of the technique used in the proof.

Theorem 4.3 (Theorem 3.1 of [208]). *For $\overline{\mathbf{u}}_0 \in L^2(\Omega)$ and $\overline{\mathbf{f}} \in L^2(\Omega \times (0, T))$, there exists at least one distributional solution to the model (4.3), (4.4).*

The theory behind this result also includes many filters, even differential filters, and many eddy viscosities $\nu_T(\mathbf{w})$ which minimally satisfy the following three consistency and growth conditions: for all $\mathbf{w} \in Y$

1. $\nu + \nu_T(\mathbf{w}) \geq C_0 > 0$,
2. $\nu_T(\mathbf{w}) \in L^\infty(0, T; L^2)$,
3. $\|\nu_T(\mathbf{w})\|_{L^\infty(0,T;L^2)} \leq C\big(1 + \|\mathbf{w}\|_{L^\infty(0,T;L^2)}\big)$.

The proof of this theorem uses Lewandowski's theory of truncated transport (see [216, 217, 208]). This theory is quite technical in detail, but simple in conception: an unbounded nonlinearity is truncated to be bounded between $-n$ and n, producing an approximate solution $\mathbf{w}_n$. The intricate mathematical details (for which we refer the reader to [216, 217, 208]) lie in extracting the limit of $\mathbf{w}_n$ as $n \to \infty$ and showing it to be a solution of the original equations in a meaningful sense.

Experiments with model (4.3) have, so far, been positive. Preliminary tests of turbulent channel flow of Iliescu and Fischer [167] indicate that model (4.3) replicates the standard turbulent statistics reasonably well. For a thorough description of the computational setting and the usual statistics of the turbulent channel flow (Fig. 12.1), one of the most popular test problems for LES validation, the reader is referred to Chap. 12.

We computed statistics of the mean velocity $\overline{u}$ (Fig. 4.1), of the off-diagonal Reynolds stresses $u'\,v'$ (Fig. 4.2), and of the root mean square of the streamwise velocity fluctuations $u'\,u'$ (Fig. 4.3). A detailed description of the quantities presented in Figs. 4.1–4.3 is given in Chap. 12. The three statistics for the LES model (4.3) were compared with the fine DNS results of Moser, Kim and Mansour [242], which were used as a benchmark. We are currently comparing model (4.3) with other EV LES models, such as Smagorinsky [277], in the numerical simulation of channel flows [40]. John has conducted extensive tests of $\nu_T = \mu_0 \, \delta \, |\mathbf{w} - \overline{\mathbf{w}}|$ as an eddy viscosity term incorporated into a mixed model, also with good results [173, 174].

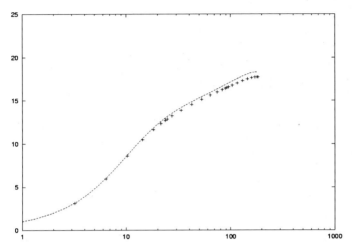

Fig. 4.1. Turbulent channel flow simulations, $Re_\tau = 180$. Statistics of the mean streamwise velocity $\langle \overline{u} \rangle$ for model (4.3) (+) and the fine DNS in [242] ($\cdot$)

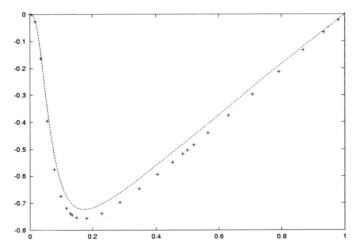

Fig. 4.2. Turbulent channel flow simulations, $Re_\tau = 180$. Statistics of the off-diagonal Reynolds stresses $\langle u' \, v' \rangle$ for model (4.3) (+) and the fine DNS in [242] ($\cdot$)

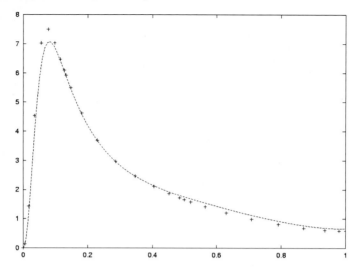

Fig. 4.3. Turbulent channel flow simulations, $Re_\tau = 180$. Statistics of the root mean square of the streamwise velocity fluctuations $\langle u'\, u' \rangle$ for model (4.3) (+) and the fine DNS in [242] ($\cdot$)

The eddy viscosity $\nu_T = \mu_0\,\delta\,|\mathbf{w} - \overline{\mathbf{w}}|$ has the simplest form and most direct connection with the physical ideas of turbulent mixing, so it is not surprising that closely related models have been independently tested in practical computations. In particular, interesting work has been done by Horiuti [156] upon scale-similarity models in general and models like the present ν_T. Sagaut and Lê [268] have tested geometric averages of ν_T and ν_{Smag}:

$$\nu = \nu_T^\theta\,\nu_{\text{Smag}}^{(1-\theta)} = C\,\delta^{2-\theta}\,|\mathbf{w} - \overline{\mathbf{w}}|^\theta\,|\nabla^s\mathbf{w}|^{1-\theta},$$

in some very challenging compressible flow problems.

What is surprising is that these models, which are simple to implement and give better results than the Smagorinsky model, have not yet replaced the Smagorinsky model in engineering calculations.

Dimensionally Equivalent Models

Not all models that are *dimensionally equivalent* can be expected to *perform analogously*. Thus, there is a real interest in exploring dimensionally equivalent versions of the model to test their differences, relative advantages and disadvantages. Surprisingly, it is an open problem to test and compare the three which come immediately to mind:

$$\nu_T = \mu_0\,\delta\,|\mathbf{w} - \overline{\mathbf{w}}|,\ \text{that is the model (4.3)}, \tag{4.5}$$

$$\nu_T = \mu_1\,\delta^2\,|\nabla^s(\mathbf{w} - \overline{\mathbf{w}})|, \tag{4.6}$$

$$\nu_T = \mu_2\,\delta^3\,|\Delta(\mathbf{w} - \overline{\mathbf{w}})|. \tag{4.7}$$

The model (4.6) was studied and tested by Hughes, Mazzei, and Jansen [160] who called it the "small-large Smagorinsky model". To date, an abstract theory for (4.6) has not (to our knowledge) been developed, but it seems attainable by using the mathematical tools of Ladyženskaya [195, 197]. The model (4.7) seems appealing computationally, but a mathematical development of it seems beyond current techniques. The Gaussian–Laplacian model of [170], which we present next, is a better candidate for a robust model.

4.2 The Gaussian–Laplacian Model

The Gaussian–Laplacian model is again an EV model based on Boussinesq's analogy presented at the beginning of this chapter. However, in contrast with the models introduced in Sect. 4.1 which are based essentially on the scale-similarity assumption, the Gaussian–Laplacian model is based on a different approach, the *approximate deconvolution*. Chapter 7 gives a detailed presentation of the approximate deconvolution. In this section, we just present the main idea in approximate deconvolution (*i.e.* use $\overline{\mathbf{u}}$ to obtain an approximation for $\mathbf{u}'$), and use it to get an approximation for k'. To do this, we will follow the presentation in [170].

Since $\mathbf{u}' = \mathbf{u} - \overline{\mathbf{u}}$, taking Fourier transforms and using the convolution theorem, we have $\widehat{\mathbf{u}'} = \widehat{\mathbf{u} - \overline{\mathbf{u}}} = \widehat{\mathbf{u}} - \widehat{g_\delta}(\mathbf{k})\,\widehat{\mathbf{u}}$ hence $\widehat{\mathbf{u}'} = (\widehat{g_\delta}(\mathbf{k})^{-1} - 1)\,\widehat{g_\delta}(\mathbf{k})\,\widehat{\mathbf{u}}$, or

$$\widehat{\mathbf{u}'}(\mathbf{k}) = \left(\frac{1}{\widehat{g_\delta}(\mathbf{k})} - 1 \right) \widehat{\overline{\mathbf{u}}}(\mathbf{k}). \tag{4.8}$$

The *key feature* of the Gaussian is its smoothing property which is equivalent to the decay of $\widehat{g_\delta}(\mathbf{k})$ as $|\mathbf{k}| \to \infty$. Thus, Taylor series expansion of $\widehat{g_\delta}(\mathbf{k})$,

$$\widehat{g_\delta}(\mathbf{k}) = 1 - \frac{\delta^2}{4\gamma} |\mathbf{k}|^2 + \dots,$$

(which have the opposite behavior) are not appropriate. The simplest approximation preserving this is the subdiagonal $(0, 1)$-Padé approximation, see Chap. 7 for details:

$$\widehat{g_\delta}(\mathbf{k}) = \frac{1}{1 + \frac{\delta^2}{4\gamma} |\mathbf{k}|^2} + O(\delta^4), \tag{4.9}$$

where γ is the constant in the definition of the Gaussian filter g_δ. The approximation (4.9), used in (4.8), gives:

$$\widehat{\mathbf{u}'}(\mathbf{k}) = \frac{\delta^2}{4\gamma} |\mathbf{k}|^2 \widehat{\overline{\mathbf{u}}}(\mathbf{k}) + O(\delta^4).$$

This gives the approximation to the kinetic energy of the turbulent fluctuations:

$$\overline{k'} = \frac{1}{2}\frac{\delta^2}{4\gamma}|\overline{\Delta\mathbf{w}}|^2 + O(\delta^4)$$

and consequently of the turbulent viscosity coefficient

$$\nu_T = \mu_3 \frac{\delta^3}{4\gamma}|g_\delta * \Delta\mathbf{w}|.$$

Here again a choice must be made regarding an outer or inner convolution, *i.e.* should the model be $|g_\delta * \Delta\mathbf{w}|^2$ (as we are inclined to believe) or $g_\delta * |\Delta\mathbf{w}|^2$ (which is also possible)? We have chosen the former. The resulting model is the Gaussian–Laplacian (GL):

$$\mathbf{w}_t + \nabla \cdot (\mathbf{w}\,\mathbf{w}^T) - \nabla q - \frac{1}{Re}\Delta\mathbf{w} - \nabla \cdot \left(\mu_3 \frac{\delta^3}{4\gamma}|g_\delta * \Delta\mathbf{w}|\nabla^s\mathbf{w}\right) = \bar{\mathbf{f}}, \quad (4.10)$$

$$\nabla \cdot \mathbf{w} = 0. \quad (4.11)$$

It is interesting to note that ν_T is active for high local fluctuations (or local curvatures) rather than gradients. In particular, for shear flows, $|\nabla\mathbf{w}|$ can be large, while $|\Delta\mathbf{w}| = 0$. Thus, the GL model (4.10) has many apparent advantages over the Smagorinsky model [277].

The eddy viscosity in (4.10) is bounded, thanks to the regularization via convolution by a Gaussian. Thus, it is possible in Theorem 4.3 to extend the Leray–Lions theory of weak solutions of the NSE to the GL model (4.10). We do this in Sect. 4.2.1.

The extra eddy viscosity term in (4.10) is called a Gaussian–Laplacian. It has other interesting mathematical properties and has been used for image smoothing in Mikula and Sgallari [236] and Catté *et al.* [57]. Interestingly enough, the structure of the Gaussian–Laplacian model (4.10) (with its explicit regularization) will come back also in the study (more specifically in the numerical implementation [175]) of the Rational model, see Chap. 7.

4.2.1 Mathematical Properties of the Gaussian–Laplacian Model

This section considers the question of existence of weak solutions to the system (4.10). Thus, we seek $(\mathbf{w}, q)$ satisfying:

$$\mathbf{w}_t + \nabla \cdot (\mathbf{w}\,\mathbf{w}^T) + \nabla q - \frac{1}{Re}\Delta\mathbf{w} - \nabla \cdot \left(\mu_3 \frac{\delta^3}{4\gamma}|g_\delta * \Delta\mathbf{w}|\nabla^s\mathbf{w}\right) = \bar{\mathbf{f}}, \quad (4.12)$$

$$\nabla \cdot \mathbf{w} = 0, \quad (4.13)$$

$$\mathbf{w}(\mathbf{x}, 0) = g_\delta * \mathbf{u}_0(\mathbf{x}), \quad \text{for } \mathbf{x} \in \Omega, \quad (4.14)$$

$$\mathbf{w}(\mathbf{x}, t) = 0, \quad \text{for } \mathbf{x} \in \partial\Omega \text{ and } t > 0. \quad (4.15)$$

The Dirichlet boundary conditions we take in (4.15) are convenient for studying the existence of solutions. The existence result we give also holds if the model is studied subject to periodic boundary conditions. It is known, however, that for modeling accuracy near $\partial\Omega$ and computational efficiency, the boundary condition (4.15) should be replaced, see Chap. 10.

Theorem 4.4. *Let $T > 0$, and Ω be a bounded domain in $\mathbb{R}^d$, $d = 2, 3$. Then, for any given $\mathbf{u}_0 \in L_\sigma^2$ and $\mathbf{f} \in L^2(0, T; L_\sigma^2)$, there exists at least one weak solution to (4.12)–(4.15) in $\Omega \times (0, T)$. This weak solution satisfies the energy inequality*

$$k(t) + \int_0^t \epsilon_{GL}(t') \, dt' \le k(0) + \int_0^t P(t') \, dt',$$

where $k(t) = \dfrac{1}{2} \int_\Omega |\mathbf{w}|^2 \, dx, \qquad P(t) = \int_\Omega \overline{\mathbf{f}} \cdot \mathbf{w} \, dx,$ *and*

$$\epsilon_{GL}(t) = \int_\Omega \frac{1}{Re} |\nabla\mathbf{w}|^2 + \mu_3 \frac{\delta^3}{4\gamma} |g_\delta * \Delta\mathbf{w}| \, |\nabla^s \mathbf{w}|^2 \, dx.$$

Remark 4.5. The model (4.12)–(4.15) without regularization is more difficult due to the unbounded coefficient $|\Delta\mathbf{w}|$ in (4.12). Appropriate mathematical tools for such problems are in their early stages of development; see Gallouët et al. [125].

Proof (of Theorem 4.4). We follow the existence proof in the NSE case, see Sect. 2.4. We shall use the Faedo–Galerkin method. Let $\mathcal{D}(\Omega) = \{\psi \in C_0^\infty(\Omega) : \nabla \cdot \psi = 0 \text{ in } \Omega\}$, L_σ^2 the completion of $\mathcal{D}(\Omega)$ in $L^2(\Omega)$, $H_{0,\sigma}^1(\Omega)$ the completion of $\mathcal{D}(\Omega)$ in $W^{1,2}(\Omega)$ and $\{\psi_r\} \subset \mathcal{D}(\Omega)$ be the orthonormal basis of L_σ^2 given in Lemma 2.3 [121]. We shall look for approximate solutions $\mathbf{w}_k$ of the form:

$$\mathbf{w}_k(\mathbf{x}, t) = \sum_{r=1}^k c_{kr}(t) \, \psi_r(\mathbf{x}) \qquad k \in \mathbb{N},$$

where the coefficients c_{kr} are required to satisfy the following system of ordinary differential equations:

$$\frac{dc_{kr}}{dt} + \sum_{i=1}^k \frac{1}{Re} (\nabla\psi_i, \nabla\psi_r) \, c_{ki} + \sum_{i,s=1}^k (\psi_i \nabla\psi_s, \psi_r) \, c_{ki} \, c_{ks}$$

$$+ \mu_3 \frac{\delta^3}{4\gamma} \sum_{i=1}^k c_{ki} \left(\sum_{j=1}^k |c_{kj}(g_\delta * \Delta\psi_r)| \nabla^s \phi_i, \nabla^s \psi_r \right) = (\overline{\mathbf{f}}, \psi_r), \qquad (4.16)$$

for $r = 1, \cdots, k$, with the initial condition

$$c_{kr}(0) = (g_\delta * \mathbf{v}_0, \psi_r).$$

Multiplying (4.16) by c_{kr}, and summing over r, we get

$$\|\mathbf{w}_k(t)\|^2 + \frac{2}{Re}\int_0^t \|\nabla\mathbf{w}_k(\zeta)\|^2 \, d\zeta + \mu_3 \frac{\delta^3}{4\gamma}\int_0^t |g_\delta * \Delta\mathbf{w}_k(\zeta)| \, |\nabla^s\mathbf{w}_k(\zeta)|^2 \, d\zeta$$
$$= 2\int_0^t (\mathbf{w}_k(\zeta), \overline{\mathbf{f}}(\zeta)) \, d\zeta + \|\mathbf{w}_k(0)\|^2 \quad \forall t \in [0, T].$$

Using the Cauchy–Schwarz inequality, Körn's inequality, and Gronwall's lemma, we get:

$$\|\mathbf{w}_k(t)\|^2 + \frac{2}{Re}\int_0^t \|\nabla\mathbf{w}_k(\zeta)\|^2 \, d\zeta \leq M \quad \forall \, t \in [0, T], \qquad (4.17)$$

with M independent of t and k. Thus,

$$|c_{kr}(t)| \leq M^{1/2} \quad \forall r = 1, \cdots, k. \qquad (4.18)$$

From the elementary theory of partial differential equations, (4.18) implies that (4.16) admits a unique solution $c_{kr} \in W^{1,2}(0, T)$ for all $k \in \mathbb{N}$ (as in Sect. 2.4.4). Using the same approach as the one in [121], from these *a priori* bounds we get the existence of $\mathbf{w} \in L^2(0, T; H^1_{0,\sigma}(\Omega))$ such that

$$\lim_{k\to\infty} (\mathbf{w}_k(t) - \mathbf{w}(t), \mathbf{w}) = 0 \text{ uniformly in } t \in [0, T], \, \forall \mathbf{w} \in L^2(\Omega), \qquad (4.19)$$

$$\lim_{k\to\infty} \int_0^T (\partial_i(\mathbf{w}_k - \mathbf{w}), \mathbf{w}) \, d\zeta = 0 \, \forall \, \mathbf{w} \in L^2(\Omega \times [0, T]), \; i = 1, \cdots, k. \, (4.20)$$

Remark 4.6. This easy energy budget, with the above weak convergence is the first, necessary step in a Faedo–Galerkin method. The next, difficult step is to prove stronger convergence, in order to show that the limit $\mathbf{w}$ satisfies the GL model (4.12)–(4.15).

Now, we shall prove the strong convergence of $\{g_\delta * \Delta\mathbf{w}_k\}$ to $g_\delta * \Delta\mathbf{w}$ in $L^2(\omega \times [0, T])$ for all[1] $\omega \subset\subset \Omega$. To show this, we need the following classical inequality; see, for instance, Lemma II.4.2, [120]):

Lemma 4.7 (Friederichs inequality). *Let C be a cube in $\mathbb{R}^d$, and let* $\mathbf{v}$ *belong to $L^2(0, T; [H^1(C)]^d)$. Then, for any $\eta > 0$, there exists $K(\eta, C) \in \mathbb{N}$ functions $\varphi_i \in L^\infty(C)$, $i = 1, \cdots, K$ such that*

$$\int_0^T \|\mathbf{v}(t)\|^2_{L^2(C)} \, dt \leq \sum_{i=1}^K \int_0^T (\mathbf{v}(t), \varphi_i)^2_C \, dt + \eta \int_0^T \|\nabla\mathbf{v}(t)\|^2_{L^2(C)} \, dt. \qquad (4.21)$$

Applying the above inequality with $\mathbf{v} = g_\delta * \Delta\mathbf{w}_k - g_\delta * \Delta\mathbf{w}$, and C a cube contained in Ω, we get

---

[1] This means for all bounded set ω such that $\overline{\omega} \subset \Omega$.

$$\int_0^t \|g_\delta * \Delta\mathbf{w}_k - g_\delta * \Delta\mathbf{w}\|_{L^2(C)}^2 \, dt$$

$$\leq \sum_{i=1}^K \int_0^T (g_\delta * (\Delta\mathbf{w}_k - \Delta\mathbf{w}), \boldsymbol{\varphi}_i)_C^2 \, dt + \eta \int_0^T \|\nabla g_\delta * (\Delta\mathbf{w}_k - \Delta\mathbf{w})\|_{L^2(C)}^2 \, dt$$

$$= -\sum_{i=1}^K \int_0^T (\nabla\mathbf{w}_k - \nabla\mathbf{w}, g_\delta * \nabla\boldsymbol{\varphi}_i)_C^2 \, dt$$

$$+\eta \int_0^T \|\nabla g_\delta * (\Delta\mathbf{w}_k - \Delta\mathbf{w})\|_{L^2(C)}^2 \, dt.$$

Using (4.20) and the fact that

$$\|\nabla(g_\delta * \Delta\mathbf{w}_k - g_\delta * \Delta\mathbf{w})\|_{L^2(C)}^2 \leq C(g_\delta, \delta) \, \|\mathbf{w}_k - \mathbf{w}\|_{L^2(C)}^2,$$

we get

$$\lim_{k\to\infty} \int_0^T \|g_\delta * \Delta\mathbf{w}_k - g_\delta * \Delta\mathbf{w}\|_{L^2(C)}^2 \, dt = 0.$$

Applying (4.21) with $\mathbf{w} = \mathbf{w}_k - \mathbf{w}$, and using (4.20), we get

$$\lim_{k\to\infty} \int_0^T \|\mathbf{w}_k(t) - \mathbf{w}(t)\|_{L^2(C)}^2 \, dt = 0. \tag{4.22}$$

Now, we shall prove that $\mathbf{w}$ is a weak solution of (4.12)–(4.15). Integrating (4.16) from 0 to $t \leq T$, we get:

$$\int_0^t -\frac{1}{Re}(\nabla\mathbf{w}_k, \nabla\boldsymbol{\psi}_r) - (\mathbf{w}_k \cdot \nabla\mathbf{w}_k, \boldsymbol{\psi}_r) \, d\zeta = -\int_0^t (\bar{\mathbf{f}}, \boldsymbol{\psi}_r) \, d\zeta$$

$$+\mu_3 \frac{\delta^3}{4\gamma} \int_0^t (|g_\delta * \Delta\mathbf{w}_k|\nabla^s\mathbf{w}_k, \nabla^s\boldsymbol{\psi}_r) \, d\zeta + \int_0^t (\mathbf{w}_k(t), \boldsymbol{\psi}_r) - (\mathbf{w}_k(0), \boldsymbol{\psi}_r) \, d\zeta. \tag{4.23}$$

From (4.19) and (4.20), we get

$$\lim_{k\to\infty} (\mathbf{w}_k(t) - \mathbf{w}(t), \boldsymbol{\psi}_r) = 0, \quad \text{and} \tag{4.24}$$

$$\lim_{k\to\infty} \int_0^t (\nabla\mathbf{w}_k(\zeta) - \nabla\mathbf{w}(\zeta), \nabla\boldsymbol{\psi}_r) \, d\zeta = 0.$$

We now focus on the nonlinear terms in (4.23) corresponding to the convective term in the NSE. Let C be a cube containing the support of $\boldsymbol{\psi}_r$. Then:

$$\left| \int_0^t (\mathbf{w}_k \cdot \nabla\mathbf{w}_k, \boldsymbol{\psi}_r) - (\mathbf{w} \cdot \nabla\mathbf{w}, \boldsymbol{\psi}_r) \, d\zeta \right|$$

$$\leq \left| \int_0^t ((\mathbf{w}_k - \mathbf{w}) \cdot \nabla\mathbf{w}_k, \boldsymbol{\psi}_r)_C \, d\zeta) \right| + \left| \int_0^t (\mathbf{w} \cdot \nabla(\mathbf{w}_k - \mathbf{w}), \boldsymbol{\psi}_r)_C \, d\zeta \right|. \tag{4.25}$$

Setting $S = \max\limits_{\mathbf{x} \in C} |\boldsymbol{\psi}_r(\mathbf{x})|$, and using (4.17), we also have:

$$\left| \int_0^t ((\mathbf{w}_k - \mathbf{w}) \cdot \nabla \mathbf{w}_k, \boldsymbol{\psi}_r)_C \, d\zeta \right|$$

$$\leq S \left(\int_0^t \|\mathbf{w}_k - \mathbf{w}\|^2_{L^2(C)} \, d\zeta \right)^{1/2} \left(\int_0^t \|\nabla \mathbf{w}_k\|^2_{L^2(C)} \, d\zeta \right)^{1/2}$$

$$\leq S \, M^{1/2} \left(\int_0^t \|\mathbf{w}_k - \mathbf{w}\|^2_{L^2(C)} \, d\zeta \right)^{1/2}.$$

Thus, using (4.22), we get

$$\lim_{k \to \infty} \left| \int_0^t ((\mathbf{w}_k - v) \cdot \nabla \mathbf{w}_k, \boldsymbol{\psi}_r)_C \, d\zeta \right| = 0. \tag{4.26}$$

We also have

$$\left| \int_0^t (\mathbf{w} \cdot \nabla(\mathbf{w}_k - \mathbf{w}), \boldsymbol{\psi}_r)_C d\zeta \right| \leq \sum_{i=1}^d \left| \int_0^t (\partial_i(\mathbf{w}_k - \mathbf{w}), \mathbf{w}_i \boldsymbol{\psi}_r)_C d\zeta \right|$$

and since $\mathbf{w}_i \boldsymbol{\psi}_r \in L^2(\Omega \times [0, T])$, (4.20) implies

$$\lim_{k \to \infty} \left| \int_0^t (\mathbf{w} \cdot \nabla(\mathbf{w}_k - \mathbf{w}), \boldsymbol{\psi}_r)_C \, d\zeta \right| = 0. \tag{4.27}$$

Relations (4.25)–(4.27) yield:

$$\lim_{k \to \infty} \left| \int_0^t (\mathbf{w}_k \cdot \nabla \mathbf{w}_k - \mathbf{w} \cdot \nabla \mathbf{w}, \boldsymbol{\psi}_r) \, d\zeta \right| = 0.$$

We now treat the Gaussian–Laplacian term as follows:

$$\int_0^t (|g_\delta * \Delta \mathbf{w}_k| \nabla^s \mathbf{w}_k - |g_\delta * \Delta \mathbf{w}| \nabla^s \mathbf{w}, \nabla^s \boldsymbol{\psi}_r) \, d\zeta$$

$$\leq \left| \int_0^t (|g_\delta * \Delta \mathbf{w}| \nabla^s (\mathbf{w}_k - \mathbf{w}), \nabla^s \boldsymbol{\psi}_r) \, d\zeta \right| \tag{4.28}$$

$$+ \left| \int_0^t (|g_\delta * \Delta \mathbf{w}_k - g_\delta * \Delta \mathbf{w}| \nabla^s \mathbf{w}_k, \nabla^s \boldsymbol{\psi}_r) \, d\zeta \right|.$$

We have

$$\left| \int_0^t (|g_\delta * \Delta \mathbf{w}| \nabla^s (\mathbf{w}_k - \mathbf{w}), \nabla^s \boldsymbol{\psi}_r) \, d\zeta \right|$$

$$\leq \left| \int_0^t (\nabla^s (\mathbf{w}_k - \mathbf{w}), |g_\delta * \Delta \mathbf{w}| \nabla^s \boldsymbol{\psi}_r) \, d\zeta \right|,$$

and, since $|g_\delta * \Delta\mathbf{w}|\nabla^s\boldsymbol{\psi}_r \in L^2(\Omega \times [0, T])$, (4.20) implies

$$\lim_{k\to\infty} \left| \int_0^t (|g_\delta * \Delta\mathbf{w}|\nabla^s(\mathbf{w}_k - \mathbf{w}), \nabla^s\boldsymbol{\psi}_r)\, d\zeta \right| = 0.$$

On the other hand, setting $S = \max_{\mathbf{x}\in C} |\nabla\boldsymbol{\psi}_r(\mathbf{x})|$, and using (4.17), we get

$$\left| \int_0^t (|g_\delta * \Delta\mathbf{w}_k - g_\delta * \Delta\mathbf{w}|\nabla^s\mathbf{w}_k, \nabla^s\boldsymbol{\psi}_r)\, d\zeta \right|$$

$$\leq C\, S \left(\int_0^t \|g_\delta * \Delta\mathbf{w}_k - g_\delta * \Delta\mathbf{w}\|_{L^2(C)}^2\, d\zeta \right)^{1/2} \left(\int_0^t \|\nabla\mathbf{w}_k\|_{L^2(C)}^2\, d\zeta \right)^{1/2}$$

$$\leq C\, S\, M^{1/2} \left(\int_0^t \|g_\delta * \Delta\mathbf{w}_k - g_\delta * \Delta\mathbf{w}\|_{L^2(C)}^2\, d\zeta \right)^{1/2}.$$

Thus, using (4.22), we get

$$\lim_{k\to\infty} \left| \int_0^t (|g_\delta * \Delta\mathbf{w}_k - g_\delta * \Delta\mathbf{w}|\nabla^s\mathbf{w}_k, \nabla^s\boldsymbol{\psi}_r)\, d\zeta \right| = 0. \qquad (4.29)$$

Relations (4.28) and (4.29) yield

$$\lim_{k\to\infty} \left| \int_0^t (|g_\delta * \Delta\mathbf{w}_k|\nabla^s\mathbf{w}_k - |g_\delta * \Delta\mathbf{w}|\nabla^s\mathbf{w}, \nabla^s\boldsymbol{\psi}_r)\, d\zeta \right| = 0. \qquad (4.30)$$

Therefore, taking the limit over $k \to \infty$ in (4.23), and using (4.24), (4.2.1), and (4.30), we get

$$\int_0^t -\frac{1}{Re}(\nabla\mathbf{w}, \nabla\mathbf{w}_r) - (\mathbf{w}\cdot\nabla\mathbf{w}, \boldsymbol{\psi}_r)\, d\zeta = -\int_0^t (\overline{\mathbf{f}}, \boldsymbol{\psi}_r)\, d\zeta$$

$$+ (\mathbf{w}(t), \boldsymbol{\psi}_r) - (\mathbf{w}(0), \boldsymbol{\psi}_r)\, d\zeta + \mu_3 \frac{\delta^3}{4\gamma} \int_0^t (|g_\delta * \Delta\mathbf{w}|\nabla^s\mathbf{w}, \nabla^s\boldsymbol{\psi}_r)\, d\zeta.$$

However, from Lemma 2.3 in [121], we know that every function $\psi \in \mathcal{D}(\Omega)$ can be uniformly approximated in $C^2(\overline{\Omega})$ by functions of the form

$$\boldsymbol{\psi}_N(\mathbf{x}) = \sum_{r=1}^N \gamma_r\, \boldsymbol{\psi}_r(\mathbf{x}) \qquad N \in \mathbb{N},\ \gamma_r \in \mathbb{R}.$$

So, writing (4.29) with $\boldsymbol{\psi}_N$ instead of $\boldsymbol{\psi}_r$, and passing to the limit as $N \to \infty$, we get the validity of (4.29) for all $\psi \in \mathcal{D}(\Omega)$. Thus, $\mathbf{w}$ is a weak solution of (4.12)–(4.15). $\square$

4.3 $k - \varepsilon$ Modeling

Together with LES models other very simple models (algebraic or involving one or two scalar equations) are used in the description of turbulent flows.

We have seen that in the Boussinesq approximation, the problem is reduced to predicting k' and l. In [260] Prandtl formulated the so called "mixing length hypothesis". By using ideas from kinetic theory of gases, he assumed ν_T proportional to the product of the scale of *mean fluctuating velocity* (scale velocity) and of the *mixing length* (scale length). The mixing length l was defined in an experimental way as a nondecreasing function of the distance from the boundary. This is known as a "zero equation model", since it involves no equations for the evolution of ν_T.

In order to overcome natural limitations of the mixing length hypothesis (it works to predict mixing layers, jets, and wakes, but not transitions from one type to another one) more sophisticated models were developed. By observing that the most significant scale for velocity fluctuations is $\sqrt{k'}$, it is possible to derive the Kolmogorov–Prandtl (4.1) expression. The $k - \varepsilon$ model then predicts k' by solving the transport equation

$$k'_t + \overline{\mathbf{u}} \cdot \nabla k' - \nabla \cdot \left[\left(\frac{1}{Re} + \frac{\nu_T}{\sigma_k} \right) \nabla k' \right] - 2\,\nu_T \nabla^s \overline{\mathbf{u}}\, \nabla \overline{\mathbf{u}} = \epsilon',$$

where ϵ' can be approximated – within K41 theory – by $c_D\, k'^{3/2}\, l$ (the constants c_D and σ_k are empirical constants).

Since the length scale (characterizing the larger eddies containing energy) is also subject to a transport process, it is reasonable to derive equations for l. By observing that an equation for l does not necessarily need the mixing length itself as a dependent variable (any combination of k' and l will be enough), several models have been derived. The most popular is the $k - \varepsilon$ model involving the turbulent energy dissipation ϵ'. The transport equation for ϵ reads

$$\epsilon'_t + \overline{\mathbf{u}} \cdot \nabla \epsilon' = -\nabla \cdot \left[\left(\frac{1}{Re} + \frac{\nu_T}{\sigma_\epsilon} \right) \nabla \epsilon' \right] - 2\, c_{1\epsilon}\, \nu_T \frac{\epsilon'}{k'} \nabla^s \overline{\mathbf{u}}\, \nabla \overline{\mathbf{u}} = c_{2\epsilon} \frac{\epsilon'^2}{k'},$$

where again $c_{1\epsilon}$, $c_{2\epsilon}$ and σ_ϵ are empirical constants. Though these models are very rough, they may be employed rather successfully after very fine tuning of the empirical constants. For the mathematical analysis of $k - \varepsilon$ methods, see Mohammadi and Pironneau [239] and Coletti [67].

We will not present further details of these models (since they are outside the primary scope of LES) but the reader should be aware of them, since they are commonly used in CFD commercial software.

4.3.1 Selective Models

Together with the dynamic procedure of Germano [129] (that we will present in Chap. 8 on scale-similarity models), other dynamic or selective methods have been introduced, especially to improve the prediction of intermittent phenomena. The fundamental idea behind these *selective models* is to modulate the subfilter-scale model so as to apply it only when the assumptions underlying the model are satisfied. One then needs to know:

(a) when and where the smallest scales of the exact solution are not resolved;
(b) where the flow is fully developed turbulence.

The assumptions that can be made are generally of a very deep and precise mathematical nature. We will briefly introduce a couple of methods, developed by the group of Cottet [75, 78, 77, 76] in recent years, since they involve precise mathematical ideas and methods.

The Anisotropic Selective Model

The starting point of this method, introduced in Cottet [75] and Cottet and Wray [78], is to consider the balance equation for the vorticity field $\boldsymbol{\omega} = \nabla \times \mathbf{u}$:

$$\boldsymbol{\omega}_t + \mathbf{u} \cdot \nabla \boldsymbol{\omega} - \frac{1}{Re} \Delta \boldsymbol{\omega} = \boldsymbol{\omega} \cdot \nabla \mathbf{u}.$$

Multiplying by $\boldsymbol{\omega}$ and integrating by parts (recall that $\nabla \cdot \boldsymbol{\omega} = 0$), we obtain

$$\frac{1}{2} \frac{d}{dt} \|\boldsymbol{\omega}\|^2 + \frac{1}{Re} \|\nabla \boldsymbol{\omega}\|^2 = \int_\Omega \boldsymbol{\omega} \cdot \nabla \mathbf{u} \, \boldsymbol{\omega} \, dx = \int_\Omega \boldsymbol{\omega} \cdot \nabla^s \mathbf{u} \, \boldsymbol{\omega} \, dx. \qquad (4.31)$$

The term on the right-hand side is the *stretching term* and the lack of suitable estimate on it can also be seen as a possible source of nonsmooth solutions for the Navier–Stokes equations.

Formula (4.31) has as an easy consequence that the enstrophy $\|\boldsymbol{\omega}\|$ may increase when the vorticity is aligned with directions corresponding with positive eigenvalues of $\nabla^s \mathbf{u}$. By denoting by $(\nabla^s \mathbf{u})^+$ the positive part of $\nabla^s \mathbf{u}$ we can also write

$$\int_\Omega \boldsymbol{\omega} \cdot \nabla \mathbf{u} \, \boldsymbol{\omega} \, dx = \int_\Omega \sum_{i,j=1}^3 \omega_i (\nabla^s \mathbf{u})_{ij} \omega_j \leq \int_\Omega \sum_{i,j=1}^3 \omega_i (\nabla^s \mathbf{u})_{ij}^+ \omega_j.$$

In this formula the *positive part* of a tensor means the tensor that is obtained after diagonalization and replacement of negative eigenvalues with zero. The idea developed in [75, 78] is to limit the enstrophy increase, by introducing an eddy viscosity tensor proportional to $(\nabla^s \mathbf{u})^+$. Since the sum of eigenvalues of $\nabla^s \mathbf{u}$ is zero ($\nabla \cdot \mathbf{u} = 0$) there is also another natural candidate $-(\nabla^s \mathbf{u})^-$, and in Cottet, Jiroveanu, and Michaux [76] there is a physical-geometric motivation for the second choice. The proposed eddy viscosity is then

$$\nu_T = -(C_S \delta)^2 (\nabla^s \mathbf{u})^-.$$

In this case one computational problem is that this requires a possibly expensive diagonalization of the matrix $\nabla^s \mathbf{u}$ at each point. Another approach that avoids diagonalization is developed in [75] (together with a possible implementation). After extensive further development [78], this model becomes

$$\nabla \cdot \boldsymbol{\tau}(\mathbf{u}) = \frac{c}{\delta^4} \int \left\{ [\mathbf{u}(\mathbf{x}) - \mathbf{u}(\mathbf{y})] \cdot \nabla \zeta \left(\frac{\mathbf{x} - \mathbf{y}}{\delta} \right) \right\}_+ [\mathbf{u}(\mathbf{x}) - \mathbf{u}(\mathbf{y})] \, d\mathbf{y},$$

where

$$\{f\}_+ := \max \{f, 0\},$$

while ζ is a filter function, with spherical symmetry satisfying a moment condition

$$\int x_k x_l \zeta(\mathbf{x}) \, d\mathbf{x} = \delta_{kl}.$$

The above method can be seen as a modified Gradient LES model, in which the energy backscatter is controlled. The advantage of this method is that it allows energy to dissipate in one or more directions while controlling the backscatter in other directions.

The Selective Smagorinsky Model

The dynamic of vorticity is twofold: the effect of the stretching term in (4.31) may increase the value of vorticity and may also change the direction of the vorticity vector. The control of the growth of vorticity and its role in the global existence of smooth solutions (for the Euler equations, too) was introduced by Beale, Kato, and Majda [18] and developed also by Beirão da Veiga [19]. In particular it can be proved that the condition

$$\boldsymbol{\omega} \in L^r(0, T; L^s(\Omega)) \quad \text{for} \quad \frac{2}{r} + \frac{3}{s} = 2, \quad 1 \leq r \leq 2, \tag{4.32}$$

implies the full regularity of the solutions to the NSE (compare it with (2.31) of Chap. 2 and see also Chap. 7, p. 172). One first idea is then to detect the regions where the vorticity is large (in the sense of L^p-norms) and to put an EV tensor that vanishes outside these regions.

A new LES-model, whose introduction and implementation can be found in [76], is based on new geometric insight into the problems of regularity for the NSE. By using some exact formulas derived in Constantin [72], Constantin and Fefferman [73] introduced a new criterion for regularity that involves only the vorticity direction (the magnitude is not relevant). This results is related to the study of bending of *vortex lines* (lines everywhere tangent to the vorticity field) and on the stretching of vortex tubes (tubes made by vortex lines). These are phenomena of pure 3D nature, since they are absent in the dynamics of 2D fluids.

Another model of Cottet, Jiroveanu, and Michaux [76] is a variant of the Smagorinsky model, in which the turbulence model is applied only in regions of intense *vortex activity*. The idea of controlling the behavior of vorticity to ensure regularity of solutions is connected with the outstanding problem of global existence of smooth solutions for the 3D Navier–Stokes equations.

The results of Constantin and Fefferman were improved in Beirão da Veiga and Berselli [22]. As a particular case they imply the following theorem:

Theorem 4.8. *Assume that there exist positive constants Ω and ρ such that in the region where the vorticity magnitude at two points x and y is larger than Ω,*

$$\exists\, C > 0 \text{ such that} \quad \sin\theta(\mathbf{x},\mathbf{y},t) \le C\,|\mathbf{x}-\mathbf{y}|^{1/2}, \quad \forall t \in [0,T],$$

where $\theta(\mathbf{x},\mathbf{y},t)$ is the angle between the vorticity vectors at the points $\mathbf{x}$ and $\mathbf{y}$, at time t. Then the NSE have a unique regular solution on $[0,T]$.

This result gives a new criterion to detect regions of turbulent behavior: to compute the angle between the vorticity at a given grid point and the average vorticity at the six closest neighboring points. (Note that everything could be done at a continuous level, but we prefer to show directly a possible numerical implementation; see also David [85].) We define

$$\boldsymbol{\omega}_m(x,t) = \frac{1}{6}\sum_{i=1}^{3}\boldsymbol{\omega}(\mathbf{x}+\Delta_i x_i, t) - \boldsymbol{\omega}(\mathbf{x}-\Delta_i x_i, t),$$

where Δ_i is the grid-width in the direction of x_i. The average angle θ_m is then defined as

$$\theta_m(\mathbf{x},t) = \arcsin\frac{\boldsymbol{\omega}(\mathbf{x},t)\times\boldsymbol{\omega}_m(\mathbf{x},t)}{|\boldsymbol{\omega}(\mathbf{x},t)|\,|\boldsymbol{\omega}_m(\mathbf{x},t)|}.$$

The next step is to define the function filter

$$\Psi(x,t) = \begin{cases} 1, & \text{if } \beta_0 \le \beta_m \le \pi - \beta_0 \\[2mm] 0, & \text{otherwise,} \end{cases}$$

where β_0 is some threshold angle (a common value of β_0 is $\pi/12$). Finally the eddy viscosity for the *selective Smagorinsky model* is expressed by

$$\nu_T = \Psi(x,t)C_S\delta^2|\nabla^s\mathbf{u}|.$$

4.4 Conclusions

Eddy viscosity models are inherently dissipative, see Fig. 3.2, and do not allow for backscatter of energy, Sect. 3.5. This dissipativity is not a significant detriment in flows in which there is a large power input and calculation is over a moderate timescale. Thus, they have proven to be very useful for calculating the statistics of turbulent flows in industrial settings. On the other hand, they are not the models of choice if accurate representation of the mean velocity and pressure is needed or for calculations over a long time interval or for problems with delicate energy balance.

The diffusivity of eddy viscosity models retards separation and transition even over moderate time intervals. Thus, one main path to improvement (explored in this chapter) is formulas for the turbulent viscosity coefficient which are more localized in space. We have seen that simple changes in the turbulent viscosity coefficient lead to large improvement in performance. Interestingly, eddy viscosity, the oldest idea in turbulence modeling, is of great current interest due to new models whose diffusivity acts only on the smallest resolved scales (and is thus localized in scale-space). The idea occurs very naturally in spectral methods (in Tadmor's spectral vanishing viscosity method [226].) Recent extensions to general variational methods in the work of Guermond [144], Hughes' Variational Multiscale method [160], and in [204] are presented in a later chapter. Properly calibrated, eddy viscosity models continue to be the workhorse of industrial turbulence calculations. Improvements in eddy viscosity models, such as development of models localized in both physical and scale-space, are of great practical importance.

Accurate solution of turbulent flows will likely be attained only as a synthesis of many good ideas, and eddy viscosity models continue to make a key contribution in practical problems. An important example of a useful synthesis occurs with, so-called, mixed models. In mixed models, an eddy viscosity term is added to a dispersive model to improve its stability properties. For example, the eddy viscosity hypothesis applies most sensibly to the Reynolds stress term $\mathbf{u'u'}^T$. Thus it is certainly physically sensible to combine eddy viscosity models for the Reynolds stress term with a dispersive model for the other terms in the expansion of the subfilter-scale stress tensor. At present, determining the right combination is an important open problem.

5

Uncertainties in Eddy Viscosity Models and Improved Estimates of Turbulent Flow Functionals

5.1 Introduction

Important decisions are made and significant designs are produced based on turbulent flow, which are simulated using various models of turbulence. Even when using the model which current practice considers best for a particular application, often the reliability of the model's predictions for the specific application is not assessed. This is particularly troublesome because solutions of turbulence models can display sensitivity with respect to the user-selected model parameters in addition to the sensitivity with respect to the upstream flow, subgrid model, and numerical realization of it (reported by Sagaut and Lê [268]).

For example, calculating the force a fluid exerts upon an immersed body, such as lift or drag, involves first solving the NSE

$$\mathbf{u}_t + \nabla \cdot (\mathbf{u}\,\mathbf{u}^T) + \nabla p - \frac{2}{Re} \nabla \cdot (\nabla^s \mathbf{u}) = \mathbf{f},$$

$$\nabla \cdot \mathbf{u} = 0,$$

with appropriate initial and boundary values. If B denotes the boundary of the immersed body, the force must be calculated on B via

$$\text{force on } B = \int_B \mathbf{n} \cdot \left[p\,\mathbb{I} - \frac{2}{Re} \nabla^s \mathbf{u} \right] dS, \tag{5.1}$$

where $\mathbf{n}$ is the outward unit normal to B. This requires accurate estimation of p and derivatives of $\mathbf{u}$ on the flow boundary – a problem harder than accurately predicting the turbulent velocity itself. The problem simplifies a bit if only time-averaged forces on B are needed. In this case, however, a well-calibrated turbulence model would likely give the most economical prediction.

The basic approach used for turbulent flows has been to replace the NSE by a turbulence model and then insert the couple velocity–pressure predicted by the turbulence model into the right-hand side of the functional such as (5.1).

Uncertainties arise immediately due to the typical sensitivity of such models to the model's input parameters. Perhaps more importantly, turbulence models approximate flow averages. Thus, all the information on fluctuations is lost in them. However, the small-scale fluctuations can have a determining role in functionals such as (5.1). Mathematically, this is because the derivatives occurring in (5.1) overweight velocity changes occurring across small distances. (It can be argued that an example is drag in the flow over a dimpled *vs.* smooth golf ball. In this case, the dimples change the flow geometry below the $O(\delta)$ length-scale, yet produce an $O(1)$ change in the drag.) Indeed, drag reduction strategies injecting small amounts of microbubbles or polymers near a body are, in part, based on the expectation that a little power input to alter the small scale flows can have a large effect upon the global drag.

We shall see that the sensitivity equation approach has the great promise of giving computable quantitative estimates of the local sensitivity of the models predicted flow field to variations in the input parameters. Thus, a sensitivity calculation will show over which regions the predicted velocities are reliable and hence believable and over which regions those predicted velocities are highly sensitive, and hence should be viewed with greater suspicion. We focus on the case of sensitivity with respect to the user-selected length scale δ. These ideas of Anitescu, Layton, and Pahlevani [8] were tested in Pahlevani [248]. The reason for this focus is that the sensitivity of the flow with respect to variations in δ can be used to improve the estimate of flow functionals, such as lift and drag, which can depend strongly upon the unknown and unresolved turbulent fluctuations!

5.2 The Sensitivity Equations of Eddy Viscosity Models

The continuous sensitivity equation approach is becoming increasingly important in computational fluid dynamics but is not yet a common tool in LES of turbulence. This section will apply the sensitivity idea to find the continuous sensitivity equation with respect to variations in the length scale δ. For a general treatment of sensitivities and applications to other flow problems, see (among many interesting works) [37, 282, 147].

Suppose a local spatial filter g_δ with radius δ has been selected. Filtering the NSE, leads to the problem of closure. One very common class of closure models is based on the Boussinesq or EV hypothesis (see Chaps. 3 and 4 for more details). The model we consider aims at finding the approximate large-scale velocity $\mathbf{w}(\mathbf{x}, t)$ and pressure $q(\mathbf{x}, t)$ satisfying

$$\mathbf{w}_t + \nabla \cdot (\mathbf{w}\,\mathbf{w}^T) + \nabla q - \nabla \cdot \left(\left[\frac{2}{Re} + \nu_T(\delta, \mathbf{w}) \right] \nabla^s \mathbf{w} \right) = \overline{\mathbf{f}}, \qquad (5.2)$$

$$\nabla \cdot \mathbf{w} = 0,$$

where $\overline{\mathbf{f}}$ is the space-filtered body force, and ν_T is the eddy viscosity coefficient, which must be specified to select the model. As an example, the Smagorinsky model [277], while currently not considered generally the best, is perhaps the most commonly used model and is given by the eddy viscosity choice

$$\nu_T = \nu_{\text{Smag}}(\delta, \mathbf{w}) := (C_S \delta)^2 |\nabla^s \mathbf{w}|, \qquad C_S \approx 0.17. \qquad (5.3)$$

For other EV models, the reader is referred to Chaps. 3 and 4. Once C_S, the Smagorinsky constant, initial and boundary conditions are specified, for a given δ, the equations (5.2) and (5.3) uniquely determine a solution $(\mathbf{w}, q)$ implicitly as a function of δ.

Definition 5.1. *Let $(\mathbf{w}, q)$ be the solution of (5.2), (5.3). The sensitivity of $(\mathbf{w}, q)$ to variations in δ is defined to be the derivatives of $(\mathbf{w}, q)$ with respect to δ,*

$$\left(\mathbf{w}_\delta := \frac{\partial \mathbf{w}}{\partial \delta}, \ q_\delta := \frac{\partial q}{\partial \delta} \right).$$

It is easy to derive continuous equations for the sensitivities by implicit differentiation of (5.2) and (5.3) with respect to δ. Doing so, gives the equations

$$\mathbf{w}_{\delta,t} + \nabla \cdot (\mathbf{w}\mathbf{w}_\delta^T + \mathbf{w}_\delta \mathbf{w}^T) - Re^{-1} \Delta \mathbf{w}_\delta + \nabla q_\delta$$
$$-\nabla \cdot \left(\left[\frac{\partial}{\partial \delta} \nu_T(\delta, \mathbf{w}) + \frac{\partial}{\partial \mathbf{w}} \nu_T(\delta, \mathbf{w}) \cdot \mathbf{w}_\delta \right] \nabla^s \mathbf{w} + \nu_T(\delta, \mathbf{w}) \nabla^s \mathbf{w}_\delta \right) = \frac{\partial \overline{\mathbf{f}}}{\partial \delta}, \quad (5.4)$$
$$\nabla \cdot \mathbf{w}_\delta = 0. \quad (5.5)$$

Remark 5.2. We have to be careful in calculating the term $\dfrac{\partial}{\partial \mathbf{w}} \nu_T(\delta, \mathbf{w}) \cdot \mathbf{w}_\delta$. It should be understood in the sense of a Gateaux derivative when (as in the Smagorinsky model (5.3)), ν_T involves differential operators. For example, by direct calculation,

$$\frac{\partial}{\partial \delta}(|\nabla \mathbf{w}|^2) = \frac{\partial}{\partial \delta}(\nabla \mathbf{w} : \nabla \mathbf{w}) = (\nabla \mathbf{w}_\delta : \nabla \mathbf{w} + \nabla \mathbf{w} : \nabla \mathbf{w}_\delta) = 2 \nabla \mathbf{w}_\delta : \nabla \mathbf{w},$$

rather than $2|\nabla \mathbf{w}|\nabla \mathbf{w}_\delta$, as the expression (5.4) would seem to suggest.

Thus, once the large eddy velocity and pressure, $(\mathbf{w}, q)$, are calculated, the corresponding sensitivities can then be found by solving a linear problem for $(\mathbf{w}_\delta, q_\delta)$, which is precisely the nonlinear LES model linearized about $(\mathbf{w}, q)$. Thus, sensitivities can be quickly and economically calculated by the same program used to calculate $(\mathbf{w}, q)$.

For the Smagorinsky model, we have $\nu_T = (C_S \delta)^2 |\nabla^s \mathbf{w}|$. Thus, the bracketed term is, by direct calculation,

$$\frac{d}{d\delta}[\nu_T(\delta, \mathbf{w}(\delta))] = \frac{\partial \nu_T}{\partial \delta} + \frac{\partial \nu_T}{\partial \mathbf{w}} \cdot \mathbf{w}_\delta,$$
$$= 2 C_S^2 \delta |\nabla^s \mathbf{w}| + (C_S \delta)^2 \left(\frac{\nabla^s \mathbf{w}}{|\nabla^s \mathbf{w}|} \right) : \nabla^s \mathbf{w}_\delta.$$

5.2.1 Calculating $\overline{\mathbf{f}}_\delta = \frac{\partial}{\partial\delta}\overline{\mathbf{f}}$

If the body forces acting on the flow vary slowly in space, $\overline{\mathbf{f}}_\delta$ is negligible. Otherwise, the right-hand side of (5.4), $\overline{\mathbf{f}}_\delta$, can play an important role in the sensitivity equation, since it incorporates information about body force fluctuations. When $\mathbf{f}$ is not smooth, the exact value of $\overline{\mathbf{f}}_\delta$ will depend on the precise filter specified. When $\overline{\mathbf{f}}$ is defined by convolutions, extending $\mathbf{f}$ by zero off the flow domain and then defining

$$\overline{\mathbf{f}} = \int_{\mathbf{R}^d} \delta^{-d} g_\delta\left(\frac{\mathbf{x}'}{\delta}\right) \mathbf{f}(\mathbf{x} - \mathbf{x}')\, d\mathbf{x},$$

where $g_\delta(\mathbf{x})$ is the chosen filter kernel, then $\overline{\mathbf{f}}_\delta$ can be calculated explicitly.

When $\overline{\mathbf{f}}$ is defined using differential filters (introduced in the pioneering work of Germano [127, 126]), a small modification is needed, which depends on the exact differential filter specified. Two interesting differential filters are defined by solving the Helmholtz problem for $\overline{\mathbf{f}}$:

$$\begin{cases} -\delta^2 \Delta \overline{\mathbf{f}} + \overline{\mathbf{f}} = \mathbf{f} & \text{in } \Omega, \\ \overline{\mathbf{f}} = \mathbf{0} & \text{on } \partial\Omega, \end{cases} \tag{5.6}$$

and by solving the shifted Stokes problem for $\overline{\mathbf{f}}$:

$$\begin{cases} -\delta^2 \Delta \overline{\mathbf{f}} + \overline{\mathbf{f}} + \nabla\lambda = \mathbf{f} & \text{in } \Omega, \\ \nabla \cdot \overline{\mathbf{f}} = \mathbf{0} & \text{in } \Omega, \\ \overline{\mathbf{f}} = \mathbf{0} & \text{on } \partial\Omega. \end{cases} \tag{5.7}$$

The second differential filter (5.7) preserves incompressibility and is thus interesting in spite of its extra cost over the first.

With the first filter $\overline{\mathbf{f}} = (-\delta^2\Delta + \mathbb{I})^{-1}\mathbf{f}$, we can differentiate implicitly with respect to δ, to derive an equation for $\overline{\mathbf{f}}_\delta$:

$$(-\delta^2\Delta + \mathbb{I})\,\overline{\mathbf{f}}_\delta = 2\delta\Delta\overline{\mathbf{f}} = \ (via \ (5.6)) \ = \left(-\frac{2}{\delta}\right)(\mathbf{f} - \overline{\mathbf{f}}).$$

Thus,

$$\overline{\mathbf{f}}_\delta = \left(-\frac{2}{\delta}\right)(\overline{\mathbf{f}'}).$$

On the contrary, the only way to calculate $\mathbf{w}_\delta$ is to solve the linear PDEs (5.4) and (5.5). Analogously, we obtain the initial condition for $\mathbf{w}_\delta$

$$\mathbf{w}_\delta(\mathbf{x}, 0) = \overline{\mathbf{u}}_{0,\delta}(\mathbf{x}) = \left(-\frac{2}{\delta}\right)(\overline{\mathbf{u}'_0})$$

and then we can try to solve the initial value problem (5.4) and (5.5).

5.2.2 Boundary Conditions for the Sensitivities

Boundary conditions for sensitivities must be specified. The most interesting and important cases are sensitivity with respect to the (modeled) upstream conditions and the (modeled) outflow conditions, see Sagaut and Lê [268]. At this point, a mathematical formulation of the former is still very unclear while there are very many options for the latter. Thus we will consider here only boundary conditions for the sensitivities at solid walls.

With a differential filter like (5.6) and (5.7), the boundary conditions for the sensitivities are clear since there is no error and no variability with respect to δ in the conditions for $\mathbf{w}$ on the wall: $\mathbf{w}_\delta = \mathbf{0}$ on the boundary.

Some slight modifications of the wall models are necessary to compute sensitivities when near wall models are used. These are usually associated with averaging by convolutions when filtering through a wall must be performed, see Chap. 10 for more details. Many near wall models/numerical boundary conditions are possible. For specificity and clarity, we treat the simplest ones considered in Chap. 10, in which the large structures that action on the boundary are modeled by no penetration and slip-with-friction conditions:

$$\mathbf{w} \cdot \mathbf{n} = 0 \qquad \text{and} \qquad \beta(\delta, Re)\,\mathbf{w} \cdot \boldsymbol{\tau}_j + \mathbf{n} \cdot \left(\frac{2}{Re} \nabla^s \mathbf{w} \right) \cdot \boldsymbol{\tau}_j = 0 \quad \text{on } \partial\Omega,$$

where $\mathbf{n}$ is the unit normal to the boundary and $\boldsymbol{\tau}_1, \boldsymbol{\tau}_2$ are a system of unit tangent vectors on the boundary. Implicit differentiation with respect to δ gives the boundary conditions for the sensitivities

$$\mathbf{w}_\delta \cdot \mathbf{n} = 0 \qquad\qquad \text{on } \partial\Omega,$$

$$\beta(\delta, Re)\,\mathbf{w}_\delta \cdot \boldsymbol{\tau}_j + \mathbf{n} \cdot \left(\frac{2}{Re} \nabla^s \mathbf{w}_\delta \right) \cdot \boldsymbol{\tau}_j = -\beta_\delta\,(\delta, Re)\,\mathbf{w} \cdot \boldsymbol{\tau}_j \qquad \text{on } \partial\Omega.$$

It is reasonable (see Chap. 10) to suppose that, at fixed Re, $\beta_\delta < 0$, since $\beta(\delta)$ increases monotonically to infinity as $\delta \to 0$ (that is, $\beta(\delta)$ decreases as δ increases, see also Fig. 2 on p. 1140 in [178]). Thus, slippage in the flow velocity $\mathbf{w}$ acts to decrease the slippage in the sensitivities when they are aligned and increase it when they are opposed.

5.3 Improving Estimates of Functionals of Turbulent Quantities

Suppose (optimistically) that the LES solution $\mathbf{w}$ implicitly defines a smooth function of δ, $\mathbf{w} = \mathbf{w}(\delta)$, with the property that

$$\mathbf{w}(\delta) \to \mathbf{u} \qquad \text{as} \qquad \delta \to 0.$$

This is a minimal analytic condition for consistency (known as "limit consistency") for an LES model which, nevertheless, has so far been proven to hold for only a few LES models, such as the Smagorinsky model, the S^4 model, the zeroth order model, and the Stolz–Adams approximate deconvolution model in Chap. 8.

Let us suppose that a functional J (for example drag, lift, *etc.*) is well-defined for LES velocity and pressure. If J is smooth (say of class C^1), then the composition

$$\delta \mapsto J(\mathbf{w}(\delta)) = \mathcal{J}(\delta)$$

defines a smooth map. The value $\mathcal{J}(\delta) = J(\mathbf{w}(\delta))$ is computable, while $J(\mathbf{u}) = J(\mathbf{w}(0)) = \mathcal{J}(0)$ is sought. Since δ is small, the linear approximation to $\mathcal{J}(0)$ is justified. The linear approximation to $\mathcal{J}(\delta)$ yields a first-order approximation to $J(\mathbf{u})$

$$J(\mathbf{u}) \approx J(\mathbf{w}(\delta)) - \delta\, J'(\mathbf{w}(\delta)) \cdot \mathbf{w}_\delta. \tag{5.8}$$

The increment $\delta\, J'(\mathbf{w}(\delta))$ incorporates effects of unresolved scales on $J(\mathbf{u})$ and is computable once the solutions sensitivities $\mathbf{w}_\delta$ are calculated.

It may happen that the functional J itself is regularized, so $J(\mathbf{u})$ is approximated by a δ-dependent approximation

$$J(\delta, \mathbf{w}(\delta)).$$

Accordingly, (5.8) is modified to

$$J(\mathbf{u}) \approx J(\delta, \mathbf{w}(\delta)) - \delta\left(J_{\mathbf{w}}(\delta, \mathbf{w}(\delta)) \cdot \mathbf{w}_\delta + J_\delta(\delta, \mathbf{w}(\delta)) \right),$$

where $J_{\mathbf{w}}$ and J_δ denote the partial Gateaux derivative of $J(\delta, \mathbf{w})$ with respect to $\mathbf{w}$ and δ, respectively. For further details on partial derivatives of functionals defined on infinite dimensional linear spaces, see Rudin [266].

We now give a couple of examples to show the context in which we can use this approach to improve estimates on functionals involving turbulent quantities, as we can find in real life applications.

Example 1: Lift, drag, and other forces on boundaries in turbulent flows.

In many applications, forces exerted by fluid on boundaries must be estimated. In this case, the functional is given by

$$J(\mathbf{u}, p, \widehat{\mathbf{a}}) := \oint_B \mathbf{n} \cdot \left[p\,\mathbb{I} - \frac{2}{Re} \nabla^s \mathbf{u} \right] \cdot \widehat{\mathbf{a}}\, dS, \tag{5.9}$$

where $\widehat{\mathbf{a}}$ is a unit vector and B is the boundary of the immersed body. If $\widehat{\mathbf{a}}$ points in the direction of motion, $J(\mathbf{u}, p, \widehat{\mathbf{a}})$ represents drag[1], while if $\widehat{\mathbf{a}}$ points in the direction of gravity, $J(\mathbf{u}, p, \widehat{\mathbf{a}})$ represents lift.

---

[1] It is interesting to note that in the 2D case the problem of shape optimization – known also as the submarine problem – leads to very interesting purely theoretical results on some basic uniqueness questions for the Stokes problem, see Šverák [292]. In fact, the problem of finding the shape that minimizes drag allowed the author to precisely find the hypotheses that give in 2D the equality between $H^1_{0,\sigma}$ and the closure of $\mathcal{V}$ in H^1_0, see Sect. 2.4.1.

Remark 5.3. For general surfaces B, the straightforward approximation of $J(\mathbf{u}, p, \widehat{\mathbf{a}})$ by $J(\mathbf{w}, q, \widehat{\mathbf{a}})$ is not so clear because (see, for example, Sagaut [267]), while ideally $\mathbf{w}(\delta) \to \mathbf{u}$ as $\delta \to 0$, $q(\delta) \to p + 1/3 \operatorname{Trace}[\boldsymbol{\tau}]$, where $\boldsymbol{\tau}$ is the subfilter-scale stress tensor

$$\tau_{ij} := \overline{\mathbf{u}_i \mathbf{u}_j} - \overline{\mathbf{u}}_i \overline{\mathbf{u}}_j.$$

Thus, $q(\delta)$ is not, for general surfaces B, a direct approximation to p and its use in (5.9) could skew the estimate of the force on a general surface B. For a wall B, the situation is clearer. Indeed, let $k(\mathbf{v}) := 1/2|\mathbf{v}|^2(\mathbf{x}, t)$ denote the kinetic energy distribution of a velocity field $\mathbf{v}$. We have

$$q(\delta) = p + (2/3)(k(\overline{\mathbf{u}}) - \overline{k(\mathbf{u})}). \tag{5.10}$$

Since $\mathbf{w}$ approximates $\overline{\mathbf{u}}$, the excess pressure contribution to q from $k(\overline{\mathbf{u}})$ is computable and thus correctable using $k(\mathbf{w})$ but that contributed by $\overline{k(\mathbf{u})}$ is not easily calculable for general surfaces B. If B is a solid wall and the averaging operator, such as (5.6) and (5.7), preserves the no-slip condition, then $k(\overline{\mathbf{u}}) = 0$ on B and $\overline{k(\mathbf{u})} = 0$ on B, too. If the averaging operator does not preserve zero boundary conditions (such as filtering by convolution with constant averaging radius), then $\overline{k(\mathbf{u})}$ does not in general vanish on B.

To proceed, for (5.9) and other functionals involving boundary pressures, there are two cases that must be considered. The first case is when

$$q(\delta)\big|_B \to p\big|_B \qquad \text{as} \quad \delta \to 0.$$

In this case, since the boundary-force functional $J(\mathbf{w}, q)$, given by (5.9) (suppressing, to simplify notation, the explicit dependence on $\widehat{\mathbf{a}}$) is a linear functional with respect to $\mathbf{u}$ and p, $J' = J$ and we obtain the corrected approximation to the force on B:

$$J(\mathbf{u}, p) \approx J(\mathbf{w}, q) - \delta J(\mathbf{w}_\delta, q_\delta) = J(\mathbf{w} - \delta \mathbf{w}_\delta, q - \delta q_\delta)$$
$$= \oint_B \mathbf{n} \cdot \left[(q - \delta q_\delta)\, \mathbb{I} - \frac{2}{Re}\, \nu \, \nabla^s (\mathbf{w} - \delta \mathbf{w}_\delta) \right] \cdot \widehat{\mathbf{a}}\, dS.$$

The second case is when $\overline{k(\mathbf{u})}$ is nonnegligible on B and its effect in $q\big|_B$ must be adjusted for. With the given information, the best available estimator of $(\overline{k(\mathbf{u})})\big|_B$ is

$$\overline{k(\mathbf{u})}\Big|_B \approx \overline{k(\mathbf{w} - \delta\, \mathbf{w}_\delta)}\Big|_B,$$

which is computable. This gives the computable approximation to the pressure on the wall, from (5.10),

$$\widetilde{p}(\delta) := q(\delta) - \frac{2}{3}\left(k(\mathbf{w}) - \overline{k(\mathbf{w} - \delta\, \mathbf{w}_\delta)} \right).$$

Let $\widetilde{p}_\delta$ denote $\dfrac{\partial}{\partial \delta}\widetilde{p}(\delta)$. Then, $\widetilde{p}_\delta$ is computable in principle from the above formula and implicit differentiation (although it is not an agreeable calculation). In this second case, the approximation to $J(\mathbf{u}, p)$ is

$$\oint_B \mathbf{n} \cdot \left[(\widetilde{p} - \delta\,\widetilde{p}_\delta)\,\mathbb{I} - \frac{2}{Re}\nabla^s(\mathbf{w} - \delta\,\mathbf{w}_\delta) \right] \cdot \widehat{\mathbf{a}}\,dS.$$

Again, we stress that at this point it is not known if $\overline{k(\mathbf{u})} \neq 0$ on B has a significant or negligible effect!

Example 2: Flow matching.

Flow matching, needs four steps: (1) a desired velocity field $\mathbf{u}^*$ is specified, (2) the flow is simulated, (3) a functional such as

$$J(\mathbf{u}) := \frac{1}{2}\int_{\Omega\times(0,T)} |\mathbf{u} - \mathbf{u}^*|^2\,dxdt$$

is calculated, and (4) the design/control parameters are used to drive $J(\cdot)$ to its minimum value. Thus, one aspect of flow matching involves getting the best estimate of $J(\cdot)$; this is challenging in the case of a turbulent flow.

Abstractly, given the LES velocity $\mathbf{w}$ and its sensitivity $\mathbf{w}_\delta$, (5.8) provides an estimate of $J(\mathbf{u})$ improving the estimate given by $J(\mathbf{w})$. Since in this case $J(\cdot)$ is quadratic, it is straightforward to calculate

$$J'(\mathbf{w})\mathbf{w}_\delta = \int_{\Omega\times(0,T)} (\mathbf{w} - \mathbf{w}^*)\cdot \mathbf{w}_\delta\,d\mathbf{x}\,dt,$$

giving the approximation

$$J(\mathbf{u}) \approx \int_{\Omega\times(0,T)} \frac{1}{2}|\mathbf{w} - \mathbf{u}^*|^2 + \delta(\mathbf{w} - \mathbf{u}^*)\cdot \mathbf{w}_\delta\,dxdt.$$

5.4 Conclusions: Are $\overline{\mathbf{u}}$ and $\overline{p}$ Enough?

It is important to keep in mind that the goal of LES is not to produce colorful animations, but rather reliable predictions of important physical quantities. Often, this means estimating functionals accurately and giving an assessment of the reliability of the LES prediction. In all cases, the flow sensitivities provide useful and possibly essential information about the quality of the simulation and predictions obtained from it. Possibly more importantly, they can be used to improve those predictions! When LES codes are designed from the start with the idea of producing *both* velocities *and* sensitivities, the increase in computational cost is negligible over just computing velocities on the same mesh [248, 38].

In addition, a key problem is that we currently do not know all the characteristic features of the filtered field (independently from the modeling error); therefore, the sensitivity analysis seems to be of primary importance.

Apart from these practical facts it should be emphasized that, from the theoretical point of view, the calculation of sensitivity poses serious mathematical problems. In order to better understand its role a more detailed analysis, involving also filters that are not of an approximate deconvolution or differential type, seems necessary. The ideas we present in this chapter should be applied to each LES model the reader tries to use, implement, understand or improve!

We have the feeling that the role sensitivity calculations will play in LES will increase. In addition, the reader should be aware that this is not the only source of error, when comparing "true" functionals and flows with computed ones. A comparison of the relative magnitude of uncertainties arising from sensitivity of the model should be done with both (1) the commutation error (see Chap. 9) and (2) the boundary effects (see Chap. 10). Sensitivity is the first accuracy condition that the practitioner should keep in mind when trying to deduce quantitative and also qualitative properties of the "true" flow, from those of the LES simulated ones.

Part III

Advanced Models

6

Basic Criteria for Subfilter-scale Modeling

6.1 Modeling the Subfilter-scale Stresses

Over long time intervals, and especially in geometrically simple domains, the closure model selected for the subfilter-scale stress (SFS) tensor

$$\tau(\mathbf{u}, \mathbf{u}) := (\overline{\mathbf{u}\, \mathbf{u}^T} - \overline{\mathbf{u}}\, \overline{\mathbf{u}}^T) \approx \mathcal{S}(\overline{\mathbf{u}}, \overline{\mathbf{u}}) \tag{6.1}$$

is extremely important for an accurate simulation.

A related formulation is to incorporate the mean normal subfilter-scale stresses into the pressure by

$$p^* := \frac{1}{3}\, \text{trace}\, \boldsymbol{\tau}(\mathbf{u}, \mathbf{u}), \qquad \boldsymbol{\tau}^*(\mathbf{u}, \mathbf{u}) := \boldsymbol{\tau}(\mathbf{u}, \mathbf{u}) - p^* \mathbb{I}.$$

Then, the closure problem is to find a tensor $\mathcal{S}^*(\overline{\mathbf{u}}, \overline{\mathbf{u}})$ with zero trace, which approximates $\boldsymbol{\tau}^*(\mathbf{u}, \mathbf{u})$. Some closure models arise naturally by approximating $\boldsymbol{\tau}(\mathbf{u}, \mathbf{u})$ and some, such as eddy viscosity models, by approximating $\boldsymbol{\tau}^*(\mathbf{u}, \mathbf{u})$. At present there are many, many SFS models that have been proposed (well surveyed in Sagaut [267]) and a "universal" model is yet to be found.

The ultimate goal is a SFS model with the property that discretizations without either explicit or implicit dissipation produce simulations with high accuracy in the large eddies over long time intervals. This goal has not yet been attained, so an intermediate goal has been to find SFS models for which $\mathcal{S}(\overline{\mathbf{u}}, \overline{\mathbf{u}})$ replicates important features of the true SFS stresses. Before presenting recent models, we will therefore summarize some important features sought in a model $\mathcal{S}(\overline{\mathbf{u}}, \overline{\mathbf{u}})$. So far, models satisfying all these "easier" conditions have been elusive!

Ignoring boundaries for the moment, once the model (6.1) is chosen, the solution of the new equation is naturally no longer the true filtered velocity and pressure but rather an approximation $\mathbf{w}$ to $\overline{\mathbf{u}}$, induced by (6.1) and satisfying:

$$\mathbf{w}_t + \nabla \cdot (\mathbf{w}\, \mathbf{w}^T) - \nu \Delta \mathbf{w} + \nabla q + \nabla \cdot \mathcal{S}(\mathbf{w}, \mathbf{w}) = \overline{\mathbf{f}}, \tag{6.2}$$

$$\nabla \cdot \mathbf{w} = 0, \tag{6.3}$$

subject to an initial condition $\mathbf{w}(\mathbf{x}, 0) = \bar{\mathbf{u}}_0(\mathbf{x})$, in Ω, and periodic boundary conditions.

6.2 Requirements for a Satisfactory Closure Model

Since there are many possible SFS models, any mathematical, physical and experimental guidance upon model selection is valuable. Much of this guidance comes from basic properties of the *true* SFS stresses τ, and the *true* averages $\bar{\mathbf{u}}$, of the *true* solution of the Navier–Stokes equations that should be preserved by $\mathcal{S}$ and $\mathbf{w}$ respectively.

We list here some relevant properties.

Condition 1: Reversibility. (Germano *et al.* [129].)
The true SFS stresses $\tau(\mathbf{v}_1, \mathbf{v}_2)$ are reversible, meaning

$$\tau(-\mathbf{v}_1, -\mathbf{v}_2) = \tau(\mathbf{v}_1, \mathbf{v}_2).$$

Thus, one important condition is that the approximate SFS stresses be reversible:

$$\mathcal{S}(-\mathbf{v}_1, -\mathbf{v}_2) = \mathcal{S}(\mathbf{v}_1, \mathbf{v}_2). \tag{6.4}$$

It's worth noting that (6.4) in a sense means that LES should seek a dispersive model rather than a dissipative model since eddy viscosity models are *irreversible*. Specifically, in eddy viscosity models $\mathcal{S}(\mathbf{v}_1, \mathbf{v}_2) = -\nu_T(\mathbf{v}_1)\nabla^s \mathbf{v}_2$. Since $\nu_T(\mathbf{v}_1) = \nu_T(-\mathbf{v}_1) \geq 0$, $\mathcal{S}(-\mathbf{v}_1, -\mathbf{v}_2) = -\mathcal{S}(\mathbf{v}_1, \mathbf{v}_2)$.

Condition 2: Realizability. (Sagaut [267] p. 54, Ghosal [134], and Vreman, Geurts, and Kuerten [308].)
If the filter kernel is nonnegative, $g(\mathbf{x}) \geq 0$ for all $\mathbf{x}$, then the true SFS stresses are positive semi-definite:

$$\xi^T \tau(\mathbf{u}, \mathbf{u})\, \xi \geq 0, \text{ for all } \xi \in \mathbb{R}^3.$$

Thus, it is natural to impose definiteness as an algebraic condition on any model sought:

$$\xi^T \mathcal{S}(\bar{\mathbf{u}}, \bar{\mathbf{u}})\, \xi \geq 0, \text{ for all } \xi \in \mathbb{R}^3. \tag{6.5}$$

Realizability is a simple and clear condition – but its significance in the final model is not well understood. This is because $\mathrm{div}\,(\tau(\mathbf{u}, \mathbf{u}))$ occurs in the model rather than $\tau(\mathbf{u}, \mathbf{u})$. Thus, any shift of $\mathcal{S}(\bar{\mathbf{u}}, \bar{\mathbf{u}})$ by a constant diagonal tensor does not change the final model.

This condition (6.5) also becomes less clear if the large scales are defined by techniques other than explicit filtering, or if the kernel changes sign, as with sharp spectral cutoff.

Condition 3: Finite kinetic energy. (Layton [203], Iliescu *et al.* [169] and John [176].)

Young's inequality for convolutions implies immediately that $\frac{1}{2}\int_\Omega |\overline{\mathbf{u}}|^2 dx \leq C \frac{1}{2}\int_\Omega |\mathbf{u}|^2 dx$, which is bounded by problem data. Since $\mathbf{w} \cong \overline{\mathbf{u}}$, it is natural, even essential, that the kinetic energy in the model does not blow up in finite time for general problem data

$$\frac{1}{2}\int_\Omega |\mathbf{w}|^2 dx \leq C^* < \infty,$$

where $C^* = C^*$ (problem data) is bounded uniformly in δ.

There are many models for which practical tests have reported stability problems which are typically "corrected" by the addition of enough *ad hoc*, extra eddy viscosity to prevent blow up. See [175, 169] for an example. Thus, if a model has the correct kinetic energy balance, such extra terms can be added to increase its *accuracy* rather than enforce stability. These considerations lead naturally to the next condition.

Condition 4: A lucid global energy balance relation. (Layton and Lewandowski [210, 204].)
The connection between the most general mathematical description of fluid flow and the physics of fluid motion is through the global energy inequality for the NSE. Define (for simplicity assume that $|\Omega| = 1$)

$$k_{NSE}(t) := \frac{1}{2}\int_\Omega |\mathbf{u}|^2\, dx, \quad \epsilon_{\text{NSE}}(t) := \int_\Omega 2\nu\, |\nabla^s \mathbf{u}|^2\, dx$$

$$\text{and } P(t) := \int_\Omega \mathbf{f} \cdot \mathbf{u}\, dx.$$

The energy inequality states

$$k_{NSE}(t) + \int_0^t \epsilon_{\text{NSE}}(t')\, dt' \leq k_{NSE}(0) + \int_0^t P(t')\, dt'.$$

The associated necessary condition is that the solution of the model (6.2), (6.3) satisfies a related global energy balance

$$k_{\text{model}}(t) + \int_0^t \epsilon_{\text{Model}}(t')\, dt' \leq k_{\text{model}}(0) + \int_0^t P_{\text{Model}}(t')\, dt', \qquad (6.6)$$

where, as $\delta \to 0$,

$$k_{\text{model}} \to k_{NSE}, \quad \epsilon_{\text{Model}} \to \epsilon, \quad \text{and } \quad P_{\text{Model}} \to P.$$

Condition 5: Modeling consistency.
In computational studies this is often called accuracy (a misnomer) and is assessed experimentally as follows. A velocity field $\mathbf{u}$ is obtained either from a moderate Reynolds number DNS or from experimental data and $\overline{\mathbf{u}}$ is explicitly calculated. Next, the modeling consistency is evaluated by calculating

$$\|\tau(\mathbf{u}, \mathbf{u}) - \mathcal{S}(\overline{\mathbf{u}}, \overline{\mathbf{u}})\|.$$

These are called *a priori* tests in the LES literature, implying that there is *no* actual LES modeling (such as, eddy viscosity) used in the numerical simulations, all the data being obtained from a DNS (more details are given in Chap. 12).

Important analytic studies of consistency can (and should) also be obtained as follows: for the fluctuating part of $\mathbf{u}$, model consistency should be expressed by the total model possessing a smoothing property.

For the mean field/smooth components of $\mathbf{u}$ a reasonable condition is that

$$\|\boldsymbol{\tau}(\mathbf{u}, \mathbf{u}) - \mathcal{S}(\overline{\mathbf{u}}, \overline{\mathbf{u}})\| \leq C(\mathbf{u})\,\delta^{\alpha} \text{ for } \mathbf{u} \text{ smooth, for some } \alpha \geq 2.$$

The reason for the restriction $\alpha \geq 2$ is that, for smooth $\mathbf{u}$, $\|\boldsymbol{\tau}(\mathbf{u}, \mathbf{u})\| \leq C(\mathbf{u})\delta^2$, so $\alpha = 2$ is minimal for consistency.

The third expression of consistency is for a Leray–Hopf weak solution $\mathbf{u}$ of the Navier–Stokes equations

$$\int_0^T \|\boldsymbol{\tau}(\mathbf{u}, \mathbf{u}) - \mathcal{S}(\overline{\mathbf{u}}, \overline{\mathbf{u}})\|^2 dt \to 0 \text{ as } \delta \to 0.$$

Condition 6: Existence of solutions for large data and long times.

It is known that global-in-time weak solutions $\mathbf{u}$ of the Navier–Stokes equations exist for large data and arbitrary Reynolds' numbers; Galdi [121]. A model for $\mathbf{w}$ approximating $\overline{\mathbf{u}} = g_\delta * \mathbf{u}$ should minimally replicate this property. (In fact, since $\overline{\mathbf{u}}$ is more regular than $\mathbf{u}$, the model for $\mathbf{w}$ should have more agreeable mathematical properties than the Navier–Stokes equations.)

Condition 7: Smoothing.

Given a weak solution $\mathbf{u}$ to the Navier–Stokes equations and a smooth filter g_δ, the true local averages satisfy

$$\overline{\mathbf{u}} \in C^\infty(\Omega), \text{ for each } t > 0.$$

Since $\mathbf{w} \cong \overline{\mathbf{u}}$, a reasonable (and minimal) condition is that the solution $\mathbf{w}$ to the model is regular enough that, for $\delta > 0$,

- *the model's weak solution $\mathbf{w}$ is a globally unique strong solution, and*
- *the model's energy inequality (6.6) is actually an energy equality.*

Condition 8: Limit consistency. (Layton and Lewandowski [204, 209].)
As $\delta \to 0, \overline{\mathbf{u}} = g_\delta * \mathbf{u} \to \mathbf{u}$, a weak solution of the NSE. Thus, two minimal conditions (the second studied in [204]) are

- *as $\delta \to 0$, there is a subsequence δ_j such that $\mathbf{w}(\delta_j) \to \mathbf{u}$, a weak solution of the NSE, and*
- *if the NSE weak solution $\mathbf{u}$ is regular enough to be unique, $\mathbf{w} \to \mathbf{u}$ as $\delta \to 0$.*

Condition 9: Verifiability (Layton and Kaya [204, 186].)
Since accuracy of a model is assessed experimentally by checking that $\|\tau(\mathbf{u}, \mathbf{u})$
$-\mathcal{S}(\overline{\mathbf{u}}, \overline{\mathbf{u}})\|$ is small, it is necessary that $\|\tau(\mathbf{u}, \mathbf{u}) - \mathcal{S}(\overline{\mathbf{u}}, \overline{\mathbf{u}})\|$ small implies that
$\|\overline{\mathbf{u}} - \mathbf{w}\|$ is small. In other words, minimally

$$\|\overline{\mathbf{u}} - \mathbf{w}\|_{L^{\infty}(0,T;L^2(\Omega))} \leq C\|\tau(\mathbf{u}, \mathbf{u}) - \mathcal{S}(\overline{\mathbf{u}}, \overline{\mathbf{u}})\|_{L^2(0,T;L^2(\Omega))}$$
$$(+ \text{ Terms that } \to 0 \text{ as } \delta \to 0).$$

Condition 10: Accuracy. (Layton and Lewandowski [209].)
For a weak solution $\mathbf{u}$ of the NSE,

$$\|\overline{\mathbf{u}} - \mathbf{w}\| \to 0 \text{ as } \delta \to 0$$

with some provable rate (at least in favorable cases).

Condition 11: Important experimental conditions.
In experiments with a minimal of algorithmic or model tuning, the model's
solution should replicate

- *the $k^{-5/3}$ energy spectrum of homogeneous, isotropic turbulence with appropriately modified k_c,*
- *statistics of turbulent channel flow (Moser, Kim, and Mansour [242], see also Fischer and Iliescu [106, 165]), and*
- *some important (and as yet not agreed upon) functionals of turbulence driven by interaction of a laminar flow with a more complex boundary.*

Condition 12: Frame invariance. (Speziale [279]).
Since the Navier–Stokes equations are themselves frame invariant, it is natural
to impose this as a reasonable condition upon any reduced system. Imposing
frame invariance gives some structure to the (very difficult) area of modeling non-Newtonian fluids. It has also given insight into conventional turbulence models; Speziale [279]. For a good exposition on frame invariance, see
the books of Sagaut [267], Pope [258], and Mohammadi and Pirroneau [239].
Frame invariance has three component parts: *translation invariance, Galilean
invariance*, and *rotation invariance*. We consider the first two on a homogeneous model:

$$\mathbf{w}_t + \nabla \cdot (\mathbf{w}\,\mathbf{w}^T) - \nu\Delta\mathbf{w} + \nabla q + \nabla \cdot \mathcal{S}(\mathbf{w}, \mathbf{w}) = \mathbf{0} \qquad \text{and} \qquad (6.7)$$
$$\nabla \cdot \mathbf{w} = 0.$$

Translation Invariance

Definition 6.1 (Translation invariance.). *Let $\mathbf{Z}$ be a fixed but arbitrary
constant vector. Let $\mathbf{y} = \mathbf{x} + \mathbf{Z}$, $\mathbf{W}(\mathbf{y}, t) = \mathbf{w}(\mathbf{x}, t)$, and $Q(\mathbf{y}, t) = q(\mathbf{x}, t)$. The
model (6.7) is translation invariant if $\mathbf{W}(\mathbf{y}, t)$ is a solution whenever $\mathbf{w}(\mathbf{x}, t)$
is a solution.*

It is easy to check that the Navier–Stokes equations (the case $\mathcal{S} = 0$ in the model (6.7)) are translation invariant.

Proposition 6.2. *Let $\mathcal{S} \equiv 0$ so (6.7) reduces to the Navier–Stokes equations. Then, (6.7) is translation invariant.*

Proof. By changing variables, we find

$$\frac{\partial \mathbf{w}}{\partial x_i} = \frac{\partial \mathbf{v}}{\partial y_i}.$$

Thus, trivially,

$$\mathbf{w}_t + \nabla_x \cdot (\mathbf{w}\,\mathbf{w}^T) - \nu \Delta_x \mathbf{w} + \nabla_x q = \mathbf{W}_t + \nabla_y \cdot (\mathbf{W}\,\mathbf{W}^T) - \nu \Delta_y \mathbf{W} + \nabla_y Q$$

and $0 = \nabla_x \cdot \mathbf{w} = \nabla_y \cdot \mathbf{W}$. □

This proposition implies that (6.7) is translation invariant provided that the model for $\boldsymbol{\tau}(\mathbf{u}, \mathbf{u})$ itself is too.

Proposition 6.3. *Suppose the averaging process is translation invariant. Then, with $\mathbf{y} = \mathbf{x} + \mathbf{Z}$, $\mathbf{Z} \in R^3$, and $\mathbf{U}(\mathbf{y}, t) = \mathbf{u}(\mathbf{y}, t)$,*

$$\nabla_x \cdot \boldsymbol{\tau}(\mathbf{u}(\mathbf{x}, t), \mathbf{u}(\mathbf{x}, t)) = \nabla_y \cdot \boldsymbol{\tau}(\mathbf{U}(\mathbf{y}, t), \mathbf{U}(\mathbf{y}, t)).$$

The model (6.7) is translation invariant if and only if whenever $\mathbf{y} = \mathbf{x} + \mathbf{Z}$, $\mathbf{Z} \in \mathbb{R}^3$

$$\nabla_x \cdot \mathcal{S}(\mathbf{w}(\mathbf{x}, t), \mathbf{w}(\mathbf{x}, t)) = \nabla_y \cdot \mathcal{S}(\mathbf{W}(\mathbf{y}, t), \mathbf{W}(\mathbf{y}, t)).$$

A sufficient condition is that whenever $\mathbf{y} = \mathbf{x} + \mathbf{Z}$, $\mathbf{Z} \in \mathbb{R}^3$,

$$\mathcal{S}(\mathbf{w}(\mathbf{x}, t), \mathbf{w}(\mathbf{x}, t)) = \mathcal{S}(\mathbf{W}(\mathbf{y}, t), \mathbf{W}(\mathbf{y}, t)).$$

Proof. Since $\nabla_x = \nabla_y$, this is clear. □

Galilean Invariance

Definition 6.4 (Galilean invariance). *Let $\mathbf{Z}$ be a fixed but arbitrary constant vector. Let $\mathbf{y} = \mathbf{x} + \mathbf{Z}\,t$ and $\mathbf{v}(\mathbf{y}, t) = \mathbf{w}(\mathbf{x}, t)$. The model (6.7) is Galilean invariant if $\mathbf{v}(\mathbf{y}, t)$ is a solution whenever $\mathbf{w}(\mathbf{x}, t)$ is a solution.*

Let us now consider the shift by a constant velocity $\mathbf{y} = \mathbf{x} + \mathbf{Z}t$, where $\mathbf{Z} \in \mathbb{R}^3$ is the fixed but arbitrary velocity vector. This corresponds to a shift of the velocity $\mathbf{w}$ by a constant $\mathbf{Z}$. Similarly to the previous case it is easy to prove the following proposition (see [239] for its proof).

Proposition 6.5. *The transformation $\mathbf{y} = \mathbf{x} + \mathbf{Z}t$ leaves the Navier–Stokes equations unchanged.*

It is important to note that the space-filtered Navier–Stokes equations are invariant under a shift $\mathbf{y} = \mathbf{x} + \mathbf{Z}\,t$ as well. This follows since, provided that averaging is exact on constants,

$$
\begin{aligned}
\nabla \cdot \tau(\mathbf{u} + \mathbf{Z}, \mathbf{u} + \mathbf{Z}) &= \nabla \cdot (\overline{(\mathbf{u} + \mathbf{Z})(\mathbf{u} + \mathbf{Z})} - \overline{(\mathbf{u} + \mathbf{Z})}\ \overline{(\mathbf{u} + \mathbf{Z})}) \\
&= \nabla \cdot [(\overline{\mathbf{u}\mathbf{u}} - \overline{\mathbf{u}}\,\overline{\mathbf{u}}) + \overline{\mathbf{u}\mathbf{Z}} + \overline{\mathbf{Z}\mathbf{u}} + \overline{\mathbf{Z}\mathbf{Z}} - \overline{\mathbf{u}}\mathbf{Z} + \overline{\mathbf{Z}\mathbf{u}} - \overline{\mathbf{Z}}\ \overline{\mathbf{Z}}] \\
&= \nabla \cdot (\overline{\mathbf{u}\mathbf{u}} - \overline{\mathbf{u}}\,\overline{\mathbf{u}}) \\
&= \nabla \cdot \tau(\mathbf{u}, \mathbf{u}).
\end{aligned}
$$

Definition 6.6. *The model (6.7) is Galilean invariant if*

$$
\nabla \cdot \mathcal{S}(\mathbf{w} + \mathbf{Z}, \mathbf{w} + \mathbf{Z}) = \nabla \cdot \mathcal{S}(\mathbf{w}, \mathbf{w})
$$

for any $\mathbf{Z} \in \mathbb{R}^3$ *and any* $\mathbf{w}$ *that is a solution of (6.7).*

We consider four examples:

Example 6.7. The Smagorinsky model [277] described in Chap. 3 is Galilean invariant.

Indeed, $\mathcal{S}(\mathbf{w}, \mathbf{w}) = (C_s \delta)^2 |\nabla^s \mathbf{w}| \nabla^s \mathbf{w}$, so that

$$
\mathcal{S}(\mathbf{w} + \mathbf{Z}, \mathbf{w} + \mathbf{Z}) = (C_s \delta)^2 |\nabla^s(\mathbf{w} + \mathbf{Z})| \nabla^s(\mathbf{w} + \mathbf{Z}) = \mathcal{S}(\mathbf{w}, \mathbf{w}),
$$

since $\mathbf{Z}$ is a constant vector.

Example 6.8. The eddy viscosity model $\nu_T = \mu\delta|\mathbf{w} - \overline{\mathbf{w}}|$ presented in Chap. 4 is Galilean invariant, provided that $\overline{\mathbf{Z}} = \mathbf{Z}$ for constant vectors $\mathbf{Z}$. Indeed,

$$
\mathcal{S}(\mathbf{w} + \mathbf{Z}, \mathbf{w} + \mathbf{Z}) = \mu\delta|(\mathbf{w} + \mathbf{Z}) - \overline{(\mathbf{w} + \mathbf{Z})}| \nabla^s(\mathbf{w} + \mathbf{Z}) = \mathcal{S}(\mathbf{w}, \mathbf{w}).
$$

Example 6.9. The Bardina Scale-similarity model [13] (described in Chap. 8) is Galilean invariant, provided that $\overline{\mathbf{Z}} = \mathbf{Z}$, $\overline{\mathbf{Z}\mathbf{w}} = \mathbf{Z}\,\overline{\mathbf{w}}$ and $\overline{\mathbf{w}\mathbf{Z}} = \overline{\mathbf{w}}\,\mathbf{Z}$ for constant vectors $\mathbf{Z}$.

Here $\mathcal{S}_{Bardina}(\mathbf{w}, \mathbf{w}) = \overline{\mathbf{w}\mathbf{w}} - \overline{\mathbf{w}}\ \overline{\mathbf{w}}$. Thus,

$$
\begin{aligned}
\mathcal{S}(\mathbf{w} + \mathbf{Z}, \mathbf{w} + \mathbf{Z}) &= \overline{(\mathbf{w} + \mathbf{Z})(\mathbf{w} + \mathbf{Z})} - \overline{(\mathbf{w} + \mathbf{Z})}\ \overline{(\mathbf{w} + \mathbf{Z})} \\
&= (\overline{\mathbf{w}\mathbf{w}} - \overline{\mathbf{w}}\ \overline{\mathbf{w}}) + \overline{\mathbf{w}\mathbf{Z}} + \overline{\mathbf{Z}\mathbf{w}} + \overline{\mathbf{Z}\mathbf{Z}} - (\overline{\mathbf{w}}\mathbf{Z} + \overline{\mathbf{Z}}\overline{\mathbf{w}} + \overline{\mathbf{Z}}\ \overline{\mathbf{Z}}) \\
&= \mathcal{S}(\overline{\mathbf{w}}, \overline{\mathbf{w}}) + (\overline{\mathbf{w}\mathbf{Z}} - \overline{\mathbf{w}}\ \overline{\mathbf{Z}}) + (\overline{\mathbf{Z}\mathbf{w}} - \overline{\mathbf{Z}}\ \overline{\mathbf{w}}) + (\overline{\mathbf{Z}\mathbf{Z}} - \overline{\mathbf{Z}}\ \overline{\mathbf{Z}}) \\
&= \mathcal{S}(\mathbf{w}, \mathbf{w}).
\end{aligned}
$$

Example 6.10. The model $\mathcal{S}(\overline{\mathbf{w}}, \overline{\mathbf{w}}) = \overline{\mathbf{w}}\,\overline{\mathbf{w}} - \mathbf{w}\,\mathbf{w}$ (Chap. 8) is Galilean invariant, provided that the averaging preserves incompressibility ($\overline{\nabla \cdot \mathbf{w}} = \nabla \cdot \mathbf{w}$), constant vectors ($\overline{\mathbf{Z}} = \mathbf{Z}$), $\overline{\mathbf{Z}\mathbf{w}} = \mathbf{Z}\,\overline{\mathbf{w}}$ and $\overline{\mathbf{w}\mathbf{Z}} = \overline{\mathbf{w}}\,\mathbf{Z}$.

This is a model in which $\mathcal{S}(\mathbf{w} + \mathbf{Z}, \mathbf{w} + \mathbf{Z}) \neq \mathcal{S}(\mathbf{w}, \mathbf{w})$ and yet the model is still Galilean invariant. Indeed,

$$
\begin{aligned}
\mathcal{S}(\mathbf{w} + \mathbf{Z}, \mathbf{w} + \mathbf{Z}) &= \overline{(\mathbf{w} + \mathbf{Z})(\mathbf{w} + \mathbf{Z})} - (\overline{\mathbf{w} + \mathbf{Z}})(\overline{\mathbf{w} + \mathbf{Z}}) \\
&= (\overline{\mathbf{ww}} - \overline{\mathbf{w}}\,\overline{\mathbf{w}}) + (\overline{\mathbf{wZ}} - \overline{\mathbf{w}}\mathbf{Z}) + (\overline{\mathbf{Zw}} - \mathbf{Z}\overline{\mathbf{w}}) + (\overline{\mathbf{Z}\,\mathbf{Z}} - \mathbf{Z}\,\mathbf{Z}) \\
&= \mathcal{S}(\mathbf{w}, \mathbf{w}) + (\overline{\mathbf{w}} - \mathbf{w})\mathbf{Z} + \mathbf{Z}(\overline{\mathbf{w}} - \mathbf{w}).
\end{aligned}
$$

Thus, since $\nabla \cdot \mathbf{w} = \nabla \cdot \overline{\mathbf{w}} = 0$,

$$
\begin{aligned}
\nabla \cdot \mathcal{S}(\mathbf{w} + \overline{\mathbf{Z}}, \mathbf{w} + \overline{\mathbf{Z}}) &= \nabla \cdot \mathcal{S}(\mathbf{w}, \mathbf{w}) + (\nabla \cdot \overline{\mathbf{w}} - \nabla \cdot \mathbf{w})\mathbf{Z} + \mathbf{Z}(\nabla \cdot \overline{\mathbf{w}} - \nabla \cdot \mathbf{w}) \\
&= \nabla \cdot \mathcal{S}(\mathbf{w}, \mathbf{w}).
\end{aligned}
$$

Example 6.11. The Rational LES model [122] as well as the Gradient LES model [212, 65] (see Chap. 7 for details) are Galilean invariant. Indeed,

$$
\begin{aligned}
\mathcal{S}(\mathbf{w} + \mathbf{Z}, \mathbf{w} + \mathbf{Z}) &= \left(\mathbb{I} - \frac{\delta^2}{4\gamma}\Delta\right)^{-1}\left(\frac{\delta^2}{2\gamma}\nabla(\mathbf{w} + \mathbf{Z})\nabla(\mathbf{w} + \mathbf{Z})\right) \\
&= \left(\mathbb{I} - \frac{\delta^2}{4\gamma}\Delta\right)^{-1}\left(\frac{\delta^2}{2\gamma}\nabla\mathbf{w}\nabla\mathbf{w}\right) \\
&= \mathcal{S}(\mathbf{w}, \mathbf{w}).
\end{aligned}
$$

Remark 6.12. In proving translation and Galilean invariance for all the LES models that we considered above, we assumed that δ (the radius of the spatial filter) is constant in space. If, however, $\delta = \delta(\mathbf{x})$, then the translation and Galilean invariance might not hold anymore.

7

Closure Based on Wavenumber Asymptotics

This chapter is devoted to the derivation and mathematical analysis of three *approximate deconvolution* LES models. A formal[1] definition of the approximation deconvolution approach might be the following:

Definition 7.1. *A deconvolution method is defined by means of an operator $\mathcal{D}$ such that if $v = g_\delta * u$, then $u = \mathcal{D}(v)$. An "order α" approximate deconvolution operator is an operator $\mathcal{D}_\alpha$ such that if $u = \mathcal{D}_\alpha(v)$ and u is smooth enough, then $g_\delta * u = v + O(\delta^\alpha)$.*

Essentially, all approximate deconvolution LES models aim at recovering (some of) the information lost in the filtering process (*i.e.* $\mathbf{u}' = \mathbf{u} - \overline{\mathbf{u}}$) by using the available approximation of the filtered flow variables (*i.e.* $\overline{\mathbf{u}}$). The approximate deconvolution methodology has a long and rich history in the LES community, starting with the pioneering work of Leonard [212], and continuing with Clark, Ferziger, and Reynolds [65], Geurts [130], Domaradzki and collaborators [93, 92], Stolz, Adams, and Kleiser [285, 288, 287, 2, 284, 289, 290, 291, 286, 3], Galdi and Layton [122], just to name a few. These methods have different names (such as approximate deconvolution or velocity estimation), but they all share the same philosophy: use an approximation for $\overline{\mathbf{u}}$ to recover an approximation for $\mathbf{u}$. This approximate deconvolution philosophy is *fundamentally different* from the eddy viscosity philosophy. The former is mathematical in nature, whereas the latter is based entirely on physical insight. Each approach has its own advantages and drawbacks. We described the eddy viscosity approach in Part II. We shall now start presenting some approximate deconvolution models. We shall continue this presentation in Chap. 8.

We present in this chapter a special class of approximate deconvolution models based on *wavenumber asymptotics*. The algorithm used in the derivation of these LES models is straightforward:

---

[1] Of course, in the mathematical development of a specific approximate deconvolution method, domains, ranges, *etc.* all must be specified.

Wavenumber Asymptotics Approximate Deconvolution

- **Step 1:** Apply the Fourier Transform to all the terms involved in the closure problem

- **Step 2:** Apply asymptotic expansion to approximate the resulting terms in the wavenumber space

- **Step 3:** Apply the Inverse Fourier Transform to the new terms to get approximations in the physical space of the original terms in the closure problem

Of special importance in this algorithm is the actual form of the LES spatial filter g_δ. This is in clear contrast to the eddy viscosity models where the spatial filter g_δ was used only implicitly (through the radius δ, for example). In the derivation of all the LES models in this chapter, we shall use the Gaussian filter introduced in Chap. 1.

The difference mentioned above is the essential reason for different terminologies for the stress tensor $\boldsymbol{\tau} = \overline{\mathbf{u}\,\mathbf{u}^T} - \overline{\mathbf{u}}\,\overline{\mathbf{u}}^T$ in the closure problem: for the most part, the LES community refers to $\boldsymbol{\tau}$ as the **subgrid-scale (SGS)** stress tensor. It is then implicitly assumed that the grid-scale h and the filter-scale δ are treated as one item for all practical purposes (in other words, there is no distinction made between h and δ). Thus, in general, when the SGS terminology is used, it is generally assumed that all the information below the grid-scale h (and therefore that below the filter-scale δ) is completely and irreversibly lost. Then, the LES modeling process in the closure problem employs *exclusively* physical insight to account for the subfilter-scale information.

At the other end of the spectrum are those in LES who refer to the stress tensor $\boldsymbol{\tau} = \overline{\mathbf{u}\,\mathbf{u}^T} - \overline{\mathbf{u}}\,\overline{\mathbf{u}}^T$ in the closure problem as the **subfilter-scale (SFS)** stress tensor. In this case, one makes implicitly a clear distinction between the filter-scale δ (the radius of the spatial filter g_δ) and the grid-scale h.

Although less popular than the SGS approach, the SFS approach has a long and rich history. The first SFS model was introduced by Leonard in his pioneering work in 1974 [212]. Subsequently, Clark, Ferziger, and Reynolds developed Leonard's model in [65]. Since then, there have been many SFS proposed and used successfully [130, 93, 92, 285, 288, 287, 2, 284, 289, 290, 291, 286, 3, 122, 39, 170, 106, 165, 166, 309, 308, 310, 55, 315, 316, 66, 67].

In this chapter, we analyze three such SFS models: the Gradient LES model of Leonard [212] and Clark, Ferziger, and Reynolds [65], the Rational LES model of Galdi and Layton [122], and the Higher-order Subfilter-scale model of Berselli and Iliescu [33]. All three models belong to a particular class of SFS models – the approximate deconvolution models based on wavenumber asymptotics. For each model, we start with a careful derivation followed by a thorough mathematical analysis.

7.1 The Gradient (Taylor) LES Model

The Gradient LES model is the first approximate deconvolution model based on wavenumber asymptotics. This model was introduced by Leonard [212] and was developed in Clark, Ferziger, and Reynolds [65]. The Gradient LES model has been used in numerous computational studies [65, 106, 165, 166, 309, 308, 310, 55, 315, 316]. When used as a stand-alone LES model, the Gradient LES model produces numerically unstable approximations. This was noted, for example, in the numerical simulation of 3D lid-driven cavity turbulent flows, where the Gradient LES model produced finite time blow-up of the kinetic energy (see Iliescu *et al.* [169]). To stabilize the numerical approximation, the Gradient LES model is usually supplemented by an eddy viscosity term, resulting in a so-called *mixed model* (see, [316]). The role and limitations of the Gradient LES model have recently been reconsidered by Geurts and Holm [132].

It is now widely accepted that the Gradient LES model, while recovering some of the subfilter-scale information, is very unstable in numerical computations (if it is not supplemented by an eddy viscosity model). In fact, we present in the next section a *mathematical* reason for the numerical instability of the Gradient LES model. We then introduce the Rational LES model of Galdi and Layton [122], which circumvents this drawback. Thorough numerical tests with the Gradient and Rational LES models for turbulent channel flows were performed in a series of papers by Iliescu and Fischer [106, 165, 166]. We present these tests in detail in Chap. 12. Further numerical tests were performed in [169, 175, 173, 176]. All the numerical tests with the two LES models confirm the mathematical improvement in the Rational LES model. In Sect. 7.3, we present a further improvement of the Rational LES model, the Higher-order Subfilter-scale Model of Berselli and Iliescu [33], which avoids the mathematically induced instability in the Gradient LES model.

Thus, although it is relatively clear that the Gradient LES model should be replaced by its improvements (the Rational or Higher-order Subfilter-scale LES models) in numerical computations, we shall still present a careful mathematical analysis for the Gradient LES model, mainly because of its relative popularity in the LES community.

7.1.1 Derivation of the Gradient LES Model

The unfiltered flow variable $\mathbf{u}$ is equal to its filtered part $\overline{\mathbf{u}} = g_\delta * \mathbf{u}$ plus "turbulent fluctuations" (defined by $\mathbf{u}' = \mathbf{u} - \overline{\mathbf{u}}$). Since convolution is transformed into a product by the Fourier transform, we obtain

$$\widehat{\overline{\mathbf{u}}}(\mathbf{k}) = \widehat{g}_\delta(\mathbf{k})\widehat{\overline{\mathbf{u}}}(\mathbf{k}) + \widehat{g}_\delta(\mathbf{k})\widehat{\mathbf{u}'}(\mathbf{k}).$$

Thus,

$$\widehat{\mathbf{u}'}(\mathbf{k}) = \left(\frac{1}{\widehat{g}_\delta(\mathbf{k})} - 1 \right) \widehat{\overline{\mathbf{u}}}(\mathbf{k}).$$

The above formula allows us to evaluate the following terms involved in the subfilter-scale stress tensor $\boldsymbol{\tau} = \overline{\mathbf{u}\,\mathbf{u}^T} - \overline{\mathbf{u}}\,\overline{\mathbf{u}}^T$:

$$
\widehat{\overline{\mathbf{u}\,\overline{\mathbf{u}}^T}}(\mathbf{k}) = \widehat{g}_\delta(\mathbf{k})\,\widehat{\overline{\mathbf{u}}}(\mathbf{k}) * \widehat{\overline{\mathbf{u}}}^T(\mathbf{k})
$$

$$
\widehat{\overline{\mathbf{u}\,\mathbf{u}'^T}}(\mathbf{k}) = \widehat{g}_\delta(\mathbf{k})\,\widehat{\overline{\mathbf{u}}}(\mathbf{k}) * \left[\left(\frac{1}{\widehat{g}_\delta(\mathbf{k})} - 1\right)\widehat{\overline{\mathbf{u}}}^T(\mathbf{k})\right]
$$

$$
\widehat{\overline{\mathbf{u}'\,\overline{\mathbf{u}}^T}}(\mathbf{k}) = \widehat{g}_\delta(\mathbf{k})\left[\left(\frac{1}{\widehat{g}_\delta(\mathbf{k})} - 1\right)\widehat{\overline{\mathbf{u}}}(\mathbf{k})\right] * \widehat{\overline{\mathbf{u}}}^T(\mathbf{k})
$$

$$
\widehat{\overline{\mathbf{u}'\,\mathbf{u}'^T}} = \widehat{g}_\delta(\mathbf{k})\left[\left(\frac{1}{\widehat{g}_\delta(\mathbf{k})} - 1\right)\widehat{\overline{\mathbf{u}}}(\mathbf{k})\right] * \left[\left(\frac{1}{\widehat{g}_\delta(\mathbf{k})} - 1\right)\widehat{\overline{\mathbf{u}}}^T(\mathbf{k})\right].
$$

$$(7.1)$$

The Gradient LES model is derived by using the Taylor expansion for $\widehat{g}_\delta(\mathbf{k})$. The expansion is done with respect to δ, up to terms that are $O(\delta^4)$:

$$
\widehat{g}_\delta(\mathbf{k}) = 1 - \frac{\delta^2}{4\gamma}|\mathbf{k}|^2 + O(\delta^4), \qquad \frac{1}{\widehat{g}_\delta(\mathbf{k})} - 1 = \frac{\delta^2}{4\gamma}|\mathbf{k}|^2 + O(\delta^4). \qquad (7.2)
$$

Substitution of (7.2) in (7.1), and application of the inverse Fourier transform yield

$$
\overline{\mathbf{u}\,\overline{\mathbf{u}}^T} = \overline{\mathbf{u}}\,\overline{\mathbf{u}}^T + \frac{\delta^2}{4\gamma}\Delta(\overline{\mathbf{u}}\,\overline{\mathbf{u}}^T) + O(\delta^4),
$$

$$
\overline{\mathbf{u}\,\mathbf{u}'^T} = -\frac{\delta^2}{4\gamma}\overline{\mathbf{u}}\,\Delta\overline{\mathbf{u}}^T + O(\delta^4),
$$

$$
\overline{\mathbf{u}'\,\overline{\mathbf{u}}^T} = -\frac{\delta^2}{4\gamma}\Delta\overline{\mathbf{u}}\,\overline{\mathbf{u}}^T + O(\delta^4),
$$

$$
\overline{\mathbf{u}'\,\mathbf{u}'^T} = O(\delta^4).
$$

Ignoring terms that are of order of δ^4, and by observing that

$$
\Delta(f\,g) = \Delta f\,g + 2\nabla f\nabla g + f\,\Delta g,
$$

we finally get

$$
\overline{\mathbf{u}\,\mathbf{u}^T} - \overline{\mathbf{u}}\,\overline{\mathbf{u}}^T = \frac{\delta^2}{2\gamma}\nabla\overline{\mathbf{u}}\nabla\overline{\mathbf{u}}^T + O(\delta^4).
$$

Recall that generally $\gamma = 6$, while the matrix $\nabla\mathbf{w}\nabla\mathbf{w}^T$ is defined by

$$
[\nabla\mathbf{w}\nabla\mathbf{w}^T]_{ij} = \sum_{l=1}^{d}\frac{\partial w^i}{\partial x_l}\frac{\partial w^j}{\partial x_l}, \qquad i,j = 1,\ldots,d.
$$

Collecting terms and simplifying, we obtain the so-called *Gradient LES model*

$$
\mathbf{w}_t + \nabla q + (\mathbf{w}\cdot\nabla)\mathbf{w} - \frac{1}{Re}\Delta\mathbf{w} + \nabla\cdot\left[\frac{\delta^2}{12}\nabla\mathbf{w}\,\nabla\mathbf{w}^T\right] = \overline{\mathbf{f}}, \qquad (7.3)
$$

$$
\nabla\cdot\mathbf{w} = 0, \qquad (7.4)
$$

for the unknown $\mathbf{w} \simeq \overline{\mathbf{u}}$. This model is also known as the Taylor LES model, since its derivation is based on a Taylor expansion in wavenumber space.

7.1.2 Mathematical Analysis of the Gradient LES Model

In this section we sketch out the mathematical theory of existence and uniqueness for the Gradient LES model. This theory is only local in time (very small time $0 \leq t < T$, where T depends upon all problem data) and for very small and very smooth data. This is consistent with computational experience for the Gradient LES model. We require some smooth sets of functions and in particular we define, for $k \geq 1$

$$H_\sigma^k := \left\{ \mathbf{v} \in [H^k(\Omega)]^d : \nabla \cdot \mathbf{v} = 0 \right\}.$$

Definition 7.2 (Compatibility condition). *We say that the initial datum* $\mathbf{w}_0 \in H_\sigma^3$ *satisfies compatibility conditions if* $\mathbf{w}_0$ *and* $\partial_t \mathbf{w}_0$ *have null trace on the boundary of* Ω. *In particular,* $\partial_t \mathbf{w}_0$ *means*

$$\partial_t \mathbf{w}_0 = -\nabla \cdot (\mathbf{w}_0 \, \mathbf{w}_0^T) + \frac{1}{Re} \nabla \cdot \nabla^s \mathbf{w}_0$$
$$- \frac{\delta^2}{12} \nabla \mathbf{w}_0 \nabla \mathbf{w}_0^T - \nabla p(\mathbf{x}, 0) + \overline{\mathbf{f}}(\mathbf{x}, 0)$$

and the value of $p(\mathbf{x}, 0)$ *is obtained by solving an elliptic problem; see [67].*

The following theorem of existence of smooth solutions, for small initial data was proven in Coletti [66] for the Gradient LES model.

Theorem 7.3. *Let us assume that*

1. *$\Omega \subset \mathbb{R}^3$ is an open, bounded, and connected set with regular boundary;*
2. *the initial condition $\mathbf{w}_0$ satisfies the compatibility condition of Definition 7.2;*
3. *$\mathbf{w}_0 \in H_\sigma^3$ with $\|\mathbf{w}_0\|_{H^3} \leq \delta^2$;*
4. *$\overline{\mathbf{f}}$ belongs to $L^2(0, T; H_\sigma^2) \cap H^1(0, T; L_\sigma^2)$, and satisfies $\|\overline{\mathbf{f}}\|_{L^2(0,T;H^2)} \leq \delta^2$ and $\|\partial_t \overline{\mathbf{f}}\|_{L^2(0,T;L^2)} \leq \delta^2$.*

Then, there exists a $\delta_0 > 0$ such that for every $\delta \in (0, \delta_0]$, the solution to (7.3) exists and is unique in $C(0, T; H_\sigma^3) \cap L^2(0, T; H_\sigma^4)$ for the velocity and $C(0, T; H_\sigma^2) \cap L^2(0, T; H_\sigma^3)$ for pressure.

This result requires both very high regularity and smallness of the data of the problem.

Regarding the existence of weak solutions we have the following result, proved in [67]. To analyze weak solutions it seems necessary (currently we do not know how to remove this limitation) to add a dissipative term of Smagorinsky type. In this way the Gradient LES model results as an improvement of the Smagorinsky model, that may include backscatter of energy. The global existence theory of Coletti [67] is based upon an assumption that the nonlinear diffusion added by the p-Laplacian is large enough (see (7.7)) to control any blow-up arising from the instabilities in the Gradient LES model of the nonlinear interaction terms.

$$\mathbf{w}_t + \nabla q + \nabla \cdot (\mathbf{w}\,\mathbf{w}^T) - \frac{2}{Re}\nabla \cdot \nabla^s \mathbf{w}$$
$$-\nabla \cdot (C_G |\nabla \mathbf{w}|\nabla \mathbf{w}) + \nabla \cdot \left[\frac{\delta^2}{12}\nabla \mathbf{w}\,\nabla \mathbf{w}^T\right] = \bar{\mathbf{f}}, \qquad (7.5)$$

$$\nabla \cdot \mathbf{w}_m = 0. \qquad (7.6)$$

We now give Coletti's existence theorem from [67] for the above mixed model (7.5) and (7.6).

Theorem 7.4. *If*

$$C_G > \frac{\delta^2}{6}, \qquad (7.7)$$

if the initial datum $\mathbf{u}_0 \in W^{1,3}_{0,\sigma}$ *and if* $\bar{\mathbf{f}} \in H^1(0,T;L^2_\sigma)$, *then a unique weak solution to the Gradient LES model (7.5) and (7.6) exists in* $H^1(0,T;L^2_\sigma) \cap L^3(0,T;W^{1,3}_{0,\sigma})$. *Furthermore, such a solution is unique.*

This theorem does not involve smallness on the data, but requires a sufficiently big constant C_G such that (7.7) is satisfied. Regarding the meaning of this assumption, see Sect. 7.1.3.

Proof (of Theorem 7.4). The proof of this theorem is similar to that of Theorem 3.9. The main point is to show that the new operator involved in the abstract formulation of the problem is monotone. In this case the operator A involved is

$$A(\mathbf{u}) = -\frac{2}{Re}\nabla \cdot \nabla^s \mathbf{u} - \nabla \cdot (C_G |\nabla \mathbf{u}|\nabla \mathbf{u}) + \nabla \cdot \left[\frac{\delta^2}{12}\nabla \mathbf{u}\,\nabla \mathbf{u}^T\right].$$

The constant C_G has to be large in order to show that the operator A is monotone. The Smagorinsky term (3-Laplacian) will then dominate the last term, see Lemma 7.5.

We start the proof of the theorem with the usual energy estimate.

Energy estimate. By multiplying (7.5) (or better use $\mathbf{w}_m$ as Test function in the weak formulation) by $\mathbf{w}_m$, we get

$$\frac{1}{2}\frac{d}{dt}\|\mathbf{w}_m\|^2 + \frac{2}{Re}\|\nabla^s \mathbf{w}_m\|^2 + C_G\|\nabla \mathbf{w}_m\|^3_{L^3}$$
$$\leq \frac{\delta^2}{12}\int_\Omega |\partial_l \mathbf{w}^i_m \partial_l \mathbf{w}^j_m \partial_j \mathbf{w}^i_m|\,d\mathbf{x} + \|\bar{\mathbf{f}}\|\,\|\mathbf{w}_m\|$$
$$\qquad (7.8)$$
$$\leq \frac{\delta^2}{12}\|\nabla \mathbf{w}_m\|^3_{L^3} + \frac{1}{2}\|\bar{\mathbf{f}}\|^2 + \frac{1}{2}\|\mathbf{w}_m\|^2.$$

In the derivation of the above estimate we used the linear algebra estimate

$$\sum_{i,j,k=1}^{d} a_{ki} b_{kj} c_{ji} \leq \|a\| \, \|b\| \, \|c\|,$$

which holds for all real, nonnegative, square matrices $a, b, c \in \mathcal{M}(d \times d, \mathbb{R})$; recall that $\| \, . \, \|$ is the usual norm $\|m\| = \sqrt{\sum_{i,j=1}^{d} m_{ij}^2}$.

Finally, provided that (7.7) holds, the first term on the right-hand side of (7.8) may be absorbed in the left-hand side to give

$$\frac{1}{2} \frac{d}{dt} \|\mathbf{w}_m\|^2 + \frac{2}{Re} \|\nabla^s \mathbf{w}_m\|^2 + \left(C_G - \frac{\delta^2}{6} \right) \|\nabla \mathbf{w}_m\|_{L^3}^3 \leq \frac{1}{2} \|\overline{\mathbf{f}}\|^2 + \frac{1}{2} \|\mathbf{w}_m\|^2.$$

This shows (with the standard procedure we introduced in the previous chapters) that

$$\sup_{0<t<T} \|\mathbf{w}_m(t)\|^2 + \int_0^T \|\nabla^s \mathbf{w}_m(\tau)\|^2 d\tau + \int_0^T \|\nabla \mathbf{w}_m(\tau)\|_{L^3}^3 d\tau \leq C, \quad (7.9)$$

for a constant C_G that depends on Re, δ, and $\overline{\mathbf{f}}$, but is independent of $m \in \mathbb{N}$.

Second a priori estimate. Now we proceed as in the analysis of the Smagorinsky–Ladyžhenskaya model and we multiply (7.5) (again to be more precise a weak formulation of the Galerkin Gradient LES model) by $\partial_t \mathbf{w}_m$ to get the following equation:

$$\|\partial_t \mathbf{w}_m\|^2 + \frac{1}{Re} \frac{d}{dt} \|\nabla^s \mathbf{w}_m\|^2 + \frac{C_G}{3} \frac{d}{dt} \|\nabla \mathbf{w}_m\|_{L^3}^3 = (\overline{\mathbf{f}}, \partial_t \mathbf{w}_m)$$
$$+ (\mathbf{w}_m \cdot \nabla \mathbf{w}_m, \partial_t \mathbf{w}_m) + \frac{\delta^2}{12} \left(\nabla \mathbf{w}_m \nabla \mathbf{w}_m^T, \partial_t \nabla \mathbf{w}_m \right). \tag{7.10}$$

The first term on right-hand side can be estimated as in Sect. 3.4, but the last term involves the time derivative of $\nabla \mathbf{w}_m$, which is not present on the left-hand side! Integrating by parts this term would lead to second-order (space) derivatives of $\mathbf{w}_m$! These two facts show that this estimate is not useful by itself: we need at least another tool to get estimates involving the same terms, on both sides of the inequalities.

In particular we estimate the last term on the right-hand side of (7.10) as follows:

$$\left| (\nabla \mathbf{w}_m \nabla \mathbf{w}_m^T, \partial_t \nabla \mathbf{w}_m) \right| \leq \int_\Omega |\nabla \mathbf{w}_m|^{1/2} |\partial_t \nabla \mathbf{w}_m| |\nabla \mathbf{w}_m|^{3/2} \, d\mathbf{x}$$

and with the Hölder inequality

$$\leq \left(\int_\Omega |\nabla \mathbf{w}_m| |\partial_t \nabla \mathbf{w}_m|^2 \, d\mathbf{x} \right)^{1/2} \left(\int_\Omega |\nabla \mathbf{w}_m|^3 \, d\mathbf{x} \right)^{1/2}$$
$$\leq \epsilon \int_\Omega |\nabla \mathbf{w}_m| |\partial_t \nabla \mathbf{w}_m|^2 \, d\mathbf{x} + C(\epsilon) \|\mathbf{w}_m\|_{L^3}^3.$$

Third *a priori* estimate. We prove another *a priori* estimate: we first differentiate (7.5) with respect to time, then we multiply by $\partial_t \mathbf{w}_m$, and we integrate by parts. We get the following system of PDEs:

$$\partial_t^2 \mathbf{w}_m + \partial_t \mathbf{w}_m \cdot \nabla \mathbf{w}_m + \mathbf{w}_m \cdot \nabla \partial_t \mathbf{w}_m - \frac{2}{Re} \nabla \cdot \nabla^s \partial_t \mathbf{w}_m$$

$$- \partial_t \nabla \cdot (C_G |\nabla \mathbf{w}_m| \nabla \mathbf{w}_m) + \nabla \cdot \frac{\delta^2}{12} \left(\partial_t \nabla \mathbf{w}_m \nabla \mathbf{w}_m^T + \nabla \mathbf{w}_m \partial_t \nabla \mathbf{w}_m^T \right)$$

$$= \partial_t \overline{\mathbf{f}} - \partial_t \nabla p.$$

After multiplication by $\partial_t \mathbf{w}_m$ the terms on the left-hand side can be treated as follows:

$$(\partial_t^2 \mathbf{w}_m, \partial_t \mathbf{w}_m) = \frac{1}{2} \frac{d}{dt} \|\partial_t \mathbf{w}_m\|^2, \qquad (-\nabla \cdot \nabla^s \partial_t \mathbf{w}_m, \partial_t \mathbf{w}_m) = \|\nabla \partial_t \mathbf{w}_m\|,$$

and consequently

$$- \int_\Omega \partial_t \nabla \cdot (C_G |\nabla \mathbf{w}_m| \nabla \mathbf{w}_m) \partial_t \mathbf{w}_m \, d\mathbf{x}$$

$$= -C_G \int_\Omega \left[\nabla \cdot \left(\frac{\partial}{\partial t} |\nabla \mathbf{w}_m| \nabla \mathbf{w}_m \right) + \nabla \cdot (C_G |\nabla \mathbf{w}_m| \partial_t \nabla \mathbf{w}_m) \right] \partial_t \mathbf{w}_m \, d\mathbf{x}$$

$$= C_G \int_\Omega \frac{(\nabla \mathbf{w}_m \partial_t \nabla \mathbf{w}_m)^2}{|\nabla \mathbf{w}_m|} \, d\mathbf{x} + C_G \int_\Omega |\nabla \mathbf{w}_m| \, |\partial_t \nabla \mathbf{w}_m|^2 \, d\mathbf{x}.$$

$$(7.11)$$

The SFS stress-tensor part (multiplied by $\delta^2/12$) is treated simply as follows:

$$\left| \int_\Omega \left[\partial_t \nabla \mathbf{w}_m \nabla \mathbf{w}_m^T + \nabla \mathbf{w}_m \partial_t \nabla \mathbf{w}_m^T \right] \partial_t \nabla \mathbf{w}_m \, d\mathbf{x} \right| \leq \int_\Omega |\partial_t \nabla \mathbf{w}_m|^2 |\nabla \mathbf{w}_m| \, d\mathbf{x}.$$

The usual nonlinear term can be estimated in the following manner: first note that since $\partial_t \mathbf{w}_m$ is divergence-free, $(\mathbf{w}_m \cdot \nabla \partial_t \mathbf{w}_m, \partial_t \mathbf{w}_m) = 0$. Then, we estimate the other term as follows:

$$|(\partial_t \mathbf{w}_m \cdot \nabla \mathbf{w}_m, \partial_t \mathbf{w}_m)| \leq \int_\Omega |\partial_t \mathbf{w}_m|^2 |\nabla \mathbf{w}_m| \, d\mathbf{x}$$

use Hölder inequality with exponents 6, 2, and 3

$$\leq \|\partial_t \mathbf{w}_m\|_{L^6} \|\partial_t \mathbf{w}_m\| \, \|\nabla \mathbf{w}_m\|_{L^3}$$

use the Sobolev embedding, together with Young inequality,

$$\leq \varepsilon \|\partial_t \nabla \mathbf{w}_m\|^2 + C(\epsilon, \Omega) \|\partial_t \mathbf{w}_m\|^2 \|\nabla \mathbf{w}_m\|_{L^3}^2.$$

The first nonlinear term on the right-hand side of (7.10) can be estimated in the same way.

By observing that the first term on the right-hand side of (7.11) is non-negative we arrive finally at the estimate

$$\frac{1}{2}\frac{d}{dt}\|\partial_t \mathbf{w}_m\|^2 + C_G \int_\Omega |\nabla \mathbf{w}_m||\partial_t \nabla \mathbf{w}_m|^2 \, d\mathbf{x} + \frac{2}{Re}\|\partial_t \nabla \mathbf{w}_m\|^2$$

(7.12)

$$\leq \frac{\delta^2}{6}\int_\Omega |\nabla \mathbf{w}_m||\partial_t \nabla \mathbf{w}_m|^2 \, d\mathbf{x} + \|\partial_t \overline{\mathbf{f}}\|^2 + \|\partial_t \mathbf{w}_m\|^2.$$

This estimate shows how taking C_G to satisfy (7.7) makes it possible to control the additional nonlinear term. In fact, if we add (7.10) (with the estimates for the right-hand side) and (7.12), we obtain

$$\frac{d}{dt}\left(\frac{1}{2}\|\partial_t \mathbf{w}_m\|^2 + \frac{2}{Re}\|\nabla \mathbf{w}_m\|^2 + \frac{C_G}{3}\|\nabla \mathbf{w}_m\|_{L^3}^3\right) + \|\partial_t \mathbf{w}_m\|^2$$

$$+ \frac{1}{2Re}\|\partial_t \nabla \mathbf{w}_m\|^2 + \left[C_G - \frac{\delta^2}{6}\right]\int_\Omega |\nabla \mathbf{w}_m||\partial_t \nabla \mathbf{w}_m|^2 \, d\mathbf{x}$$

$$\leq c\left[\|\nabla \mathbf{w}_m\|_{L^3}^2 + \|\partial_t \mathbf{w}_m\|\, \|\nabla \mathbf{w}_m\|_{L^3}^3 + \|\overline{\mathbf{f}}\|^2 + \|\partial_t \overline{\mathbf{f}}\|^2\right].$$

Now we can use the Gronwall lemma (recall (7.9) to show that the right-hand side satisfies the required hypotheses) to show that $\mathbf{w}_m$ is bounded uniformly (with respect to m) in the space

$$H^1(0, T; L^2) \cap L^3(0, T; W_0^{1,3}).$$

Having this bound, the proof follows as in Theorem 3.9, provided we know that the operator $A(\mathbf{v})$ is monotone. We prove this in the following lemma:

Lemma 7.5. *If condition* (7.7) *is satisfied, then*

$$\int_\Omega (A(\mathbf{v}) - A(\mathbf{w}))(\nabla \mathbf{v} - \nabla \mathbf{w}) \, d\mathbf{x} \geq \frac{2}{Re}\|\nabla^s \mathbf{v} - \nabla^s \mathbf{w}\|^2 \qquad \forall\, \mathbf{u},\, \mathbf{v} \in W_0^{1,3}.$$

Essentially this lemma states that if C_G is "big enough", then the additional term does not influence the good properties of the Smagorinsky (or better 3-Laplacian) operator analyzed in Sect. 3.4.

Proof. The proof of this lemma follows by using the same technique as for Proposition 3.18.

Let us define the operator $A : W_0^{1,3}(\Omega) \to W^{-1,3/2}(\Omega)$

$$\langle A\mathbf{u}, \mathbf{v}\rangle := \frac{2}{Re}\int_\Omega \nabla^s \mathbf{u}\nabla^s \mathbf{v}\, d\mathbf{x} + C_G \int_\Omega |\nabla \mathbf{u}|\nabla \mathbf{u}\nabla \mathbf{v}\, d\mathbf{x} - \frac{\delta^2}{12}\int_\Omega \nabla \mathbf{u}\nabla \mathbf{u}^T \nabla \mathbf{v}\, d\mathbf{x}.$$

We consider the function

$$f(s) = \langle A(s\mathbf{w}_1 + (1-s)\mathbf{w}_2), \mathbf{w}_1 - \mathbf{w}_2\rangle, \qquad s \in [0, 1].$$

The function f is monotone increasing if and only if the operator A is monotone. To prove such a result, we use the technique of Proposition 3.22 to estimate the part arising from the 3-Laplacian and we consider the additional term as a perturbation.

If we set $\mathbf{w}^s = s\mathbf{w}_1 + (1-s)\mathbf{w}_2$, we can write

$$\mathcal{B} = \int_\Omega \langle A(\mathbf{w}_1) - A(\mathbf{w}_2), \mathbf{w}_1 - \mathbf{w}_2 \rangle \, dx = \int_\Omega \left\langle \left[\int_0^1 \frac{d}{ds} A(\mathbf{w}^s) \, ds \right], \mathbf{w}_1 - \mathbf{w}_2 \right\rangle dx.$$

We can split $\mathcal{B}$ in a natural way as $\mathcal{B} = \mathcal{B}_1 + \mathcal{B}_2 + \mathcal{B}_3$, where the first term involves the "Laplacian", the second term the "3-Laplacian", while the last term represents the SFS stress tensor appearing in the Gradient LES model.

The first term of $\mathcal{B}$ is easily calculated:

$$\mathcal{B}_1 = \frac{2}{Re} \int_\Omega \|\nabla^s \mathbf{w}_1 - \nabla^s \mathbf{w}_2\|^2 \, dx.$$

By calculating explicitly the derivative with respect to the parameter s and by recalling that $|\nabla \mathbf{w}| = \sqrt{\nabla \mathbf{w} \nabla \mathbf{w}^T} = \sqrt{\sum_{i,j=1}^d (\nabla \mathbf{w})_{ij}^2}$, we get the following expression for $\mathcal{B}_2$:

$$\mathcal{B}_2 = C_G \int_\Omega \int_0^1 \sum_{i,j,k,l} \frac{(\nabla \mathbf{w}^s)_{ij}}{|\nabla \mathbf{w}^s|} (\nabla \mathbf{w}^s)_{kl} (\nabla \mathbf{w}_1 - \nabla \mathbf{w}_2)_{ij} (\nabla \mathbf{w}_1 - \nabla \mathbf{w}_2)_{kl} \, dx \, ds$$

$$+ C_G \int_\Omega \int_0^1 |\nabla \mathbf{w}^s| \, |\nabla \mathbf{w}_1 - \nabla \mathbf{w}_2|^2 \, dx \, ds.$$

Note that the second term is obviously nonnegative, while the first term is nonnegative since

$$\sum_{i,j,k,l} (\nabla \mathbf{w}^s)_{ij} (\nabla \mathbf{w}^s)_{kl} (\nabla \mathbf{w}_1 - \nabla \mathbf{w}_2)_{ij} (\nabla \mathbf{w}_1 - \nabla \mathbf{w}_2)_{kl}$$

$$= \left[\sum_{i,j} (\nabla \mathbf{w}^s)_{ij} (\nabla \mathbf{w}_1 - \nabla \mathbf{w}_2)_{ij} \right]^2.$$

For $\mathcal{B}_3$ we easily get the following expression:

$$\mathcal{B}_3 = -\frac{\delta^2}{12} \int_\Omega \int_0^1 \sum_{i,k,l} (\nabla \mathbf{w}_1 - \nabla \mathbf{w}_2)_{ik} (\nabla \mathbf{w}^s)_{il} (\nabla \mathbf{w}_1 - \nabla \mathbf{w}_2)_{lk} \, dx \, ds$$

$$- \frac{\delta^2}{12} \int_\Omega \int_0^1 \sum_{i,k,l} (\nabla \mathbf{w}_1 - \nabla \mathbf{w}_2)_{il} (\nabla \mathbf{w}^s)_{ik} (\nabla \mathbf{w}_1 - \nabla \mathbf{w}_2)_{lk} \, dx \, ds.$$

It follows, by using the Hölder inequality, that

$$|\mathcal{B}_3| \leq \frac{\delta^2}{6} \int_\Omega |\nabla \mathbf{w}^s| \, |\nabla \mathbf{w}_1 - \nabla \mathbf{w}_2|^2 \, dx \leq \frac{\delta^2}{6} \|\nabla \mathbf{w}^s\|_{L^3} \|\nabla \mathbf{w}_1 - \nabla \mathbf{w}_2\|_{L^3}^2.$$

This finally shows that if Condition (7.7) is satisfied, then $\mathcal{B}_2 + \mathcal{B}_3 \geq 0$.   □

Remark 7.6. Theorem 7.4 requires the additional Smagorinsky term to be

$$-\nabla \cdot (C_G |\nabla \mathbf{w}| \nabla \mathbf{w}).$$

This model is not frame-invariant. The question whether it is possible to replace ∇ with its symmetric (and more natural) counterpart ∇^s is still unsolved. We believe that monotonicity fails in Lemma 7.5 with this substitution.

7.1.3 Numerical Validation and Testing of the Gradient LES Model

The assumption (7.7) on the size of the constant C_G in Theorem 7.4 implies that the Smagorinsky term dominates the actual Gradient LES model term. This is in clear contradiction with the very assumption in the derivation of the Gradient LES model: indeed, the wavenumber asymptotic analysis was essentially based on the idea that one keeps all terms formally $O(\delta^2)$ and drops terms formally $O(\delta^4)$. Thus, to be consistent with the derivation of the Gradient LES model, one should consider in the mixed Gradient LES model (7.5), (7.6) a Smagorinsky term with a constant $C = O(\delta^4)$ and *not* $C = O(\delta^2)$ as in assumption (7.7).

This mathematical discrepancy between the derivation of the Gradient LES model and the assumptions that seem needed for proving existence and uniqueness of weak solutions is recovered at a numerical level as well: numerical computations with the Gradient LES model are very unstable. For example, for the 3D lid-driven cavity problem, the kinetic energy of the Gradient LES model blew-up in finite time [169]. To stabilize it, the Gradient LES model is used with an $O(\delta^2)$ Smagorinsky term in actual numerical simulations of turbulent flows, a the so-called *mixed model* (see [316]).

The Gradient LES model has been used in numerous computational studies [65, 106, 165, 166, 309, 308, 310, 55, 315, 316]. Thorough numerical tests with the Gradient LES model for turbulent channel flows were performed in a series of papers by Iliescu and Fischer [106, 165, 166]. We present these tests in detail in Chap. 12. Further numerical tests for the mixing layer were performed by John in [175, 173, 176]. All the numerical tests confirm the mathematical instability of the Gradient LES model. This illustrates one of the principles that we have tried to highlight throughout the book:

> *In LES computations of turbulent flows, mathematical analysis, physical insight, and numerical experience should permanently complement and guide each other.*

The evolution of the Gradient LES model into the Rational LES model (its improvement presented in the next section) is the perfect illustration of this principle; *cfr.* the parable by F. Bacon:

> The men of experiment are like the ant, they only collect and use; the reasoners resemble spiders, who make cobwebs out of their own substance.

But the bee takes the middle course: it gathers its material from the flowers of the garden and field, but transforms and digests it by a power of its own. Not unlike this is the true business of philosophy (science); for it neither relies solely or chiefly on the powers of the mind, nor does it take the matter which it gathers from natural history and mechanical experiments and lay up in the memory whole, as it finds it, but lays it up in the understanding altered and digested. Therefore, from a closer and purer league between these two faculties, the experimental and the rational (such as has never been made), much may be hoped (*Novum Organum*, 1620.)

7.2 The Rational LES Model (RLES)

In this section we present a careful derivation and a thorough mathematical analysis for another approximate deconvolution LES model derived through wavenumber asymptotics: the Rational LES model of Galdi and Layton [122]. The Rational LES model is an approximate deconvolution model whose derivation is based on an $O(\delta^2)$ asymptotic wavenumber expansion similar to that in the derivation of the Gradient LES model. The essential difference in the derivation of the two LES models is the approximation used in the wavenumber space: Taylor series for the Gradient LES model and a rational (Padé) approximation for the Rational LES model. As we have seen in the previous section, the main drawback of the Gradient LES model is its numerical instability in practical computations. The same instability is reflected in the mathematical analysis of the Gradient LES model by the need for an extra eddy viscosity term. The Padé approximation used in the derivation of the Rational LES model is *stable* and *consistent* with the original filtering by a Gaussian. This is reflected both in the mathematical analysis (there is no need for an extra eddy viscosity term) and in the numerical experiments where the Rational LES model is much more stable and accurate than the Gradient LES model [169, 106, 165, 166, 173, 175, 176].

Consider the periodic or pure Cauchy problem and let $\mathbf{k}$ be the dual variable of the Fourier transform. Recall that the Fourier transform of a Gaussian is again a Gaussian:

$$\mathcal{F}(g_\delta)(\mathbf{k}) = \widehat{g}_\delta(\mathbf{k}) = \mathrm{e}^{-\frac{\delta}{4\gamma}|\mathbf{k}|^2}.$$

The Fourier transform of $\overline{\mathbf{u}} = g_\delta * \mathbf{u}$ yields $\widehat{\overline{\mathbf{u}}} = \widehat{g}_\delta(\mathbf{k})\widehat{\mathbf{u}}(\mathbf{k})$ so that, proceeding formally,

$$\widehat{\mathbf{u}} = \frac{1}{\widehat{g}_\delta(\mathbf{k})}\widehat{\overline{\mathbf{u}}}(\mathbf{k}). \qquad (7.13)$$

At first sight, for the Gaussian filter this relation could be inverted and the closure problem solved exactly. Indeed, this would give a deconvolution operator $\mathbf{u} = \mathcal{D}(\overline{\mathbf{u}})$ and we could write $\overline{\mathbf{u}\,\mathbf{u}^T} = \overline{\mathcal{D}(\overline{\mathbf{u}})\mathcal{D}(\overline{\mathbf{u}}^T)}$ exactly. However, this exact solution $\mathbf{u} = \mathcal{D}(\overline{\mathbf{u}})$ is an illusion. For example, stable inversion of (7.13) in L^2 is possible only when

$$\left| \frac{1}{\widehat{g_\delta}(\mathbf{k})} \right| \text{ is bounded.}$$

Unfortunately, $|\widehat{g_\delta}(\mathbf{k})| \to 0$ (exponentially fast) as $|\mathbf{k}| \to \infty$ and this necessary condition fails. This is known as a "small divisor problem." Since $|\widehat{g_\delta}(\mathbf{k})| \to 0$ exponentially fast, (7.13) cannot be stably inverted for data $\widehat{\overline{\mathbf{u}}}(\mathbf{k})$ in any Sobolev space $H^s(\mathbb{R}^d)$ either for the same reason: $\frac{|\mathbf{k}|^s}{|\widehat{g_\delta}(\mathbf{k})|}$ is not bounded as $|\mathbf{k}| \to \infty$.

Even though no information is (in some sense) lost in (7.13), the relation (7.13) cannot be *stably* inverted because of the small divisor problem and the information lost in filtering cannot be recovered. (Since the Gaussian is the heat kernel, exact deconvolution is equivalent to solving the heat equation backwards in time stably, a well-known ill-posed problem.) Thus, it seems that inverting (7.13) in a *useful* sense depends on approximating (7.13) and inverting it *inexactly*, *i.e.* in losing information!

Given (7.13) and the above considerations, it is clear that an *approximation* to $\widehat{g_\delta}(\mathbf{k})$ in (7.13) yields an approximate deconvolution method which gives a closure model. The property of the Gaussian which is fundamental to LES is its smoothing property. Smoothing in $\mathbf{x}$ is equivalent to decay at infinity in $\mathbf{k}$. Thus, the key property that must be preserved under deconvolution is smoothing. In wavenumber space this means decay at ∞ of $|\widehat{g_\delta}(\mathbf{k})|$:

$$|\widehat{g_\delta}(\mathbf{k})| \to 0 \text{ (exponentially fast) as } |\mathbf{k}| \to \infty.$$

One early approximation to $\widehat{g_\delta}(\mathbf{k})$ is given by its Taylor polynomial

$$\widehat{g_\delta}(\mathbf{k}) = 1 - \frac{\delta^2}{4\gamma}|\mathbf{k}|^2 + O(\delta^4).$$

We have seen in Sect. 7.1.1 (and we shall see again in the numerical experiments in Chap. 12) that this approximation leads to amplification of high wavenumbers and finite time blow up.

Clearly, $|\widehat{g_\delta}(\mathbf{k})| \to 0$ at infinity but $(1 - \frac{\delta^2}{4\gamma}|\mathbf{k}|^2) \to \infty$ at $\mathbf{k} \to \infty$. Thus, the Taylor approximation will lead to an anti-smoothing model.

The simplest approximation to the exponential that preserves the smoothing property (decay at infinity in $|\mathbf{k}|$) is the subdiagonal $(0, 1)$ Padé approximation (*e.g.* Pozzi [259]):

$$\widehat{g_\delta}(\mathbf{k}) = \frac{1}{1 + \frac{\delta^2}{4\gamma}|\mathbf{k}|^2} + O(\delta^4). \tag{7.14}$$

The approximation (7.14) in (7.13) gives

$$\widehat{\overline{\mathbf{u}}} = \left(\frac{1}{1 + \frac{\delta^2}{4\gamma}|\mathbf{k}|^2} \right)^{-1} \widehat{\mathbf{u}} + O(\delta^4).$$

Inversion gives the Padé based approximate deconvolution formula ([122]):

$$\mathcal{D}_\delta := \left(-\frac{\delta^2}{4\gamma}\Delta + \mathbb{I}\right) : H^2(\mathbb{R}^d) \to L^2(\mathbb{R}^d) \text{ by}$$

$$\mathbf{u} = \mathcal{D}_\delta(\overline{\mathbf{u}}) + O(\delta^4) = \left(-\frac{\delta^2}{4\gamma}\Delta + \mathbb{I}\right)\overline{\mathbf{u}} + O(\delta^4). \tag{7.15}$$

Since $\mathbf{u}' = \mathbf{u} - \overline{\mathbf{u}}$, rearrangement also gives the approximation for $\mathbf{u}'$ in terms of $\overline{\mathbf{u}}$:

$$\mathbf{u}' = -\frac{\delta^2}{4\gamma}\Delta\overline{\mathbf{u}} + O(\delta^4). \tag{7.16}$$

This approximate deconvolution formulation can also be used to build the Rational LES model presented below. In performing this modeling it is useful to estimate the sizes of the individual terms to be modeled. To this end, consider the filtered nonlinear term. As $\mathbf{u} = \overline{\mathbf{u}} + \mathbf{u}'$, we have

$$\overline{\mathbf{u}\,\mathbf{u}^T} = \overline{\overline{\mathbf{u}}\,\overline{\mathbf{u}}^T} + \overline{\overline{\mathbf{u}}\,\mathbf{u}'^T} + \overline{\mathbf{u}'\,\overline{\mathbf{u}}^T} + \overline{\mathbf{u}'\,\mathbf{u}'^T}. \tag{7.17}$$

Lemma 7.7. *In (7.17), for smooth* $\mathbf{u}$

$$\overline{\overline{\mathbf{u}}\,\overline{\mathbf{u}}^T} = O(1),$$

$$\overline{\overline{\mathbf{u}}\,\mathbf{u}'^T} + \overline{\mathbf{u}'\,\overline{\mathbf{u}}^T} = O(\delta^2), \text{ and}$$

$$\overline{\mathbf{u}'\,\mathbf{u}'^T} = O(\delta^4).$$

Proof. That $\overline{\overline{\mathbf{u}}\,\overline{\mathbf{u}}^T} = O(1)$ is a direct calculation. The remainder follows from $\mathbf{u} - \overline{\mathbf{u}} = O(\delta^2)$ for smooth $\mathbf{u}$ (*e.g.* [158]). Indeed, for smooth $\mathbf{u}$

$$\|\overline{\overline{\mathbf{u}}\,\mathbf{u}'^T + \mathbf{u}'\,\overline{\mathbf{u}}^T}\| \le C(\overline{\mathbf{u}})\|\mathbf{u}'\| = C(\overline{\mathbf{u}})\,\|\mathbf{u} - \overline{\mathbf{u}}\| \le C(\mathbf{u})\,\delta^2$$

and

$$\|\overline{\mathbf{u}'\,\mathbf{u}'^T}\| = \|\overline{(\mathbf{u} - \overline{\mathbf{u}})(\mathbf{u} - \overline{\mathbf{u}})^T}\| \le C(\mathbf{u})\,\delta^4.$$

$\square$

Using the Padé approximation (7.15) in the individual terms in (7.17), collecting and simplifying the result and discarding terms of $O(\delta^4)$ gives an LES model (proposed in [122] and studied in [106, 165, 175, 29]) now known as the *Rational LES Model*:

$$\mathbf{w}_t + \nabla \cdot (\mathbf{w}\,\mathbf{w}) - \frac{1}{Re}\Delta\mathbf{w} + \nabla \cdot \left(\frac{\delta^2}{2\gamma}A(\nabla\mathbf{w}\nabla\mathbf{w}^T)\right) = \overline{\mathbf{f}}, \tag{7.18}$$

$$\nabla \cdot \mathbf{w} = 0, \tag{7.19}$$

with the initial condition $\mathbf{w}(\mathbf{x}, 0) = \overline{\mathbf{u}}_0(\mathbf{x})$. For mathematical convenience, we consider periodic boundary conditions for the Rational LES model. A thorough discussion of boundary conditions in LES is given in Part IV. We also discuss the challenge of equipping the Rational LES model with boundary conditions and propose a couple of solutions in Chap. 12.

In (7.18),

$$A\phi := \left(-\frac{\delta^2}{4\gamma}\Delta + \mathbb{I}\right)^{-1}\phi.$$

The Rational LES model (7.18) omits all $O(\delta^4)$ terms, including the turbulent fluctuations in (7.17), $\nabla \cdot \overline{(\mathbf{u}'\,\mathbf{u}'^T)}$. These are widely believed to be very important in the physics of turbulence around the cut-off frequency. Thus, it is both mathematically and physically sensible to append to (7.18) and (7.19) an eddy viscosity model for this term:

$$-\nabla \cdot \overline{(\mathbf{u}'\,\mathbf{u}'^T)} \approx -\nabla \cdot (\nu_T(\overline{\mathbf{u}},\delta)\nabla^s\overline{\mathbf{u}}),$$

where, in view of Lemma 7.7, *asymptotic consistency* requires that (formally)

$$\nu_T(\overline{\mathbf{u}},\delta) \doteq O(\delta^4) \quad \text{as} \quad \delta \to 0.$$

Adding this term results in the *Mixed Rational LES model*, given by

$$\mathbf{w}_t + \nabla \cdot (\mathbf{w}\,\mathbf{w}^T) - \nabla \cdot \left[\left(\frac{2}{Re} + \nu_T(\mathbf{w},\delta)\right)\nabla^s\mathbf{w}\right]$$
$$+\nabla \cdot \left[\frac{\delta^2}{2\gamma}A\left(\nabla\mathbf{w}\,\nabla\mathbf{w}^T\right)\right] = \overline{\mathbf{f}} \text{ in } \Omega \times (0,T), \quad (7.20)$$
$$\nabla \cdot \mathbf{w} = 0 \text{ in } \Omega \times (0,T),$$
$$\mathbf{w}(\mathbf{x},0) = \overline{\mathbf{u}}_0(\mathbf{x}) \text{ in } \Omega \text{ and periodic boundary conditions.}$$

These models have been tested in the work of Iliescu *et al.* [169], Fischer and Iliescu [106, 165, 166], and John [175]. The Rational LES model performs very well in reproducing the important statistics of turbulent channel flow [106, 165, 166] and in very interesting tests in shear flows [175]. The Mixed Rational model with the $O(\delta^3)$ eddy viscosity of [170]

$$\nu_T = \mu\delta|\mathbf{w} - \overline{\mathbf{w}}| \quad (= O(\delta^3) \text{ formally})$$

was reported the best performer of all models tested.

7.2.1 Mathematical Analysis for the Rational LES Model

The Rational LES Model (7.18) and (7.19) is quite complex and, so far, a complete mathematical foundation for the model for large data and long time intervals is still an open problem. In [29] existence was proved for small data/small time.

Interestingly, exquisitely careful calculations by John [175] (see also [169]) seem to indicate that without the addition of a small eddy viscosity model for the (formally $O(\delta^4)$) neglected turbulent fluctuation term $-\nabla \cdot \overline{(\mathbf{u}'\,\mathbf{u}'^T)}$, the kinetic energy of the Rational LES model can blow-up in finite time. Thus, the above analysis might be reasonably sharp. (Its blow-up time is much longer than that of the Gradient LES model.) On the other hand, the Mixed Rational LES model (7.20) was very well behaved in these calculations for very small amounts of eddy viscosity.

In this section we prove the existence of strong solutions for the Rational LES model (7.18) and (7.19). The main result is the existence of such solutions without extra and dominating dissipative terms, as are required in other LES models [66, 67, 164].

Functional Setting

We restrict out analysis to the space periodic setting that decouples the boundary effect with the modeling of the equations. Since we shall consider the problem in the space-periodic setting, we recall the basic function spaces needed to deal with this functions, and we shall follow the notation of Temam [297]. In particular, these turn out to be special cases of the Sobolev spaces introduced in previous chapters. We denote by $H^m_{per}(Q)$, $m \in \mathbb{N}$, the space of functions that are in $H^m_{loc}(\mathbb{R}^3)$ (i.e. $u_{|\mathcal{O}} \in H^m(\mathcal{O})$ for every bounded set $\mathcal{O}$) and that are periodic with period $\mathcal{L} > 0$:

$$u(\mathbf{x} + \mathcal{L}\,\mathbf{e}_i) = u(\mathbf{x}), \qquad i = 1, 2, 3,$$

where $\{\mathbf{e}_1, \mathbf{e}_2, \mathbf{e}_3\}$ represents the canonical basis of $\mathbb{R}^3$, and $Q =]0, \mathcal{L}[^3$ is a cube of side length $\mathcal{L}$.

In the case $m = 0$, $H^0_{per}(Q)$ coincides simply with the Lebesgue space $L^2(Q)$. For an arbitrary $m \in \mathbb{N}$, $H^m_{per}(Q)$ is a Hilbert space and the functions in $H^m_{per}(Q)$ are easily characterized by their Fourier series expansion

$$H^m_{per}(Q) = \left\{ u = \sum_{\mathbf{k} \in \mathbb{Z}^3} c_{\mathbf{k}} e^{\frac{2i\pi \mathbf{k} \cdot \mathbf{x}}{\mathcal{L}}}, \quad \bar{c}_{\mathbf{k}} = c_{-\mathbf{k}}, \quad \sum_{\mathbf{k} \in \mathbb{Z}^3} (1 + |\mathbf{k}|)^{2m} |c_{\mathbf{k}}|^2 < \infty \right\}.$$

$$(7.21)$$

The definition (7.21) allows us to consider also[2] $m \in \mathbb{R}$. We set

$$H^m = \left\{ u \in H^m_{per}(Q) \text{ of type (7.21), such that } c_0 = 0 \right\}.$$

For $m \in \mathbb{R}$, H^m is a Hilbert space if embedded with the following norm (note that the norm involve "fractional derivatives"):

$$\|u\|^2_{H^m} = \sum_{\mathbf{k} \in \mathbb{Z}^3} |\mathbf{k}|^{2m} |c_{\mathbf{k}}|^2;$$

furthermore the spaces H^m and H^{-m} are in duality.

---

[2] In the general case of nonperiodic functions, the definition of Sobolev spaces with real index is much more involved, see Adams [4].

We now define the proper spaces involved in the theory of the Navier–Stokes equations. They are the periodic specialization of the spaces L^2_σ and $H^1_{0,\sigma}$ introduced in Chap. 2.

Two spaces frequently used in the theory of Navier–Stokes equations are

$$H = \left\{ \mathbf{u} \in [H^0]^3,\ \nabla \cdot \mathbf{u} = 0 \right\} \quad \text{and} \quad V = \left\{ \mathbf{u} \in [H^1]^3,\ \nabla \cdot \mathbf{u} = 0 \right\}. \quad (7.22)$$

Note that they are subspaces of $[H^0]^3$ and $[H^1]^3$, defined by the constraint

$$\mathbf{k} \cdot \mathbf{c_k} = 0.$$

Next, we summarize other properties of these function spaces: if $\Gamma_i = \partial Q \cap \{x_i = 0\}$, $\Gamma_{i+3} = \partial Q \cap \{x_i = \mathcal{L}\}$, and if $\mathbf{u} \in V$, then $\mathbf{u}_{|\Gamma_{j+3}} = \mathbf{u}_{|\Gamma_j}$. Let G be the orthogonal complement of H in $[H^0]^3$ (this means that $[H^0]^3 = H \oplus G$). We have the following characterization of G:

$$G = \left\{ \mathbf{u} \in [L^2(Q)]^3 :\ \mathbf{u} = \nabla q,\ q \in H^1_{per}(Q) \right\}.$$

This is an explicit realization of the Helmholtz decomposition.

Next, we need to define properly the **Stokes operator** associated with the space-periodic functions. Given $\mathbf{f} \in H^{-1} = (H^1)'$, we solve

$$\begin{cases} -\Delta \mathbf{u} + \nabla p = \mathbf{f} \text{ in } Q, \\ \nabla \cdot \mathbf{u} = 0 \text{ in } Q. \end{cases} \quad (7.23)$$

We observe that if $\mathbf{f}$ belongs to H (in particular $\sum_{\mathbf{k} \in \mathbb{Z}^3} \mathbf{k} \cdot \mathbf{f_k} = 0$, where $\mathbf{f_k}$ are the Fourier coefficients of $\mathbf{f}$), then the Fourier coefficients $\{\mathbf{u_k}, p_\mathbf{k}\}$ of the solution of (7.23) are given by

$$\mathbf{u_k} = -\frac{\mathbf{f_k}\, \mathcal{L}^2}{4\pi^2 |\mathbf{k}|^2} \quad \text{and} \quad p_\mathbf{k} = 0, \qquad \mathbf{k} \in \mathbb{Z}^3 \backslash \{0,0,0\},$$

while $(\mathbf{u_0}, p_0) = (\mathbf{0}, 0)$. We can properly define a one-to-one mapping $\mathbf{f} \to \mathbf{u}$ from H onto

$$\mathcal{D}(A) = \{ \mathbf{u} \in H,\ \Delta \mathbf{u} \in H \} = H^2 \cap H.$$

Its inverse from $\mathcal{D}(A)$ onto H is the Stokes operator denoted by A and, in fact,

$$A\mathbf{u} = -\Delta \mathbf{u}, \qquad \forall \mathbf{u} \in \mathcal{D}(A).$$

Remark 7.8. In the absence of boundaries (in this case, the space-periodic setting) the Stokes and the Laplace operator coincide, apart from the domain of definition.

If $\mathcal{D}(A)$ is endowed with the norm induced by L^2, then A becomes an isomorphism between $\mathcal{D}(A)$ and H. It follows that the norm $\|A\mathbf{u}\|$ on $\mathcal{D}(A)$ is equivalent to the norm induced by H^2. It is well known that A is an unbounded, positive, linear, and self-adjoint operator on H. Furthermore, the operator A^{-1} is linear continuous and compact. Hence A^{-1} possesses a sequence of eigenfunctions $\{\mathcal{W}_l\}_{l \in \mathbb{N}}$ that form an orthonormal basis of H,

$$
\begin{cases}
A\mathcal{W}_l = \lambda_l \mathcal{W}_l, \quad \mathcal{W}_l \in \mathcal{D}(A), \\
0 < \lambda_1 \le \lambda_2 \le \lambda_3 \ldots, \quad \text{and} \quad \lambda_l \to \infty \text{ for } l \to \infty.
\end{cases}
\tag{7.24}
$$

We can also define fractional powers A^α, $\alpha \in \mathbb{R}$: if $\mathbf{v} = \sum_{l=1}^\infty v_l \mathcal{W}_l$, then

$$
A^\alpha \mathbf{v} = \sum_{l=1}^\infty \lambda_l^\alpha v_l \mathcal{W}_l \qquad \forall \mathbf{v} \in \mathcal{D}(A^\alpha),
$$

where $\mathcal{D}(A^\alpha) \subset H = \{\mathbf{v} \in H : \sum_l \lambda_l^{2\alpha} |v_l|^2 < \infty\}$. If we set $V_\alpha = \mathcal{D}(A^{\alpha/2})$, then

$$
V_\alpha = \{\mathbf{v} \in H^\alpha,\ \nabla \cdot \mathbf{v} = 0\}.
$$

All the norms that appear in the sequel are clearly evaluated on $Q =]0, \mathcal{L}[^3$.

Proof of the Existence and Uniqueness Theorems

In this section we prove the existence and uniqueness of a particular class of solutions for (7.18) and (7.19).

Definition 7.9. *We say that* $\mathbf{w}$ *is a strong solution to system* (7.18) *and* (7.19) *if*

$$
\mathbf{w} \in L^\infty(0, T; V) \cap L^2(0, T; \mathcal{D}(A)), \quad \partial_t \mathbf{w} \in L^2(0, T; H)
\tag{7.25}
$$

and $\mathbf{w}$ *satisfies, for each* $\phi \in V$,

$$
\frac{d}{dt}(\mathbf{w}, \phi) + \frac{1}{Re}(\nabla \mathbf{w}, \nabla \phi) + ((\mathbf{w} \cdot \nabla)\,\mathbf{w}, \phi)
$$

$$
\tag{7.26}
$$

$$
- \left(\left(\mathbb{I} - \frac{\delta^2}{24}\Delta \right)^{-1} \left[\frac{\delta^2}{12} \nabla \mathbf{w} \nabla \mathbf{w}^T \right], \nabla \phi \right) = (\bar{\mathbf{f}}, \phi).
$$

Since $\mathbf{w}$ *satisfies* (7.25), *then* $\mathbf{w} \in C([0, T]; V)$ *and, by interpolation, the condition* $\mathbf{w}(\mathbf{x}, 0) = \mathbf{w}_0(\mathbf{x})$ *makes sense.*

The main result we prove is the following [29]:

Theorem 7.10. *Let* $\mathbf{w}_0 \in V$ *and* $\bar{\mathbf{f}} \in L^2(0, T; H)$. *Then there exists a strictly positive* $T^* = T^*(\delta, \mathbf{w}_0, Re, \bar{\mathbf{f}})$ *such that there exists a strong solution to* (7.18) *and* (7.19) *in* $[0, T^*)$. *A lower bound for* T^* *depending on* $\delta, \|\nabla \mathbf{w}_0\|, Re,$ *and* $\|\bar{\mathbf{f}}\|_{L^2(0,T;L^2)}$ *is obtained in* (7.35).

Remark 7.11. The strong solutions we define for the Rational LES model have the same regularity as the strong solutions of the NSE we introduced in Chap. 2. Furthermore, we shall show that also the life-span of the solution satisfies an estimate that is completely analogous to that known for the NSE.

Proof (of Theorem 7.10). We consider the Faedo–Galerkin approximation of problem (7.18), (7.19). As usual, we look for approximate functions

$$\mathbf{w}_m(\mathbf{x}, t) = \sum_{k=1}^{m} g_m^i(t) \mathcal{W}_i(\mathbf{x}),$$

satisfying for $l = 1, \ldots, m$,

$$\frac{d}{dt}(\mathbf{w}_m, \mathcal{W}_l) + \frac{1}{Re}(\nabla \mathbf{w}_m, \nabla \mathcal{W}_l) + ((\mathbf{w}_m \cdot \nabla) \mathbf{w}_m, \mathcal{W}_l)$$

$$\tag{7.27}$$

$$- \left(\left(\mathbb{I} - \frac{\delta^2}{24} \Delta \right)^{-1} \left[\frac{\delta^2}{12} \nabla \mathbf{w}_m \nabla \mathbf{w}_m^T \right], \nabla \mathcal{W}_i \right) = (\overline{\mathbf{f}}, \mathcal{W}_l),$$

$$\mathbf{w}_m(\mathbf{x}, 0) = P_m(\mathbf{w}_0(\mathbf{x})).$$

The operator P_m denotes, as usual, the orthogonal projection $P_m : H \to \text{Span}\langle \mathcal{W}_1, \ldots, \mathcal{W}_m \rangle$.

Remark 7.12. The first *a priori* estimate fails for the Rational LES model. In this case to obtain a useful estimate it is necessary to use suitable test functions. Multiplication by $\mathbf{w}_m$ as in the previous cases does not lead to an estimate that can help to find *a priori* estimates. In fact, if we multiply by $\mathbf{w}_m$ the additional nonlinear term and integrate by parts, we obtain

$$\int_Q \left(\mathbb{I} - \frac{\delta^2}{24} \Delta \right)^{-1} \left[\frac{\delta^2}{12} \nabla \mathbf{w}_m \nabla \mathbf{w}_m^T \right] \nabla \mathbf{w}_m \, d\mathbf{x}.$$

(1) This term has no definite sign, since the Rational LES model may allow backscatter of energy, as demonstrated numerically by Iliescu and Fischer [166]. These numerical results are presented in detail in Chap. 12.
(2) Even if there is smoothing due to the inverse of an elliptic operator, it does not seem possible to prove that the absolute value of the above nonlinear term is bounded by $c\|\nabla \mathbf{w}_m\|^2$, for some c. This would allow us to absorb the resulting term on the left-hand side.

To obtain *a priori* estimates, we need a different technique: we multiply (7.27) by $\mathcal{A} \mathbf{w}_m$, defined by

$$\mathcal{A} \mathbf{w}_m := \mathbf{w}_m + \frac{\delta^2}{24} A \mathbf{w}_m,$$

and use suitable integration by parts to get

$$\frac{1}{2}\frac{d}{dt}\left(\|\mathbf{w}_m\|^2 + \frac{\delta^2}{24}\|\nabla\mathbf{w}_m\|^2\right) + \frac{1}{Re}\left(\|\nabla\mathbf{w}_m\|^2 + \frac{\delta^2}{24}\|A\mathbf{w}_m\|^2\right) = (\overline{\mathbf{f}}, \mathcal{A}\mathbf{w}_m)$$

$$-((\mathbf{w}_m \cdot \nabla)\nabla\mathbf{w}_m, \mathcal{A}\mathbf{w}_m) + \left(\left(\mathbb{I} - \frac{\delta^2}{24}\Delta\right)^{-1}\left[\frac{\delta^2}{12}\nabla\mathbf{w}_m\nabla\mathbf{w}_m^T\right], \nabla\mathcal{A}\mathbf{w}_m\right).$$

The first term on the right-hand side can be estimated simply by the Schwartz inequality

$$|(\overline{\mathbf{f}}, \mathcal{A}w_m)| \leq |(\overline{\mathbf{f}}, \mathbf{w}_m)| + \frac{\delta^2}{24}|(\overline{\mathbf{f}}, Aw_m)|$$

(7.28)

$$\leq \frac{1}{6\,Re}\left(\|\nabla\mathbf{w}_m\| + \frac{\delta^2}{24}\|A\mathbf{w}_m\|^2\right) + c\|\overline{\mathbf{f}}\|^2.$$

We also use the fact that $A\mathcal{W}_m = \lambda_m\mathcal{W}_m$ to increase the L^2-norm of $\mathbf{w}_m$ with that in V. The second term can be estimated by observing that, as usual, $((\mathbf{w}_m \cdot \nabla)\mathbf{w}_m, \mathbf{w}_m) = 0$ and by using the following classical inequality (see, for instance, Prodi [261])

$$|((\mathbf{u} \cdot \nabla)\mathbf{v}, \mathbf{w})| \leq c\|\nabla\mathbf{u}\|\|\nabla\mathbf{v}\|^{1/2}\|A\mathbf{v}\|^{1/2}\|\mathbf{w}\|, \qquad (7.29)$$

that holds $\forall\,\mathbf{u} \in V$, $\forall\,\mathbf{v} \in \mathcal{D}(A)$, and $\forall\,\mathbf{w} \in H$. Thus, we obtain

$$|((\mathbf{w}_m \cdot \nabla)\nabla\mathbf{w}_m, \mathcal{A}\mathbf{w}_m)| \leq \frac{c\,\delta^2}{24}\|\nabla\mathbf{w}_m\|^{3/2}\|A\mathbf{w}_m\|^{3/2}$$

(7.30)

$$\leq \frac{1}{Re}\frac{\delta^2}{24}\|A\mathbf{w}_m\|^2 + \frac{c\,\delta^2\,Re^3}{24}\|\nabla\mathbf{w}_m\|^6.$$

Concerning the last term, not present in the previous chapters, we use the following identity: given a linear, self-adjoint, and unbounded operator B acting from $\mathcal{D}(B) \subseteq X$ into the Hilbert space $(X, (\,.\,,\,.\,))$, we have

$$(Bx, y) = (x, By) \qquad \forall\,x, y \in \mathcal{D}(B). \qquad (7.31)$$

In particular, if $B = \mathcal{A}^{-1}$, we have

$$(\mathcal{A}^{-1}x, \mathcal{A}y) = (x, y).$$

We observe that, since we are working in the space periodic setting, if $\mathcal{W}_k$ is in the domain of A, its partial derivatives also belong to the same subspace of H. We have then, by using (7.31),

$$\left| \left(\left(\mathbb{I} - \frac{\delta^2}{24}\Delta \right)^{-1} \left[\frac{\delta^2}{12}\nabla \mathbf{w}_m \nabla \mathbf{w}_m^T \right], \nabla \mathcal{A}\,\mathbf{w}_m \right) \right|$$

$$= \left| \left(\mathcal{A}^{-1} \left[\frac{\delta^2}{12}\nabla \mathbf{w}_m \nabla \mathbf{w}_m^T \right], \mathcal{A}\nabla \mathbf{w}_m \right) \right| = \frac{\delta^2}{12} |(\nabla \mathbf{w}_m \nabla \mathbf{w}_m, \nabla \mathbf{w}_m)|$$

$$\le \frac{\delta^2}{12}\|\nabla \mathbf{w}_m \nabla \mathbf{w}_m\|\,\|\nabla \mathbf{w}_m\| \le \frac{\delta^2}{12}\|\nabla \mathbf{w}_m\|_{L^4}^2\|\nabla \mathbf{w}_m\|.$$

Now, by using the classical interpolation[3] inequality

$$\|\mathbf{u}\|_{L^4} \le c\|\mathbf{u}\|^{1/4}\|\nabla\mathbf{u}\|^{3/4} \qquad \forall \mathbf{u} \in V, \tag{7.32}$$

we obtain:

$$\left| \left(\left(\mathbb{I} - \frac{\delta^2}{24}\Delta \right)^{-1} \left[\frac{\delta^2}{12}\nabla \mathbf{w}_m \nabla \mathbf{w}_m^T \right], \nabla \mathcal{A}\mathbf{w}_m \right) \right|$$

$$\le \frac{c\,\delta^2}{12}\|\nabla \mathbf{w}_m\|^{3/2}\|A\mathbf{w}_m\|^{3/2}. \tag{7.33}$$

$$\le \frac{1}{12\,Re}\frac{\delta^2}{12}\|A\mathbf{w}_m\|^2 + \frac{c\,\delta^2\,Re^3}{12}\|\nabla \mathbf{w}_m\|^6.$$

By collecting estimates (7.28)–(7.30), and (7.33), we get

$$\frac{d}{dt}\left(\|\mathbf{w}_m\|^2 + \frac{\delta^2}{24}\|\nabla \mathbf{w}_m\|^2 \right) + \frac{1}{Re}\|\nabla \mathbf{w}_m\|^2$$

$$\tag{7.34}$$

$$+ \frac{1}{Re}\frac{\delta^2}{24}\|A\mathbf{w}_m\|^2 \le c\|\overline{\mathbf{f}}\|^2 + c\,\delta^2\,Re^3\|\nabla \mathbf{w}_m\|^6.$$

The Gronwall lemma (provided $\overline{\mathbf{f}}$ belongs to $L^2(0,T;H)$) and the same results of existence for systems of ODEs we used in Chap. 2 imply that there exists $T^* > 0$ such that there exists a solution $\mathbf{w}_m$ to (7.27) in $[0,T^*)$ and

$$\{\mathbf{w}_m\} \text{ is uniformly bounded in } L^\infty(0,T^*;V) \cap L^2(0,T^*;\mathcal{D}(A)).$$

We now investigate more carefully the question related to the lower bounds on the life-span of solutions. The results of this section are similar to those used to derive (2.29). They use essentially the same technique and extend that estimate to the case of a nonvanishing external force.

---

[3] This is an adaptation, in the periodic setting, of the Ladyžhenskaya's inequality we introduced in Chap. 2.

A lower bound on the time T^* can be deduced as follows: let us set $y(t) = \|\mathbf{w}_m\| + \delta^2/24\,\|\nabla\mathbf{w}_m\|^2$. Then we study (recall (7.34)) the differential inequality

$$\frac{dy}{dt} \le c_1\|\bar{\mathbf{f}}\|^2 + \frac{c_2\,Re^3}{\delta^4}y^3.$$

Dividing both sides by $(1+y)^3 \ge 1$, we obtain

$$\frac{dy}{dt}\frac{1}{(1+y)^3} \le c_1\|\bar{\mathbf{f}}\|^2 + \frac{c_2\,Re^3}{\delta^4}.$$

This equation can be explicitly integrated to get

$$1 + y(t) \le \frac{1 + y(0)}{\sqrt{1 - (1+y(0))^2\left[c_1\int_0^t \|\bar{\mathbf{f}}(\tau)\|^2\,d\tau + \frac{c_2 Re^3}{\delta^4}t\right]}}.$$

Consequently, a condition that bounds T^* from below is the following:

$$c_1\int_0^{T^*} \|\bar{\mathbf{f}}(\tau)\|^2\,d\tau + \frac{c_2 Re^3}{\delta^4}T^* \le \frac{1}{(1+\|\nabla\mathbf{w}_0\|^2)^2}. \tag{7.35}$$

Remark 7.13. The same result can be written also as follows: there exists $\epsilon = \epsilon(T, \bar{\mathbf{f}}, \delta, Re) > 0$ such that if $\|\nabla\mathbf{w}_0\| < \epsilon$ and $\|\bar{\mathbf{f}}\|_{L^2(0,T;L^2)} < \epsilon$, then $\{\mathbf{w}_m\}$ is uniformly bounded in

$$L^\infty(0, T; V) \cap L^2(0, T; \mathcal{D}(A)). \tag{7.36}$$

Remark 7.14. The result of existence is given for a fixed averaging radius δ. In fact, the basic theory of differential inequalities implies that, if all the other quantities ($\mathbf{w}_0$, Re, and $\bar{\mathbf{f}}$) are fixed, then the *life-span* of $\mathbf{w}_m$ is, in the worst case, *at least*, $O(\delta^4)$. This limitation can be overcome and the life-span is independent of δ [30]. Later in this section, we shall state this result, together with several others whose proofs are a little bit more technical. We shall refer to the bibliography of that section for more details.

Remark 7.15. (A simple observation on energy estimates.) To prove an energy-type estimate (absolutely necessary for an existence theory, a stable computation, etc.), we multiply the equation of the model by $B\mathbf{w}$, where B is some operator. Generally, we obtain an estimate of the form:

$$\frac{d}{dt}(\text{model's kinetic energy}) + (\text{model's energy dissipation})$$

$$\le \text{data} + (\text{cubic terms arising from the model's nonlinearity})$$

Cubic terms can never be subsumed into quadratic terms (for large data and long time). Thus, the only hope for a global theory is to construct the operator B (often B is an approximate deconvolution operator) for which

$$(\text{model's nonlinearity}(\mathbf{w}), B(\mathbf{w})) = 0.$$

This is the mathematical and physical approach of the above analysis, the analysis of Chap. 8, and other work.

In view of applying the Aubin–Lions' compactness Lemma 3.11 that we used earlier, we need an estimate of the time derivative of $\mathbf{w}_m$. By comparison, *i.e.* by isolating the $\partial_t \mathbf{w}_m$ on the left-hand side, we have to estimate the following quantity

$$|(\partial_t \mathbf{w}_m, \mathcal{W}_j)| \leq |((\mathbf{w}_m \cdot \nabla) \mathbf{w}_m, \mathcal{W}_j)| + \frac{1}{Re} |(A\mathbf{w}_m, \mathcal{W}_j)|$$

$$+ \left| \left(\nabla \cdot \left(\mathbb{I} - \frac{\delta^2}{24} \Delta \right)^{-1} \left[\frac{\delta^2}{12} \nabla \mathbf{w}_m \nabla \mathbf{w}_m^T \right], \mathcal{W}_j \right) \right| + |(\mathbf{f}, \mathcal{W}_j)| .$$

Some care is needed to estimate the highly nonlinear term, while the others are treated in a standard way (see, for instance, Galdi [121] and Temam [297] for the space-periodic setting). This "bad" one can be estimated as follows:

$$\left| \left(\nabla \cdot \left(\mathbb{I} - \frac{\delta^2}{24} \Delta \right)^{-1} \left[\frac{\delta^2}{12} \nabla \mathbf{w}_m \nabla \mathbf{w}_m^T \right], \mathcal{W}_j \right) \right|$$

$$\leq \left\| \nabla \cdot \left(\mathbb{I} - \frac{\delta^2}{24} \Delta \right)^{-1} \left[\frac{\delta^2}{12} \nabla \mathbf{w}_m \nabla \mathbf{w}_m \right] \right\| \|\mathcal{W}_j\| \qquad (7.37)$$

$$\leq \left\| \left(\mathbb{I} - \frac{\delta^2}{24} \Delta \right)^{-1} \left[\frac{\delta^2}{12} \nabla \mathbf{w}_m \nabla \mathbf{w}_m^T \right] \right\|_{H^1} \|\mathcal{W}_j\| .$$

At this point we need the Sobolev embedding theorem, that we simply state. Its proof (with weaker hypotheses) can be found in Adams [4].

Proposition 7.16 (Sobolev embedding). *Let Ω be a bounded smooth subset of $\mathbb{R}^d$ and let $f \in W^{1,p}(\Omega)$, for $p < d$. Then, the number p^* is well-defined by the relation*

$$\frac{1}{p^*} = \frac{1}{p} - \frac{1}{d},$$

and there exists a positive constant $C = C(\Omega, p, q)$ (independent of f) such that

for each $p \leq q \leq p^*$ $\|f\|_{L^q} \leq C(\Omega, p, q) \|f\|_{W^{1,p}} \quad \forall f \in W^{1,p}(\Omega).$ (7.38)

This also means that for $p \leq q \leq p^*$ we have $W^{1,p} \hookrightarrow L^q$ or, in other words, that $W^{1,p}$ is continuously embedded into L^q.

Remark 7.17. The exponent p^* is known as the Sobolev exponent. Note that in Chap. 2 we used a particular case of this result. In general, if we set $q = p^*$ in inequality (7.38), then the estimate can be improved to

$$\|f\|_{L^{p^*}} \leq C(\Omega, p)\|\nabla f\|_{L^p} \quad \forall f \in W^{1,p}(\Omega).$$

This means that the L^{p^*}-norm can be controlled just by the $W^{1,p}$-semi-norm of f rather than with its complete norm. In particular, the latter is satisfied also if $f \in L^1(K)$, for each compact set $K \subset \Omega$, and just $\nabla f \in L^p$.

A very fast way to guess the latter result may be that of *scaling invariance*. This technique may be used to find very quickly interesting results, even if it is not a proof. Let us show that if the latter inequality holds, then the only possible exponent in the left-hand side is p^*. To verify this, let us suppose that $\Omega = \mathbb{R}^d$ and set

$$f_\lambda(\mathbf{x}) = f(\lambda \mathbf{x}).$$

Thus, by applying the rule of change of variables in multiple integrals, the inequality $\|f\|_{L^q} \leq C\|\nabla f\|_{L^p}$ becomes (if it is true!)

$$\|f\|_{L^q} \leq C\lambda^{\left(1 + \frac{d}{q} - \frac{d}{p}\right)}\|\nabla f\|_{L^p}.$$

This shows that necessarily $q = p^*$, if we want to have invariance with respect to λ. This invariance is needed if we want the Sobolev embedding to hold with a constant C independent of the function f.

It is clear that by induction (on the integer index k) it is possible to show that

$$\text{for each } p \leq q \leq p^*, \quad \|f\|_{W^{k-1,q}} \leq C(\Omega, k, p, q)\|f\|_{W^{k,p}}, \quad \forall f \in W^{k,p}(\Omega).$$

We use will use this result in the following special case

$$W^{2,3/2} \hookrightarrow W^{1,2}.$$

To estimate the expression in (7.37) we shall also need some classical results of *elliptic regularity*. For elliptic regularity we essentially need the following result that (in all its generality) dates back to the work of Agmon, Douglis, and Nirenberg [5, 6].

Proposition 7.18. *If $f \in L^p(\Omega)$, where $\Omega \subset \mathbb{R}^d$ is a smooth open set, and $1 < p < +\infty$, then the variational solution u of the following boundary-value problem*

$$\begin{cases} u - \Delta u = f & \text{in } \Omega, \\ u = 0 & \text{on } \partial\Omega, \end{cases}$$

belongs to $W^{2,p}$. In particular there exists $c > 0$, independent of u and f such that

$$\|u\|_{W^{2,p}} \leq c\|f\|_{L^p}.$$

Remark 7.19. The above proposition still holds if the Laplacian is replaced by a more general elliptic operator, such as the Stokes operator (see Cattabriga [56] in this case). Complete details on the precise assumptions on the operator and on the regularity of the boundary $\partial\Omega$ can be found in the classical monograph by Nečas [245]. Note that it is essential to require p being different from 1 and $+\infty$. In the limit cases it is known that Proposition 7.18 may be false. Other results concerning the Hölder regularity $C^{2,\alpha}$ for u if $f \in C^{0,\alpha}$ can be found again in [245].

By using the above proposition, the term in (7.37) is then bounded by

$$c\left\|\left(1 - \frac{\delta^2}{24}\Delta\right)^{-1}\left[\frac{\delta^2}{12}\nabla\mathbf{w}_m\nabla\mathbf{w}_m^T\right]\right\|_{W^{2,3/2}}\|\mathcal{W}_j\| \le c\|\nabla\mathbf{w}_m\nabla\mathbf{w}_m^T\|_{L^{3/2}}\|\mathcal{W}_j\|.$$

Next, we use the convex-interpolation inequality (2.26). The latter, together with the Sobolev embedding $H^1(Q) \hookrightarrow L^6(Q)$, implies that the term in (7.37) may be bounded by

$$c\|\nabla\mathbf{w}_m\|_{L^3}^2\|\mathcal{W}_j\| \le c\|\nabla\mathbf{w}_m\| \, \|\nabla\mathbf{w}_m\|_{L^6}\|\mathcal{W}_j\| \le c\|\nabla\mathbf{w}_m\| \, \|A\mathbf{w}_m\|\|\mathcal{W}_j\|.$$

Multiplying (7.27) by $dg_m^i(t)/dt$, summing over $i = 1,\dots,m$, and using the last inequality (together with well-known estimates for the other terms), we obtain

$$\frac{1}{2Re}\frac{d}{dt}\|\nabla\mathbf{w}_m\|^2 + \|\partial_t\mathbf{w}_m\|^2 \le c(1 + \|\nabla\mathbf{w}_m\|^2)\|A\mathbf{w}_m\|^2 + c\|\nabla\mathbf{w}_m\|^6 + c\|\bar{\mathbf{f}}\|^2.$$

The last differential inequality, together with (7.36), shows that $\partial_t\mathbf{w}_m$ is uniformly bounded in $L^2(0,T^*;L^2)$.

We can now use the Aubin–Lions' Lemma 3.11 with

$$p = 2, \quad X_1 = \mathcal{D}(A), \quad X_2 = V, \quad \text{and} \quad X_3 = H.$$

Thus, it is possible to extract from $\{\mathbf{w}_m\}$ a subsequence (relabeled for notational convenience again as $\{\mathbf{w}_m\}$) such that

$$\begin{cases} \mathbf{w}_m \overset{*}{\rightharpoonup} \mathbf{w} \text{ in } L^\infty(0,T^*;V) \\[2mm] \mathbf{w}_m \rightharpoonup \mathbf{w} \text{ in } L^2(0,T^*;\mathcal{D}(A)) \\[2mm] \mathbf{w}_m \to \mathbf{w} \text{ in } L^2(0,T^*;V) \text{ and a.e. in } (0,T) \times Q. \end{cases} \tag{7.39}$$

With this convergence it is easy to pass to the limit in (7.27) and to prove that $\mathbf{w}$ is a solution to (7.18), (7.19). In fact, without loss of generality, using the same subsequence, we have also

$$\partial_t\mathbf{w}_m \rightharpoonup \partial_t\mathbf{w} \quad \text{in } L^2(0,T^*;L^2).$$

By using a classical interpolation argument (see Lions and Magenes [222]), the function $\mathbf{w}$ belongs also to $C(0, T^*; V)$. The only difficult part is to show that $\mathbf{w}$ satisfies (7.26), and hence it is a strong solution. The passage to the limit is done in a standard way (the same as for the Navier–Stokes equations (see, for instance, Chap. 2) for all terms appearing in (7.27), except for

$$\left(\left(\mathbb{I} - \frac{\delta^2}{24}\Delta\right)^{-1}\left[\frac{\delta^2}{12}\nabla\mathbf{w}_m\nabla\mathbf{w}_m^T\right], \nabla\mathcal{W}_k\right).$$

To pass to the limit in the above expression, we recall (see, for instance, Lemma 6.7, Chap. 1, in Lions [221]) that if ∇f belongs to both $L^\infty(0, T^*; L^2)$ and $L^2(0, T^*; H^1)$, then Hölder inequality implies that

$$\nabla f \in L^4(0, T^*; L^3).$$

Thus, $\nabla\mathbf{w}_m\nabla\mathbf{w}_m^T$ is bounded in $L^2(0, T^*; L^{3/2})$, and the third relation in (7.39) implies that

$$\nabla\mathbf{w}_m\nabla\mathbf{w}_m^T \rightharpoonup \nabla\mathbf{w}\nabla\mathbf{w} \quad \text{in } L^2(0, T^*; L^{3/2}).$$

This implies that $\forall \phi \in C_{per}^\infty(Q)$

$$\left(\left(\mathbb{I} - \frac{\delta^2}{24}\Delta\right)^{-1}\left[\frac{\delta^2}{12}\nabla\mathbf{w}_m\nabla\mathbf{w}_m^T\right], \nabla\phi\right)$$

$$= \left(\frac{\delta^2}{12}\nabla\mathbf{w}_m\nabla\mathbf{w}_m, \left(\mathbb{I} - \frac{\delta^2}{24}\Delta\right)^{-1}\nabla\phi\right) \rightarrow \left(\frac{\delta^2}{12}\nabla\mathbf{w}\nabla\mathbf{w}, \left(\mathbb{I} - \frac{\delta^2}{24}\Delta\right)^{-1}\nabla\phi\right)$$

in $L^2(0, T^*)$. The proof concludes with a density argument. $\square$

The solutions we have proved to exist are rather regular. It is also possible to prove a uniqueness result.

Theorem 7.20. *Under the same hypotheses as in Theorem 7.10, there exists at most one strong solution to* (7.18), (7.19).

Proof. As usual let us suppose that we have two solutions $\mathbf{w}_1$ and $\mathbf{w}_2$ relative to the same external force $\overline{\mathbf{f}}$ and the same initial datum $\mathbf{w}_0$. Furthermore, let us suppose that both the solutions exist in some interval $[0, T]$. By using a technique we have adopted several times, we subtract the equation satisfied by $\mathbf{w}_2$ from that one satisfied by $\mathbf{w}_1$, and we multiply the equations by $A\mathbf{w}$, where $\mathbf{w} := \mathbf{w}_1 - \mathbf{w}_2$. All the terms can be treated as in the proof of uniqueness for strong solutions to the NSE. The only term that requires some care is the additional one corresponding to the SFS stress tensor:

$$\left(\nabla \cdot \left(\mathbb{I} - \frac{\delta^2}{24}\Delta\right)^{-1}\frac{\delta^2}{12}\left[\nabla\mathbf{w}_1\nabla\mathbf{w}_1^T - \nabla\mathbf{w}_2\nabla\mathbf{w}_2^T\right], A\mathbf{w}\right).$$

By adding and subtracting $\nabla \cdot \left(\mathbb{I} - \frac{\delta^2}{24}\Delta \right)^{-1} \left[\frac{\delta^2}{12} \nabla \mathbf{w}_1 \nabla \mathbf{w}_2^T \right]$, we get

$$\left(\nabla \cdot \left(\mathbb{I} - \frac{\delta^2}{24}\Delta \right)^{-1} \frac{\delta^2}{12} \left[\nabla \mathbf{w}_1 \nabla \mathbf{w}^T - \nabla \mathbf{w} \nabla \mathbf{w}_2^T \right] , A\mathbf{w} \right). \qquad (7.40)$$

The first term in (7.40) can be estimated as follows:

$$I_1 = \left| \left(\nabla \cdot \left(\mathbb{I} - \frac{\delta^2}{24}\Delta \right)^{-1} \left[\frac{\delta^2}{12} \nabla \mathbf{w}_1 \nabla \mathbf{w}^T \right] , A\mathbf{w} \right) \right|$$

$$\leq \left\| \left(\mathbb{I} - \frac{\delta^2}{24}\Delta \right)^{-1} \left[\frac{\delta^2}{12} \nabla \mathbf{w}_1 \nabla \mathbf{w}^T \right] \right\|_{H^2} \|\nabla A\mathbf{w}\|_{H^{-2}}$$

$$\leq c \left\| \nabla \mathbf{w}_1 \nabla \mathbf{w}^T \right\| \|\nabla \mathbf{w}\| \leq c \|\nabla \mathbf{w}_1\|_{L^4} \|\nabla \mathbf{w}\|_{L^4} \|\nabla \mathbf{w}\|.$$

By using again the interpolation inequality (7.32), we obtain

$$I_1 \leq c \|\nabla \mathbf{w}_1\|_{L^4} \|\nabla \mathbf{w}\|^{5/4} \|A\mathbf{w}\|^{3/4} \leq \frac{1}{8Re} \|A\mathbf{w}\|^2 + c \|\nabla \mathbf{w}_1\|_{L^4}^{8/5} \|\nabla \mathbf{w}\|^2.$$

The other term leads to a very similar expression

$$I_2 = \left| \left(\nabla \cdot \left(\mathbb{I} - \frac{\delta^2}{24}\Delta \right)^{-1} \left[\frac{\delta^2}{12} \nabla \mathbf{w} \nabla \mathbf{w}_2^T \right] , A\mathbf{w} \right) \right|$$

$$\leq \frac{1}{8Re} \|A\mathbf{w}\|^2 + c \|\nabla \mathbf{w}_2\|_{L^4}^{8/5} \|\nabla \mathbf{w}\|^2.$$

For the sake of completeness, we shall estimate the other nonlinear term

$$I_3 = \left| \left((\mathbf{w}_1 \cdot \nabla) \mathbf{w}_1 - (\mathbf{w}_2 \cdot \nabla) \mathbf{w}_2, A\mathbf{w} \right) \right|$$

(which is also present in the Navier–Stokes equations) essentially as in the proof of Theorem 2.21. Again, by adding and subtracting the term $(\mathbf{w}_2 \cdot \nabla) \mathbf{w}_1$, we obtain

$$I_3 = \left| \left((\mathbf{w} \cdot \nabla) \mathbf{w}_1 - (\mathbf{w}_2 \cdot \nabla) \mathbf{w}, A\mathbf{w} \right) \right|.$$

Using again estimate (7.29), we obtain

$$\left| \left((\mathbf{w}_2 \cdot \nabla) \mathbf{w}, A\mathbf{w} \right) \right| \leq \|\nabla \mathbf{w}_2\| \|\nabla \mathbf{w}\|^{1/2} \|A\mathbf{w}\|^{3/2}$$

$$\leq \frac{1}{8Re} \|A\mathbf{w}\|^2 + c \|\nabla \mathbf{w}_2\|^4 \|\nabla \mathbf{w}\|^2.$$

The other term is easier to handle, since

$$I_4 = |((\mathbf{w} \cdot \nabla) \mathbf{w}_1, A\mathbf{w})| \leq \|\mathbf{w}\|_{L^4} \|\mathbf{w}_1\|_{L^4} \|A\mathbf{w}\|$$

$$\leq c\|\mathbf{w}_1\|_{L^4} \|A\mathbf{w}\| \, \|\nabla \mathbf{w}\| \quad \text{by the Sobolev embedding theorem}$$

$$\leq \frac{1}{8Re}\|A\mathbf{w}\|^2 + c\|\nabla \mathbf{w}_1\|_{L^4}^2 \|\nabla \mathbf{w}\|^2 \quad \text{by the Young inequality.}$$

Collecting all the above estimates, we obtain

$$\frac{d}{dt}\|\nabla \mathbf{w}\|^2 + \frac{1}{Re}\|A\mathbf{w}\|^2$$

$$\leq c \left(\|\nabla \mathbf{w}_1\|_{L^4}^{8/5} + \|\nabla \mathbf{w}_2\|_{L^4}^{8/5} + \|\nabla \mathbf{w}_2\|^4 + \|\nabla \mathbf{w}_1\|_{L^4} \right) \|\nabla \mathbf{w}\|^2.$$

We recall that $\mathbf{w}(\mathbf{x},0) = \mathbf{w}_1(\mathbf{x},0) - \mathbf{w}_2(\mathbf{x},0) = \mathbf{0}$. By (7.36), we get

$$\left(\|\nabla \mathbf{w}_1\|_{L^4}^{8/5} + \|\nabla \mathbf{w}_2\|_{L^4}^{8/5} + \|\nabla \mathbf{w}_2\|^4 + \|\nabla \mathbf{w}_1\|_{L^4} \right) \in L^1(0,T).$$

Since $\|\mathbf{u}\| \leq 1/\lambda_1 \|\nabla \mathbf{u}\|$, for each $\mathbf{u} \in V$, the Gronwall lemma directly implies that $\mathbf{w} \equiv \mathbf{0}$ in V. $\square$

7.2.2 On the Possible Breakdown of Strong Solutions for the Rational LES Model

In this section we introduce some criteria for the breakdown (and also for the continuation) of strong solutions, and we report some numerical results recently obtained. We compare these criteria with others in the literature. We shall use them in interpreting the numerical simulations.

The results of this section follow closely the guidelines of Leray's proof of the theorem in the epoch of irregularity cited in Chap. 2. For simplicity (although it is easy to include a smooth external force), we set $\bar{\mathbf{f}} = \mathbf{0}$. We start with the following theorem:

Theorem 7.21. *Let $\mathbf{w}$ be a strong solution in the time interval $[0, \overline{T})$. If it cannot be continued in (7.25) to $t = \overline{T}$, then we have*

$$\lim_{t \to \overline{T}^-} \|\nabla \mathbf{w}(t)\| = +\infty. \tag{7.41}$$

Furthermore, we have the following blow-up estimate:

$$\|\nabla \mathbf{w}(t)\| \geq \frac{C\delta}{Re^{3/4}} \frac{1}{(\overline{T} - t)^{1/4}}, \quad t < \overline{T}. \tag{7.42}$$

Proof. We observe that if $\overline{\mathbf{f}} = \mathbf{0}$, the estimate (7.35) of the life span of the strong solution such that $\mathbf{w}(\mathbf{x}, 0) = \mathbf{w}_0(\mathbf{x})$ can be replaced by the more explicit

$$T^* \geq \frac{C\delta^4}{Re^3 \|\nabla \mathbf{w}_0\|^4},$$

as can easily be seen by using the same technique as that on p. 163. We now prove (7.41) by contradiction. Let us assume that (7.41) does not hold. Then, there would exist an increasing sequence of numbers in $(0, \overline{T})$ $\{t_k\}_{k \in \mathbb{N}}$ (such that $t_k \uparrow \overline{T}$) and a positive number M such that

$$\|\nabla \mathbf{w}(t_k)\| \leq M.$$

Since $\mathbf{w}(t_k) \in H^1$, by using Theorem 7.10 we may construct a solution $\overline{\mathbf{w}}$ with initial datum $\mathbf{w}(t_k)$ in a time interval $[t_k, t_k + T^*)$, where

$$T^* \geq \frac{C}{\|\nabla \mathbf{w}(t_k)\|^4} \geq \frac{C}{M^4} := T^0.$$

By using the uniqueness Theorem 7.20, we have $\mathbf{w} \equiv \overline{\mathbf{w}}$ in $[t_k, t_k + T^0)$. We may now select $k_0 \in \mathbb{N}$ such that $t_{k_0} + T^0 > \overline{T}$ to contradict the assumption on the boundedness of $\|\nabla \mathbf{w}(t)\|$. This proves (7.41).

To obtain the estimate on the growth of $\|\nabla \mathbf{w}(t)\|$, we argue as in the proof of Theorem 7.10. We multiply (7.18) by $A\mathbf{w}$, and we get that $Y(t) := \|\nabla \mathbf{w}(t)\|^2$ satisfies, in the time interval $[0, \overline{T})$,

$$\frac{dY(t)}{dt} \leq \frac{c\,Re^3}{\delta^4} [Y(t)]^3.$$

Integrating the above equation, we find

$$\frac{1}{\|\nabla w(t)\|^4} - \frac{1}{\|\nabla \mathbf{w}(\tau)\|^4} \leq \frac{c\,Re^3}{\delta^4}(\tau - t) \qquad 0 < t < \tau < \overline{T}.$$

Letting $\tau \to \overline{T}$, and recalling (7.41), we obtain (7.42). □

In the case of the Rational LES model it seems difficult to obtain regularity of solutions that satisfy condition (7.43) on the velocity. In fact, we recall that a well-known result of Leray–Prodi–Serrin states that if a weak solution $\mathbf{u}$ of the NSE satisfies

$$\mathbf{u} \in L^r(0, T; L^s(\Omega)) \quad \text{for} \quad \frac{2}{r} + \frac{3}{s} = 1, \quad \text{and} \quad s \in (3, \infty], \tag{7.43}$$

then (see p. 57) it is unique and smooth. For full details, see Serrin [275].

By using Theorem 7.21, we can prove, for the Rational LES model, the following blow-up criteria, involving appropriate $L^r(0, T; L^s)$-norms of $\nabla \mathbf{w}$.

Theorem 7.22. *Let* **w** *be a strong solution to* (7.18), (7.19), *and suppose that there exists a time* $\overline{T}$ *such that the solution cannot be continued in the class* (7.25) *to* $T = \overline{T}$. *Assume that* $\overline{T}$ *is the first such time. Then*

$$\int_0^{\overline{T}} \|\nabla \mathbf{w}(\tau)\|_{L^\beta}^\alpha \, d\tau = +\infty,$$

(7.44)

$$for \quad \frac{2}{\alpha} + \frac{3}{\beta} = 2, \quad 1 \le \alpha < \infty, \quad 3/2 < \beta \le \infty.$$

Remark 7.23. Condition (7.44) is the same as that involved in the study of the breakdown (or the global regularity) of the 3D NSE; see Beirão da Veiga [19] for the Cauchy problem (also in $\mathbb{R}^n$) and Berselli [27] for the initial boundary value problem (recall condition on p. 120). In the limit case $\beta = \infty$, condition (7.44) is related to the Beale–Kato–Majda [18] criterion for the 3D Euler equations.

Proof (of Theorem 7.22). The proof is done by contradiction. We assume that

$$\int_0^{\overline{T}} \|\nabla \mathbf{w}(\tau)\|_{L^\beta}^\alpha \, d\tau \le C < \infty$$

(7.45)

and use estimates similar to those derived in the existence theorem. Let us suppose that $[0, \overline{T})$ is the maximal interval of existence of the unique strong solution starting from $\mathbf{w}_0$ at time $t = 0$. We multiply (7.18) by (recall Remark 7.8)

$$\mathcal{A} \mathbf{w} = \mathbf{w} + \frac{\delta^2}{24} A \mathbf{w} = \mathbf{w} - \frac{\delta^2}{24} \Delta \mathbf{w},$$

and we obtain, with suitable integrations by parts,

$$\frac{1}{2} \frac{d}{dt} \left(\|\mathbf{w}_m\|^2 + \frac{\delta^2}{24} \|\nabla \mathbf{w}\|^2 \right) + \frac{1}{Re} \left(\|\nabla \mathbf{w}\|^2 + \frac{\delta^2}{24} \|A \mathbf{w}\|^2 \right)$$

(7.46)

$$\le \frac{\delta^2}{24} |((\mathbf{w} \cdot \nabla) \nabla \mathbf{w}, \Delta \mathbf{w})| + \left| \left\langle \mathcal{A}^{-1} \left[\frac{\delta^2}{12} \nabla \mathbf{w} \nabla \mathbf{w}^T \right], \nabla \mathcal{A} \mathbf{w} \right\rangle_{V,V'} \right|,$$

where $\langle \, . \, , . \, \rangle_{V,V'}$ denotes the pairing between V and its topological dual V'. The first term on the right-hand side can be estimated with an integration by parts. We have, in fact,

$$\int_Q (\mathbf{w} \cdot \nabla) \mathbf{w} \, \Delta \mathbf{w} \, dx = - \sum_{i,j,k=1}^3 \int_Q \frac{\partial w_j}{\partial x_k} \frac{\partial w_i}{\partial x_j} \frac{\partial w_i}{\partial x_k} \, dx$$

$$- \sum_{i,j,k=1}^3 \int_Q w_j \frac{\partial^2 w_i}{\partial x_j \partial x_k} \frac{\partial w_i}{\partial x_k} \, dx.$$

(7.47)

The term

$$\sum_{i,j,k=1}^{3} \int_Q w_j \frac{\partial^2 w_i}{\partial x_j \partial x_k} \frac{\partial w_i}{\partial x_k} \, dx = \sum_{i,j,k=1}^{3} \frac{1}{2} \int_Q w_j \frac{\partial}{\partial x_j} \left(\frac{\partial w_i}{\partial x_k} \right)^2 dx$$

is identically zero, as can be seen with another integration by parts, since $\nabla \cdot \mathbf{w} = 0$.

The other term on the right-hand side of (7.47) can be estimated in the following manner, for $3/2 < \beta \leq \infty$:

$$\left| \sum_{i,j,k=1}^{3} \int_Q \frac{\partial w_j}{\partial x_k} \frac{\partial w_i}{\partial x_j} \frac{\partial w_i}{\partial x_k} \, dx \right| \leq c \|\nabla \mathbf{w}\|_{L^{2\beta'}}^2 \|\nabla \mathbf{w}\|_{L^\beta} \quad \text{for} \quad \frac{1}{\beta} + \frac{1}{\beta'} = 1.$$

Then, we use the interpolation inequality (2.26) (observe that $1 \leq \beta' < 3$, and if $\beta' = 1$ there is nothing to do), together with the Sobolev embedding $H^1(Q) \subset L^6(Q)$, to obtain

$$\left| \sum_{i,j,k=1}^{3} \int_Q \frac{\partial w_j}{\partial x_k} \frac{\partial w_i}{\partial x_j} \frac{\partial w_i}{\partial x_k} \, dx \right| \leq c \|\nabla \mathbf{w}\|^{\frac{2\beta-3}{\beta}} \|\Delta \mathbf{w}\|^{\frac{3}{\beta}} \|\nabla \mathbf{w}\|_{L^\beta}.$$

By using Young's inequality with exponents $x = 2\beta/3$, $x' = 2\beta/(2\beta - 3)$, we obtain

$$\left| \sum_{i,j,k=1}^{3} \int_Q \frac{\partial w_j}{\partial x_k} \frac{\partial w_i}{\partial x_j} \frac{\partial w_i}{\partial x_k} \, dx \right| \leq \frac{1}{4 \, Re} \|\Delta \mathbf{w}\|^2 + c \|\nabla \mathbf{w}\|_{L^\beta}^{\frac{2\beta}{2\beta-3}} \|\nabla \mathbf{w}\|^2. \quad (7.48)$$

The other term in (7.46) can be estimated as follows:

$$\left\langle \mathcal{A}^{-1} \left[\frac{\delta^2}{12} \nabla \mathbf{w} \nabla \mathbf{w}^T \right], \mathcal{A} \, \nabla \mathbf{w} \right\rangle_{V,V'} = \frac{\delta^2}{12} (\nabla \mathbf{w} \nabla \mathbf{w}^T, \nabla \mathbf{w}),$$

and the latter can be treated as in (7.48).

The above estimates lead to

$$\frac{d}{dt} \left(\|\mathbf{w}_m\|^2 + \frac{\delta^2}{24} \|\nabla \mathbf{w}\|^2 \right) \leq c \|\nabla \mathbf{w}\|_{L^\beta}^\alpha \|\nabla \mathbf{w}\|^2, \quad \text{where} \quad \alpha = \frac{2\beta}{2\beta - 3},$$

and hence α, β are as in (7.44). The Gronwall lemma, together with (7.45), imply that

$$\nabla \mathbf{w} \in L^\infty(0, \overline{T}; L^2).$$

The latter condition implies (from Theorem 7.21) that the solution $\mathbf{w}$ can be uniquely continued beyond $\overline{T}$, and this contradicts the maximality of the existence interval $[0, \overline{T})$. $\square$

Remark 7.24. The same techniques may be used to prove that there exists $\eta > 0$ such that, if

$$\sup_{0<t<\overline{T}} \|\nabla \mathbf{w}(t)\|_{L^{3/2}} < \eta,$$

then the strong solution exists up to $\overline{T}$. The constant η does not depend on $\mathbf{w}$ but only on Re, δ, and $\mathcal{L}$. The proof easily follows by observing that

$$\left| \sum_{i,j,k=1}^{3} \int_Q \frac{\partial w_j}{\partial x_k} \frac{\partial w_i}{\partial x_j} \frac{\partial w_i}{\partial x_k} \, dx \right| \leq c \|\nabla \mathbf{w}\|_{L^6}^2 \|\nabla \mathbf{w}\|_{L^{3/2}} \leq c \|A\mathbf{w}\|^2 \|\nabla \mathbf{w}\|_{L^{3/2}}.$$

Consequently, in

$$\frac{1}{2} \frac{d}{dt} \left(\|\mathbf{w}\|^2 + \frac{\delta^2}{24} \|\nabla \mathbf{w}\|^2 \right) + \frac{1}{Re} \left(\|\nabla \mathbf{w}\|^2 + \frac{\delta^2}{24} \|A\mathbf{w}\|^2 \right) \leq c \|A\mathbf{w}\|^2 \|\nabla \mathbf{w}\|_{L^{3/2}}$$

we can apply the Gronwall lemma to deduce a bound for $\|\nabla \mathbf{w}\|_{L^\infty(0,\overline{T};L^2)}$, provided

$$\eta < \frac{\delta^2}{c \, 24 Re}.$$

Remark 7.25. From the Sobolev embedding theorem, we have $W^{1,p} \subset L^{p^*}$, for $1 \leq p < 3$. Consequently, if $\nabla \mathbf{w}$ belongs to $L^\alpha(0, T; L^\beta)$ (with α, and β as in Theorem 7.22, $\beta < 3$), then (2.31) is satisfied.

By using some classical results on elliptic systems and on singular integrals, we can also introduce breakdown criteria involving the vorticity $\omega = \text{curl} \, \mathbf{w}$. We start by observing that, for a divergence-free function $\mathbf{w}$, we have $-\Delta \mathbf{w} = \text{curl}(\text{curl} \, \mathbf{w})$, and the following estimate holds:

$$\|\nabla \mathbf{w}\|_{L^p} \leq c_p \|\text{curl} \, \mathbf{w}\|_{L^p} \qquad \text{for} \quad 1 < p < \infty, \tag{7.49}$$

with c_p a positive constant depending only on p. The above estimate follows by observing that the Biot–Savart law implies

$$\mathbf{w}(\mathbf{x}) = \int G(\mathbf{x} - \mathbf{x}') \, \text{curl} \, \mathbf{w}(\mathbf{x}') \, d\mathbf{x}', \tag{7.50}$$

where $G(\mathbf{y})$ is given explicitly by

$$G(\mathbf{y}) = \nabla \left[\frac{1}{4\pi} \lim_{N \to \infty} \sum_{\mathbf{k} \in \mathbb{Z}^3, \, |\mathbf{k}| \leq N} \left(\frac{1}{|\mathbf{y} + L\mathbf{k}|} + \frac{1}{|\mathbf{k} \, L|} \right) \right].$$

By taking the gradient of (7.50) (with respect to the variable $\mathbf{x}$), we obtain that

$$\nabla \mathbf{w} = P(\text{curl} \, \mathbf{w}),$$

with P a (linear) singular operator of Calderón–Zygmund type. The estimate (7.49) follows by using the properties of such operators; see Stein [283].

Using estimate (7.49) and Theorem 7.22, one can easily prove the following result:

Corollary 7.26. *Let* $\mathbf{w}$ *be a strong solution to* (7.18), (7.19) *in the time interval* $[0, \overline{T})$. *If it cannot be continued in* (7.25) *to* $t = \overline{T}$, *then*

$$\int_0^{\overline{T}} \|\operatorname{curl}\mathbf{w}(\tau)\|_{L^\beta}^\alpha \, d\tau = \infty \quad \text{for} \quad \frac{2}{\alpha} + \frac{3}{\beta} = 2, \ 1 < \alpha < \infty, \ 3/2 < \beta < \infty.$$

This breakdown criterion is very interesting from a physical point of view. In fact, if $\beta = 2$, and consequently $\alpha = 4$, we have the blow-up criterion involving the so-called *enstrophy*, that is, the L^2-norm of the vorticity field:

$$\int_0^{\overline{T}} \|\operatorname{curl}\mathbf{w}(\tau)\|^4 \, d\tau = +\infty.$$

Energy Budget and Existence Proof

For the Rational LES model it will be interesting to see if it is possible to have an energy balance and existence of weak solutions, provided we add a Smagorinsky dissipative term (a mixed Rational LES model). This question has been analyzed by Iliescu [164] in the space-periodic case. In [164] it is shown that for the Rational LES model it is enough to add the following term on the left-hand side:

$$-\nabla \cdot (C\delta^2 |\nabla\mathbf{w}|^{2\mu}\nabla\mathbf{w}) \quad \text{for } \mu \geq \frac{1}{10}.$$

The nonlinear term derived from the turbulent stress-tensor can be estimated by using results of elliptic regularity to yield the following energy estimate:

$$\frac{1}{2}\frac{d}{dt}\|\mathbf{w}\|^2 + \frac{1}{Re}\|\nabla\mathbf{w}\|^2 + C\|\nabla\mathbf{w}\|_{L^{2+2\mu}}^{2+2\mu}(1 - c\|\nabla\mathbf{w}\|^{1-\beta}) \leq c_1\|\overline{\mathbf{f}}\|^2,$$

where $\beta \in (0, 1)$. Then, by using appropriate smallness of the data ($\|\mathbf{w}_0\| < \epsilon$ and $\|\overline{\mathbf{f}}\|_{L^2(0,T;L^2)} \leq \epsilon$) it is possible to prove the global bound for $\mathbf{w}$

$$\sup_{0<t<T} \|\mathbf{w}(t)\|^2 + \int_0^T \|\nabla\mathbf{w}_m(\tau)\|^2 d\tau + \int_0^T \|\nabla\mathbf{w}_m(\tau)\|_{L^{2+2\mu}}^{2+2\mu} d\tau \leq C \quad t \in [0, T],$$

for a constant C that depends on $Re, \delta, \mathbf{w}_0$, and $\overline{\mathbf{f}}$.

The energy balance together with estimates similar to those obtained for the Gradient LES model allow one to prove the existence of weak solutions, at least for small enough data. Again the crucial point is to show that the operator

$$A(\mathbf{u}) = -\frac{1}{Re}\Delta\mathbf{u} - \nabla \cdot (C\delta^2|\nabla\mathbf{u}|^{2\mu}\nabla\mathbf{u}) + \nabla \cdot \left(\mathbb{I} - \frac{\delta^2}{24}\Delta\right)^{-1}\left[\frac{\delta^2}{12}\nabla\mathbf{u}\,\nabla\mathbf{u}^T\right]$$

is monotone, in order to pass to the limit along Galerkin sequences.

It is interesting to note that due to the regularizing effect of the operator $(\mathbb{I} - \frac{\delta^2}{24}\Delta)^{-1}$, here a dissipative term much weaker than that appearing in the Gradient LES model is required. Recall that in that case the existence results have been proved for $\mu = 1/2$ and they clearly hold also for $\mu \geq 1/2$.

The results in [164] have recently been extended to the nonperiodic case by using also the theory of locally-strongly-monotone operators [30].

Miscellaneous Results for the Rational LES Model

In this section we briefly review some of the recent results that have been proved for the Rational LES model. These results require more involved mathematical results, so we quote them without proofs. These results seem interesting since they prove many of the experimental observations for general flow problems.

First, we collect in one theorem several results proved in Berselli and Grisanti [31] and in Barbato, Berselli, and Grisanti [12]. In these papers the authors proved full regularity of the strong solutions and also consistency results, *i.e.* the convergence of the solution to the Rational LES model to the strong solutions of the NSE as the averaging radius δ goes to zero. (In the sequel we denote by $\mathbf{w}^\delta$ the solution to (7.18), (7.19) corresponding to a given $\delta > 0$.)

Theorem 7.27. *Let $\mathbf{w}_0^\delta \in H^1$ and $\overline{\mathbf{f}} \in L^2(0,T;L^2)$. Then the following results hold:*

(a) *The life-span of a strong solution to the Rational LES model depends on $\|\nabla\mathbf{w}_0^\delta\|_{L^2}$, Re, and $\overline{\mathbf{f}}$, but it is independent of δ.*

(b) *If $\|\nabla\mathbf{w}_0^\delta\|_{L^2}$ is small enough, then a unique strong solution exists on $[0,+\infty)$.*

(c) *If furthermore $\mathbf{w}_0 \in C^\infty(Q)$, and if $\mathbf{w}_\delta$ is a strong solution to the Rational LES model in $[0,T^*[$, then*

$$\mathbf{w}^\delta \in C^\infty(]0,T^*[\times Q).$$

(d) *Let $\mathbf{w}^\delta$ be a strong solution to (7.18)-(7.19) and $\mathbf{u}$ be a solution to the NSE, in the common time interval $[0,T]$. Suppose that both initial data are smooth (say $\mathbf{w}^\delta(\mathbf{x},0)$ and $\mathbf{u}(\mathbf{x},0)$ belong to H^2) and that*

$$\exists c_1 > 0 \quad \text{such that} \quad \|\mathbf{w}^\delta(\mathbf{x},0) - \mathbf{u}(\mathbf{x},0)\|_{L^2} \leq c_1\delta^2.$$

Then we have, for some $c_2 > 0$,

$$\sup_{t\in[0,T]} \|\mathbf{w}^\delta(\mathbf{x},t) - \mathbf{u}(\mathbf{x},t)\|_{L^2} \leq c_2\,\delta^2.$$

If, in addition,

$$\exists\, c_3 > 0 \quad such\ that \quad \|\mathbf{w}^\delta(\mathbf{x},0) - \mathbf{u}(\mathbf{x},0)\|_{H^1} \le c_3\,\delta,$$

then we have, for some $c_4 > 0$,

$$\sup_{t\in[0,T]} \|\mathbf{w}^\delta(\mathbf{x},t) - \mathbf{u}(\mathbf{x},t)\|_{H^1} \le c_4\delta.$$

7.2.3 Numerical Validation and Testing of the Rational LES Model

In [12] we discovered that some classes of exact solution to the NSE (3D generalization of Ross Ethier and Steinman [265] of the 2D Taylor solutions) are also classical solutions to both the Gradient and the Rational LES models. These exact solutions are:

$$\begin{cases} u_1 = [A\sin(\pi z) + C\cos(\pi y)]\,e^{-\pi^2 t/Re} \\[2mm] u_2 = [B\sin(\pi x) + A\cos(\pi z)]\,e^{-\pi^2 t/Re} \\[2mm] u_3 = [C\sin(\pi y) + B\cos(\pi x)]\,e^{-\pi^2 t/Re} \\[2mm] p = -[BC\cos(\pi x)\sin(\pi y) + AB\sin(\pi x)\cos(\pi z) \\[2mm] \qquad\quad + AC\sin(\pi z)\cos(\pi y)]\,e^{-2\pi^2 t/Re}, \end{cases} \qquad (7.51)$$

where A, B, and C are arbitrary constants.

Straightforward calculations show how the SFS stress tensor $\boldsymbol{\tau} = \overline{\mathbf{u}\,\mathbf{u}^T} - \overline{\mathbf{u}}\,\overline{\mathbf{u}}^T$ vanishes identically when evaluated on these solutions. The fact that the NSE and the Rational LES model share similar solutions also suggests that the two models have some common mathematical structure.

The family of solutions (7.51) is very simple since it involves only one Fourier mode. Although *they certainly do not represent turbulence*, they can be useful in debugging and validating complex codes. For a detailed description of numerical validation and testing of LES models (including the Rational and Gradient LES models), the reader is referred to Chap. 12.

Note also that the flow in (7.51) is the viscous counterpart of the classical ABC/Arnold–Beltrami–Childress flow, that has been studied by Arnold [9], Beltrami [23], and Childress [60] in connection with problems of stability and breakdown of smooth solutions for the Euler equations.

In [28] the well-known class of Taylor–Green solutions is analyzed in the context of eddy viscosity LES models. Recall that the so called "Taylor–Green vortex" is widely used as a test case since with its symmetries it can be implemented in a rather efficient way. The solution starting from this vortex is interesting for the complexity of the small scales generated [44] and also for the detection of possible singularities in 3D fluids, see the review in Majda and Bertozzi [227].

To perform physically meaningful calculations the author followed the classical approach of Green and Taylor [138] and considered the flow developing from the very simple initial datum:

$$
\begin{cases}
u_1(x_1, x_2, x_3, 0) = A\cos(ax_1)\sin(bx_2)\sin(cx_3), \\[2mm]
u_2(x_1, x_2, x_3, 0) = B\sin(ax_3)\cos(bx_2)\sin(cx_3), \\[2mm]
u_3(x_1, x_2, x_3, 0) = C\sin(ax_1)\sin(bx_2)\cos(cx_3),
\end{cases}
\tag{7.52}
$$

where, to satisfy the divergence-free constraint, it is necessary to require that $aA + bB + cC = 0$.

This initial datum, with only one frequency (in each space direction), may generate a complex flow. In [28] is shown that the period doubling of this flow is well reproduced by the Rational LES model. Other results regarding the mean value of the pressure show how the Rational LES model seem to be validated, while the Gradient LES model exhibits some divergences. Recall that this kind of analysis was introduced by Taylor and Green in [138] with the "philosophical idea" that:

> "It appears that nothing but a complete solution of the equations of motion in some special case will suffice to illustrate the process of grinding down of large eddies into smaller ones..."

In [28] it is also shown that symmetries of the flow are still present if the simulation is performed with the Rational LES model (instead of the full 3D NSE as in the study of Orszag [247].) This suggests how to reduce the computational time if the simulation is done by means of a spectral Galerkin method in a periodic box.

Numerical Validation and Testing

A careful numerical validation and testing of the Rational LES model was started by Iliescu *et al.* [169] for the 3D lid-driven cavity problem. The authors investigated the behavior of the total kinetic energy for the Gradient and Rational LES models. In all the numerical tests, the Rational LES model was much more stable than the Gradient LES model.

Further numerical tests for the two LES models were carried out by John [173, 175, 176] for the mixing layer test case. Both LES models were equipped with the same extra eddy viscosity term. It is interesting to note that the Rational LES model with the $O(\delta^3)$ eddy viscosity term of Iliescu and Layton [170] (see Chap. 4),

$$
\nu_T = \mu\,\delta\,\|\mathbf{w} - \overline{\mathbf{w}}\|,
$$

outperformed the other LES models.

Iliescu and Fischer started a thorough comparison of the Rational and Gradient LES models in the numerical simulation of turbulent channel flows [106, 165, 166]. These results are presented in detail in Chap. 12 and they also converge to the same conclusion: the Rational LES model outperforms (in terms of numerical stability and accuracy) the Gradient LES model.

7.3 The Higher-order Subfilter-scale Model (HOSFS)

In the previous section we derived the Rational LES model by using an asymptotic wavenumber expansion of order $O(\delta^2)$ of the terms appearing in the SFS stress tensor $\boldsymbol{\tau} = \overline{\mathbf{u}\,\mathbf{u}^T} - \overline{\mathbf{u}}\,\overline{\mathbf{u}}^T$. Since we dropped all terms formally $O(\delta^4)$, the Rational LES model does not include the turbulent fluctuations term $\overline{\mathbf{u}'\mathbf{u}'^T}$ which is believed to be important in the physics of turbulent flow. One way to overcome this drawback is to consider a higher-order asymptotic expansion in the wavenumber space. By using such an approximation, Berselli and Iliescu introduced in [33] the *Higher-order Subfilter-scale (HOSFS)* model. In this section, we present a careful mathematical analysis of the HOSFS model. We note that the mathematical techniques involved in the proof of existence of strong solutions for the HOSFS model, although essentially following the same path as in the mathematical analysis of the Rational LES model, will require more powerful tools.

7.3.1 Derivation of the HOSFS Model

By following the same technique as in the derivation of the Rational LES model, we approximate the Fourier transform of the Gaussian filter $\widehat{g_\delta}(\mathbf{k})$ by using a higher-order (0,2)-Padé rational approximation:

$$e^{-\beta x} = \frac{1}{1 + \beta x + \frac{1}{2}\beta^2 x^2} + O(x^3).$$

Thus, we get

$$\widehat{g_\delta}(\mathbf{k}) = \frac{1}{1 + \frac{\delta^2}{24}|\mathbf{k}|^2 + \frac{\delta^4}{1152}|\mathbf{k}|^4} + O\left(\delta^6|\mathbf{k}|^6\right). \tag{7.53}$$

By using a Taylor series approximation, we also get

$$\frac{1}{\widehat{g_\delta}(\mathbf{k})} = 1 + \frac{\delta^2}{24}|\mathbf{k}|^2 + \frac{\delta^4}{1152}|\mathbf{k}|^4 + O\left(\delta^6|\mathbf{k}|^6\right). \tag{7.54}$$

Thus, by using (7.54) in the usual expression

$$\widehat{\mathbf{u}}(\mathbf{k}) = \frac{1}{\widehat{g_\delta}(\mathbf{k})}\,\widehat{\overline{\mathbf{u}}}(\mathbf{k}),$$

and dropping all terms formally of order $O(\delta^6)$ and higher, we get (in index[4] notation)

$$
\begin{aligned}
\mathbf{u}_i &= \overline{\mathbf{u}}_i - \frac{\delta^2}{24}\nabla^2\overline{\mathbf{u}}_i + \frac{\delta^4}{1152}\nabla^4\overline{\mathbf{u}}_i + 2\,\frac{\delta^4}{1152}\nabla_k^2\nabla_l^2\overline{\mathbf{u}}_i \\
&= \overline{\mathbf{u}}_i - \frac{\delta^2}{24}\Delta\overline{\mathbf{u}}_i + \frac{\delta^4}{1152}\Delta^2\overline{\mathbf{u}}_i,
\end{aligned}
\tag{7.55}
$$

where

$$
\Delta\overline{\mathbf{u}} = \nabla^2\overline{\mathbf{u}} := \sum_{j=1}^{3}\frac{\partial^2\overline{\mathbf{u}}}{\partial\mathbf{x}_j^2}, \qquad \nabla^4\overline{\mathbf{u}} := \sum_{j=1}^{3}\frac{\partial^4\overline{\mathbf{u}}}{\partial\mathbf{x}_j^4},
$$

$$
\nabla_k^2\nabla_l^2\overline{\mathbf{u}} := \sum_{k,l=1;k\neq l}^{3}\frac{\partial^4\overline{\mathbf{u}}}{\partial\mathbf{x}_k^2\partial\mathbf{x}_l^2}, \qquad \Delta^2\overline{\mathbf{u}} = \nabla^4\overline{\mathbf{u}} + 2\nabla_k^2\nabla_l^2\overline{\mathbf{u}}.
$$

By using (7.55), we get

$$
\begin{aligned}
\mathbf{u}_p\mathbf{u}_q &= \overline{\mathbf{u}}_p\overline{\mathbf{u}}_q - \frac{\delta^2}{24}\overline{\mathbf{u}}_p\nabla^2\overline{\mathbf{u}}_q - \frac{\delta^2}{24}\nabla^2\overline{\mathbf{u}}_p\overline{\mathbf{u}}_q + \frac{\delta^4}{1152}\overline{\mathbf{u}}_p\nabla^4\overline{\mathbf{u}}_q + \frac{\delta^4}{1152}\nabla^4\overline{\mathbf{u}}_p\overline{\mathbf{u}}_q \\
&\quad + 2\frac{\delta^4}{1152}\nabla_k^2\nabla_l^2\overline{\mathbf{u}}_p\overline{\mathbf{u}}_q + 2\frac{\delta^4}{1152}\overline{\mathbf{u}}_p\nabla_k^2\nabla_l^2\overline{\mathbf{u}}_q + \frac{\delta^4}{576}\nabla^2\overline{\mathbf{u}}_p\nabla^2\overline{\mathbf{u}}_q.
\end{aligned}
$$

Now with the aid of (7.53), we get

$$
\begin{aligned}
\overline{\mathbf{u}_p\mathbf{u}_q} &= \left(\frac{\delta^4}{1152}\Delta^2 - \frac{\delta^2}{24}\Delta + \mathbb{I}\right)^{-1}(\mathbf{u}_p\mathbf{u}_q) \\
&\approx \overline{\mathbf{u}}_p\overline{\mathbf{u}}_q + \left(\frac{\delta^4}{1152}\Delta^2 - \frac{\delta^2}{24}\Delta + \mathbb{I}\right)^{-1} \\
&\quad \left[\frac{\delta^2}{24}\Delta(\overline{\mathbf{u}}_p\overline{\mathbf{u}}_q) - \frac{\delta^4}{1152}\Delta^2(\overline{\mathbf{u}}_p\overline{\mathbf{u}}_q) - \frac{\delta^2}{24}\overline{\mathbf{u}}_p\nabla^2\overline{\mathbf{u}}_q - \frac{\delta^2}{24}\nabla^2\overline{\mathbf{u}}_p\overline{\mathbf{u}}_q \right. \\
&\quad + \frac{\delta^4}{1152}\overline{\mathbf{u}}_p\nabla^4\overline{\mathbf{u}}_q + \frac{\delta^4}{1152}\nabla^4\overline{\mathbf{u}}_p\overline{\mathbf{u}}_q + 2\frac{\delta^4}{1152}\nabla_k^2\nabla_l^2\overline{\mathbf{u}}_p\overline{\mathbf{u}}_q \\
&\quad \left. + 2\frac{\delta^4}{1152}\overline{\mathbf{u}}_p\nabla_k^2\nabla_l^2\overline{\mathbf{u}}_q + \frac{\delta^4}{576}\nabla^2\overline{\mathbf{u}}_p\nabla^2\overline{\mathbf{u}}_q\right].
\end{aligned}
\tag{7.56}
$$

We need to expand some of the terms in this formula. After a simple calculation, we get

$$
\nabla^4(\overline{\mathbf{u}}_p\overline{\mathbf{u}}_q) = \nabla^4\overline{\mathbf{u}}_p\overline{\mathbf{u}}_q + 4\nabla^3\overline{\mathbf{u}}_p\nabla\overline{\mathbf{u}}_q + 6\nabla^2\overline{\mathbf{u}}_p\nabla^2\overline{\mathbf{u}}_q + 4\nabla\overline{\mathbf{u}}_p\nabla^3\overline{\mathbf{u}}_q + \overline{\mathbf{u}}_p\nabla^4\overline{\mathbf{u}}_q,
$$

and

$$
\begin{aligned}
\nabla_k^2\nabla_l^2(\overline{\mathbf{u}}_p\overline{\mathbf{u}}_q) &= \nabla_k^2\nabla_l^2\overline{\mathbf{u}}_p\overline{\mathbf{u}}_q + 2\nabla_k\nabla_l^2\overline{\mathbf{u}}_p\nabla_k\overline{\mathbf{u}}_q + \nabla_l^2\overline{\mathbf{u}}_p\nabla_k^2\overline{\mathbf{u}}_q + 2\nabla_k^2\nabla_l\overline{\mathbf{u}}_p\nabla_l\overline{\mathbf{u}}_q \\
&\quad + 4\nabla_k\nabla_l\overline{\mathbf{u}}_p\nabla_k\nabla_l\overline{\mathbf{u}}_q + 2\nabla_l\overline{\mathbf{u}}_p\nabla_k^2\nabla_l\overline{\mathbf{u}}_q + \nabla_k^2\overline{\mathbf{u}}_p\nabla_l^2\overline{\mathbf{u}}_q \\
&\quad + 2\nabla_k\overline{\mathbf{u}}_p\nabla_k\nabla_l^2\overline{\mathbf{u}}_q + \overline{\mathbf{u}}_p\nabla_k^2\nabla_l^2\overline{\mathbf{u}}_q.
\end{aligned}
$$

---

[4] In this section we prefer to use the index notation since some formulas may be misunderstood, if written in a compact way.

Thus, by replacing the two relations in (7.56), we get the higher-order SFS (HOSFS) model:

$$
\begin{aligned}
\tau_{pq} = {}& \left(\frac{\delta^4}{1152} \Delta^2 - \frac{\delta^2}{24} \Delta + \mathbb{I} \right)^{-1} \left[\frac{\delta^2}{12} \nabla \overline{u}_p \nabla \overline{u}_q - \frac{\delta^4}{1152} (4 \nabla^3 \overline{u}_p \nabla \overline{u}_q \right. \\
& + 4 \nabla^2 \overline{u}_p \nabla^2 \overline{u}_q + 4 \nabla \overline{u}_p \nabla^3 \overline{u}_q) - 2 \frac{\delta^4}{1152} (4 \nabla_k \nabla_l^2 \overline{u}_p \nabla_k \overline{u}_q + 2 \nabla_l^2 \overline{u}_p \nabla_k^2 \overline{u}_q \\
& + \left. 4 \nabla_k \nabla_l \overline{u}_p \nabla_k \nabla_l \overline{u}_q + 4 \nabla_l \overline{u}_p \nabla_k^2 \nabla_l \overline{u}_q) \right] \\
= {}& \left(\frac{\delta^4}{1152} \Delta^2 - \frac{\delta^2}{24} \Delta + \mathbb{I} \right)^{-1} \left[\frac{\delta^2}{12} \nabla \overline{u}_p \nabla \overline{u}_q - \frac{\delta^4}{288} (\nabla^3 \overline{u}_p \nabla \overline{u}_q + \nabla^2 \overline{u}_p \nabla^2 \overline{u}_q \right. \\
& + \nabla \overline{u}_p \nabla^3 \overline{u}_q) - \frac{\delta^4}{288} (2 \nabla_k \nabla_l^2 \overline{u}_p \nabla_k \overline{u}_q + \nabla_l^2 \overline{u}_p \nabla_k^2 \overline{u}_q \\
& + 2 \nabla_k \nabla_l \overline{u}_p \nabla_k \nabla_l \overline{u}_q + 2 \nabla_l \overline{u}_p \nabla_k^2 \nabla_l \overline{u}_q) \Big] \\
= {}& \left(\frac{\delta^4}{1152} \Delta^2 - \frac{\delta^2}{24} \Delta + \mathbb{I} \right)^{-1} \left[\frac{\delta^2}{12} \nabla \overline{u}_p \nabla \overline{u}_q - \frac{\delta^4}{288} (\widetilde{\nabla^3} \overline{u}_p \widetilde{\nabla} \overline{u}_q + \widetilde{\Delta} \overline{u}_p \widetilde{\Delta} \overline{u}_q \right. \\
& + \left. \widetilde{\nabla} \overline{u}_p \widetilde{\nabla^3} \overline{u}_q) \right],
\end{aligned}
\tag{7.57}
$$

where, for simplicity of exposition, we have used the following notation:

$$
\begin{aligned}
\widetilde{\nabla^3} \overline{u}_p \widetilde{\nabla} \overline{u}_q &:= \nabla^3 \overline{u}_p \nabla \overline{u}_q + 2 \nabla_k \nabla_l^2 \overline{u}_p \nabla_k \overline{u}_q \\
\widetilde{\Delta} \overline{u}_p \widetilde{\Delta} \overline{u}_q &:= \nabla^2 \overline{u}_p \nabla^2 \overline{u}_q + \nabla_l^2 \overline{u}_p \nabla_k^2 \overline{u}_q + 2 \nabla_k \nabla_l \overline{u}_p \nabla_k \nabla_l \overline{u}_q \\
\widetilde{\nabla} \overline{u}_p \widetilde{\nabla^3} \overline{u}_q &:= \nabla \overline{u}_p \nabla^3 \overline{u}_q + 2 \nabla_l \overline{u}_p \nabla_k^2 \nabla_l \overline{u}_q.
\end{aligned}
$$

Remark 7.28. We note that the need for a higher-order approximation (to $O(\delta^6)$) to account for $\overline{u'u'^T}$ was first advocated in [164] and independently in [175]. A similar approach was also used in De Stefano, Denaro, and Riccardi [86] and Katopodes, Street, and Ferziger [182].

7.3.2 Mathematical Analysis of the HOSFS Model

As usual, in our analysis we shall uncouple the problem of wall modeling and boundary conditions from the interior closure problem, by using the space-periodic setting. We shall use the same function spaces and notation introduced in Sect. 7.2.1.

As we shall see, the nonlinear term now requires smoother functions to be correctly evaluated and estimated.

Definition 7.29. *We say that the vector* $\mathbf{w}$ *is a strong solution to the model with the HOSFS stress tensor (7.57) if it has the following regularity:*

$$
\mathbf{w} \in L^\infty(0, T; \mathcal{D}(A)) \cap L^2(0, T; \mathcal{D}(A^{3/2})) \qquad \partial_t \mathbf{w} \in L^2(0, T; \mathcal{D}(A^{1/2}))
$$

and satisfies

$$\frac{d}{dt}(\mathbf{w}, \phi) + \frac{1}{Re}(\nabla\mathbf{w}, \nabla\phi) + (\nabla \cdot (\mathbf{w}\,\mathbf{w}), \phi)$$

$$- \left(\left(\mathbb{I} - \frac{\delta^2}{24}\Delta + \frac{\delta^4}{1152}\Delta^2 \right)^{-1} \left[\frac{\delta^2}{12}\nabla\mathbf{w}\nabla\mathbf{w} \right.\right.$$

$$\left.\left. - \frac{\delta^4}{288}(\widetilde{\nabla}^3\mathbf{w}\widetilde{\nabla}\mathbf{w} + \widetilde{\Delta}\mathbf{w}\widetilde{\Delta}\mathbf{w} + \widetilde{\nabla}\mathbf{w}\widetilde{\nabla}^3\mathbf{w}) \right], \nabla\phi \right) = (\overline{\mathbf{f}}, \phi),$$

for each $\phi \in C^\infty$, *with* $\nabla \cdot \phi = 0$.

Note that since in the HOSFS model we have terms that involve higher-order derivatives, we need rather smooth functions to give meaning to these terms. Recall that, on the other hand, in the Gradient and Rational LES models, there are just terms involving products of first-order derivatives. This is the main reason for the increase of regularity stated in the above definition.

Theorem 7.30. *Assume* $\mathbf{w}_0 \in \mathcal{D}(A)$ *and* $\overline{\mathbf{f}} \in L^2(0, T; \mathcal{D}(A^{1/2}))$. *Then there exists a strictly positive* $T^* = T^*(\mathbf{w}_0, Re, \overline{\mathbf{f}})$ *such that there exists a strong solution to the HOSFS model in* $[0, T^*)$.

For details of the proof, see [33]. Here we only sketch the essential steps in the proof of the local existence of smooth solutions for the HOSFS model, by showing the *a priori* estimates that can be derived.

   Again, the life span could in principle depend also on δ but it is possible to show that it is independent of it.

Proof (of Theorem 7.30). We construct a solution by solving problems for approximate Faedo–Galerkin functions $\mathbf{w}_m(\mathbf{x}, t) = \sum_{k=1}^m g_m^i(t)\mathcal{W}_i(\mathbf{x})$, satisfying for $k = 1, \ldots, m$,

$$\frac{d}{dt}(\mathbf{w}_m, \mathcal{W}_k) + \frac{1}{Re}(\nabla\mathbf{w}_m, \nabla\mathcal{W}_k) + (\nabla \cdot (\mathbf{w}_m\,\mathbf{w}_m), \mathcal{W}_k)$$

$$- \left(\left(\mathbb{I} - \frac{\delta^2}{24}\Delta + \frac{\delta^4}{1152}\Delta^2 \right)^{-1} \left[\frac{\delta^2}{12}\nabla\mathbf{w}_m\nabla\mathbf{w}_m - \frac{\delta^4}{288}(\widetilde{\nabla}^3\mathbf{w}_m\widetilde{\nabla}\mathbf{w}_m \right.\right. \quad (7.58)$$

$$\left.\left. + \widetilde{\Delta}\mathbf{w}_m\widetilde{\Delta}\mathbf{w}_m + \widetilde{\nabla}\mathbf{w}_m\widetilde{\nabla}^3\mathbf{w}_m) \right], \nabla\mathcal{W}_k \right) = (\overline{\mathbf{f}}, \mathcal{W}_k).$$

We use, as test function in (7.58), the function $A^2\mathbf{w}_m$ (namely, we multiply by $\lambda_k^2\mathcal{W}_k$ and perform summation over k). We obtain

$$\int_Q \partial_t \mathbf{w}_m A^2 \mathbf{w}_m \, d\mathbf{x} = \frac{1}{2} \frac{d}{dt} \|A\mathbf{w}_m\|^2,$$

$$\frac{1}{Re} \int_Q A\mathbf{w}_m A^2 \mathbf{w}_m \, dx = \frac{1}{Re} \|A^{3/2}\mathbf{w}_m\|^2.$$

The usual nonlinear term can be estimated as follows (see Lemma 10.4 in Constantin and Foiaş [74] and the definition of fractional powers of A previously given):

$$|A\, B(\mathbf{u}, \mathbf{v})| \le c\|A\mathbf{u}\| \, \|A^{3/2}\mathbf{v}\|,$$

where $B(\mathbf{u}, \mathbf{v})$ is the usual nonlinear operator defined by $B(\mathbf{u}, \mathbf{v}) = P(\mathbf{u} \cdot \nabla \mathbf{v})$. Consequently,

$$\int_Q B(\mathbf{w}_m, \mathbf{w}_m)\, A^2 \mathbf{w}_m \, d\mathbf{x} \le c\|A\mathbf{w}_m\| \, \|A^{3/2}\mathbf{w}_m\| \, \|A\mathbf{w}_m\|$$

$$\le c\|A\mathbf{w}_m\|^2 \, \|A^{3/2}\mathbf{w}_m\|.$$

The most delicate term is the nonlinear term deriving from the LES modeling. For simplicity, we define the following fourth-order linear differential operator:

$$L := \mathbb{I} - \frac{\delta^2}{24}\Delta + \frac{\delta^4}{1152}\Delta^2.$$

The operator L acts on $\mathcal{D}(A^2)$, with values in H. With this notation, the term to be estimated (the extra stress-tensor) becomes

$$\nabla \cdot L^{-1}\left(\frac{\delta^2}{12}\nabla\mathbf{w}_m \nabla\mathbf{w}_m - \frac{\delta^4}{288}(\widetilde{\nabla}^3\mathbf{w}_m \widetilde{\nabla}\mathbf{w}_m + \widetilde{\Delta}\mathbf{w}_m \widetilde{\Delta}\mathbf{w}_m + \widetilde{\nabla}\mathbf{w}_m \widetilde{\nabla}^3\mathbf{w}_m)\right).$$

Multiplying by $A^2\mathbf{w}_m$ and integrating by parts over Q, we obtain

$$\left(L^{-1}\left[\frac{\delta^2}{12}\nabla\mathbf{w}_m \nabla\mathbf{w}_m - \frac{\delta^4}{288}(\widetilde{\nabla}^3\mathbf{w}_m \widetilde{\nabla}\mathbf{w}_m + \widetilde{\Delta}\mathbf{w}_m \widetilde{\Delta}\mathbf{w}_m + \widetilde{\nabla}\mathbf{w}_m \widetilde{\nabla}^3\mathbf{w}_m)\right]\right.$$
$$\left. \nabla A^2\mathbf{w}_m\right).$$

It is enough to estimate terms with the higher-order derivatives, since the first one is easier. We have

$$\left|\left(L^{-1}(\widetilde{\nabla}^3\mathbf{w}_m \widetilde{\nabla}\mathbf{w}_m), \nabla A^2\mathbf{w}_m\right)\right| = \|L^{-1}(\widetilde{\nabla}^3\mathbf{w}_m \widetilde{\nabla}\mathbf{w}_m)\|_{H^4} \|\nabla A^2\mathbf{w}_m\|_{H^{-4}}.$$

Furthermore, by recalling that the bi-Laplacian acts as an isomorphism between L^2 and H^4, we obtain

$$\|\widetilde{\nabla}^3\mathbf{w}_m \widetilde{\nabla}\mathbf{w}_m\| \, \|\nabla\mathbf{w}_m\| \le C \, \|\nabla^3\mathbf{w}_m\| \, \|\nabla\mathbf{w}_m\|_{L^\infty} \|\nabla\mathbf{w}_m\|.$$

At this point we turn the H^s-bound into an L^∞-bound. We can do this with the aid of a result of Morrey.

Proposition 7.31 (Morrey). *Let $u \in W^{1,p}(\mathbb{R}^d)$, with $p > d$. Then $u \in L^\infty(\mathbb{R}^d)$ and*

$$\|u\|_{L^\infty} \leq C\|u\|_{W^{1,p}}$$

with C a constant independent of u. Furthermore, the following inequality holds:

$$|u(\mathbf{x}) - u(\mathbf{y})| \leq C|\mathbf{x} - \mathbf{y}|^\alpha \|\nabla u\|_{W^{1,p}} \qquad \alpha = 1 - \frac{d}{p}.$$

Remark 7.32. First, the reader may note that the case $p > d$ was not covered by the Sobolev embedding Theorem (7.38). Further, as a simple corollary, we can deduce that

$$\text{if} \quad \frac{1}{p} - \frac{m}{d} < 0, \text{ then } \quad W^{m,p}(\mathbb{R}^d) \hookrightarrow L^\infty(\mathbb{R}^d).$$

The same results hold if the functions are periodic or defined on a smooth enough domain $\Omega \subset \mathbb{R}^d$ (see Adams [4]).

In the sequel we will need also this simple interpolation inequality that is a particular case of much more general results on interpolation in Sobolev spaces (see Bergh and Löfström [25]).

Proposition 7.33. *Let $u \in H^r(\mathbb{R}^d) \cap H^s(\mathbb{R}^d)$. Then u belongs to $H^t(\mathbb{R}^d)$, for each $r \leq t \leq s$, and the following estimate holds:*

$$\|u\|_{H^t} \leq C\|u\|_{H^r}^\theta \|u\|_{H^r}^{1-\theta}, \quad \theta \text{ being defined by } t = \theta r + (1-\theta)s,$$

for a constant C independent of u.

Then, using the Sobolev embedding $H^{3/2+\epsilon} \hookrightarrow L^\infty$, and with the above result of interpolation of H^s-spaces, we obtain

$$\|\nabla \mathbf{w}_m\|_{L^\infty} \leq \|\mathbf{w}_m\|_{H^{5/2+\epsilon}} \leq \|\mathbf{w}_m\|_{H^2}^{\frac{1}{2}-\epsilon} \|\mathbf{w}_m\|_{H^3}^{\frac{1}{2}+\epsilon}, \quad \text{for each } \epsilon \in (0, 1/2).$$

By recalling the Young inequality, we have

$$\|\widetilde{\nabla}^3 \mathbf{w}_m \widetilde{\nabla} \mathbf{w}_m\| \|\nabla \mathbf{w}_m\| \leq C \|\mathbf{w}_m\|_{H^3}^{\frac{1}{2}+\epsilon} \|\mathbf{w}_m\|_{H^2}^{\frac{3}{2}-\epsilon} \leq \frac{\eta}{2} \|\mathbf{w}_m\|_{H^3}^2 + \frac{C^2}{2\eta} \|\mathbf{w}_m\|_{H^2}^2.$$

We also have

$$\left| \left(L^{-1}(\widetilde{\Delta} \mathbf{w}_m \widetilde{\Delta} \mathbf{w}_m), \nabla A^2 \mathbf{w}_m \right) \right| = \|L^{-1}(\widetilde{\Delta} \mathbf{w}_m \widetilde{\Delta} \mathbf{w}_m)\|_{H^4} \|\nabla A^2 \mathbf{w}_m\|_{H^{-4}}$$

and

$$\|L^{-1}(\widetilde{\Delta} \mathbf{w}_m \widetilde{\Delta} \mathbf{w}_m)\|_{H^4} \|\nabla A^2 \mathbf{w}_m\|_{H^{-4}} \leq C \|\nabla^2 \mathbf{w}_m\|_{L^4}^2 \|\nabla \mathbf{w}_m\|$$

$$\leq C \|\nabla^3 \mathbf{w}_m\|^{3/2} \|\nabla^2 \mathbf{w}_m\|^{3/2}.$$

They finally imply

$$\left|\left(L^{-1}(\widetilde{\Delta}\mathbf{w}_m\widetilde{\Delta}\mathbf{w}_m), \nabla A^2\mathbf{w}_m\right)\right| \leq \frac{\eta}{2}\|\mathbf{w}_m\|_{H^3}^2 + \frac{C^4 3^3}{2^5\eta^2}\|\mathbf{w}_m\|_{H^2}^6.$$

The last term in (7.58) is estimated in the obvious way:

$$\left|(\overline{\mathbf{f}}, A^2\mathbf{w}_m)\right| = \left|\left(A^{1/2}\overline{\mathbf{f}}, A^{3/2}\mathbf{w}_m\right)\right| \leq \frac{\eta}{2}\|A^{3/2}\mathbf{w}_m\|^2 + \frac{1}{2\eta}\|A^{1/2}\overline{\mathbf{f}}\|^2,$$

where we used the identity

$$(A\mathbf{u}, \mathbf{v}) = (A^\epsilon\mathbf{u}, A^{1-\epsilon}\mathbf{v}),$$

which holds for $\mathbf{u}, \mathbf{v} \in \mathcal{D}(A)$. We finally obtain

$$\frac{1}{2}\frac{d}{dt}\|A\mathbf{w}_m\|^2 \frac{1}{2Re}\|A^{3/2}\mathbf{w}_m\|^2 \leq C(Re, \delta)(\|A\mathbf{w}_m\|^4 + \|A\mathbf{w}_m\|^6 + \|A^{1/2}\overline{\mathbf{f}}\|^2).$$

The last estimate implies (by using classical existence results for ordinary differential equations) that there exists a unique solution $\mathbf{w}_m$ to (7.58), in some time interval $[0, T^*)$, for a strictly positive T^*, and that

$$\mathbf{w}_m \in L^\infty(0, T^*; H^2) \cap L^2(0, T^*; H^3). \tag{7.59}$$

Let us now turn to an estimate for the time derivative. Multiplying (7.58) by $\partial_t A\mathbf{w}_m$ and integrating by parts, we obtain

$$\|\partial_t\nabla\mathbf{w}_m\|^2 + \frac{1}{2\,Re}\frac{d}{dt}|A\mathbf{w}_m|^2 \leq |((\mathbf{w}_m \cdot \nabla)\mathbf{w}_m, A\partial_t\mathbf{w}_m)|$$

$$+ \left|\left(L^{-1}\left[\frac{\delta^2}{12}\nabla\mathbf{w}_m\nabla\mathbf{w}_m\right.\right.\right. \tag{7.60}$$

$$\left.\left.\left. - \frac{\delta^4}{1152}(\widetilde{\nabla}^3\mathbf{w}_m\widetilde{\nabla}\mathbf{w}_m + \widetilde{\Delta}\mathbf{w}_m\widetilde{\Delta}\mathbf{w}_m + \widetilde{\nabla}\mathbf{w}_m\widetilde{\nabla}^3\mathbf{w}_m)\right], \nabla A\partial_t\mathbf{w}_m\right)\right|.$$

To estimate the right-hand side we proceed as follows: We start with the term

$$\left|\left(L^{-1}(\widetilde{\nabla}^3\mathbf{w}_m\widetilde{\nabla}\mathbf{w}_m), \nabla A\partial_t\mathbf{w}_m\right)\right|$$

$$\leq \|L^{-1}(\widetilde{\nabla}^3\mathbf{w}_m\widetilde{\nabla}\mathbf{w}_m)\|_{W^{3,6/5}}\|\nabla A\partial_t\mathbf{w}_m\|_{W^{-3,6}}.$$

By recalling the Sobolev embedding $H^1 \hookrightarrow L^6$, we can easily bound the second term by

$$\|\partial_t\mathbf{w}_m\|_{L^6} \leq C\|\partial_t\nabla\mathbf{w}_m\|,$$

while the first one needs the following treatment:

$$\|L^{-1}(\tilde{\nabla}^3\mathbf{w}_m\tilde{\nabla}\mathbf{w}_m)\|_{W^{3,6/5}} \leq C \,\|\nabla^3\mathbf{w}_m\nabla\mathbf{w}_m\|_{W^{-1,6/5}}.$$

This is handled as follows:

$$\|\nabla^3\mathbf{w}_m\nabla\mathbf{w}_m\|_{W^{-1,6/5}} = \sup_{\psi\neq 0}\frac{\langle\nabla^3\mathbf{w}_m\nabla\mathbf{w}_m,\psi\rangle}{\|\psi\|_{W^{1,6}}}.$$

An integration by parts, together with the periodicity of the functions, implies (we replace the duality with the integral, since the functions $\mathbf{w}_m$ are smooth)

$$\left|\int_Q \nabla^3\mathbf{w}_m\nabla\mathbf{w}_m\psi\,d\mathbf{x}\right| = \left|\int_Q \nabla^2\mathbf{w}_m\nabla^2\mathbf{w}_m\psi\,d\mathbf{x} + \int_Q \nabla^2\mathbf{w}_m\nabla\mathbf{w}_m\nabla\psi\,d\mathbf{x}\right|$$

with the Hölder inequality

$$\leq \|\nabla^2\mathbf{w}_m\nabla^2\mathbf{w}_m\|_{L^{6/5}}\|\psi\|_{L^6} + \|\nabla^2\mathbf{w}_m\nabla\mathbf{w}_m\|\|\nabla\psi\|_{L^6}$$
$$\leq (\|\nabla^2\mathbf{w}_m\nabla^2\mathbf{w}_m\|_{L^{6/5}} + \|\nabla^2\mathbf{w}_m\nabla\mathbf{w}_m\|_{L^{6/5}})\|\psi\|_{W^{1,6}}.$$

The terms involving $\mathbf{w}_m$ may be bounded as follows:

$$\|\nabla^2\mathbf{w}_m\nabla^2\mathbf{w}_m\|_{L^{6/5}} \leq \|\nabla^2\mathbf{w}_m\|^{3/2}\|\nabla^3\mathbf{w}_m\|^{1/2} + \|\nabla^2\mathbf{w}_m\|_{L^{6/5}}\|\nabla\mathbf{w}_m\|_{L^\infty}$$
$$\leq \|\nabla^2\mathbf{w}_m\|^{3/2}\|\nabla^3\mathbf{w}_m\|^{1/2} + \|\nabla^2\mathbf{w}_m\|\|\nabla^3\mathbf{w}_m\|.$$
$$(7.61)$$

In the derivation of (7.61) we used the embedding $H^3 \hookrightarrow L^\infty$ and the interpolation inequality

$$\|f\|_{L^{12/5}} \leq \|f\|_{L^2}^{3/4}\|f\|_{L^6}^{1/4} \leq C\|f\|^{3/4}\|\nabla f\|^{1/4},$$

that derives from the convex-interpolation inequality and the Sobolev embedding. The same method shows how to estimate the last term appearing in (7.60). The term

$$\left|\left(L^{-1}(\tilde{\Delta}\mathbf{w}_m\tilde{\Delta}\mathbf{w}_m),\nabla A\partial_t\mathbf{w}_m\right)\right| = \left|\nabla\left(L^{-1}(\tilde{\Delta}\mathbf{w}_m\tilde{\Delta}\mathbf{w}_m),A\partial_t\mathbf{w}_m\right)\right|$$

may be bounded as follows:

$$\|\nabla(L^{-1}(\tilde{\Delta}\mathbf{w}_m\tilde{\Delta}\mathbf{w}_m))\|_{W^{2,6/5}}\,\|\nabla A\partial_t\mathbf{w}_m\|_{W^{-2,6}},$$

which in turn is bounded by

$$C\|\nabla^2\mathbf{w}_m\|_{L^{12/5}}^2\,\|\partial_t\mathbf{w}_m\|_{L^6} \leq C\|\nabla^3\mathbf{w}_m\|^{1/2}\|\nabla^2\mathbf{w}_m\|^{3/2}\,\|\partial_t\nabla\mathbf{w}_m\|.$$

The term involving the lower derivative can be handled similarly.

Using the Young inequality, we obtain

$$\|\partial_t\nabla\mathbf{w}_m\|^2 + \frac{1}{Re}\frac{d}{dt}\|A\mathbf{w}_m\|^2 \leq c(\|A\mathbf{w}_m\|^2 + \|A\mathbf{w}_m\|^3)\|A^{3/2}\mathbf{w}_m\|^2. \quad (7.62)$$

By recalling the bound previously obtained in (7.59), we can integrate (7.62) with respect to time over $[0, T^*)$ to obtain

$$\int_0^{T^*} \|\partial_t \nabla \mathbf{w}_m\|^2 \le C,$$

which gives the desired bound on the time derivative.

With the bounds we can extract from $\{\mathbf{w}_m\}$ a subsequence (relabeled as $\{\mathbf{w}_m\}$) such that

$$\begin{cases} \mathbf{w}_m \overset{*}{\rightharpoonup} \mathbf{w} \text{ in } L^\infty(0, T^*; H^2) \\ \mathbf{w}_m \rightharpoonup \mathbf{w} \text{ in } L^2(0, T^*; H^3) \\ \mathbf{w}_m \to \mathbf{w} \text{ in } L^2(0, T^*; H^2) \quad \text{and a.e. in } (0, T) \times Q. \end{cases}$$

The argument is based on the classical Aubin–Lions lemma, and the reasoning follows the guidelines of the previous section.

Regarding the term with $\widetilde{\nabla}^3 \mathbf{w}_m \widetilde{\nabla} \mathbf{w}_m$, we observe that since $\partial_t \nabla \mathbf{w}_m$ converges weakly in $L^2(0, T^*; H^1)$, then $\mathbf{w}_m$ converges weakly in $L^\infty(0, T^*; H^1)$. Consequently, this implies that $\forall \phi \in C^\infty_{per}(Q)$

$$\left(\left(\mathbb{I} - \frac{\delta^2}{24} \Delta + \frac{\delta^4}{1152} \Delta^2 \right)^{-1} \left[\widetilde{\nabla}^3 \mathbf{w}_m \widetilde{\nabla} \mathbf{w}_m \right], \nabla \phi \right)$$

$$= \left(\widetilde{\nabla}^3 \mathbf{w}_m \widetilde{\nabla} \mathbf{w}_m, \left(\mathbb{I} - \frac{\delta^2}{24} \Delta + \frac{\delta^4}{1152} \Delta^2 \right)^{-1} \nabla \phi \right)$$

$$\to \left(\widetilde{\nabla}^3 \mathbf{w} \widetilde{\nabla} \mathbf{w}, \left(\mathbb{I} - \frac{\delta^2}{24} \Delta + \frac{\delta^4}{1152} \Delta^2 \right)^{-1} \nabla \phi \right)$$

in $L^2(0, T^*)$. The convergence of the terms involving $\widetilde{\Delta} \mathbf{w}_m \widetilde{\Delta} \mathbf{w}_m$ can be obtained by observing that the classical results of interpolation show that $\widetilde{\Delta} \mathbf{w}_m \in L^\infty(0, T^*; L^2) \cap L^2(0, T^*; H^1)$. This implies, by the Hölder inequality, that $\widetilde{\Delta} \mathbf{w}_m \in L^4(0, T^*; L^3)$. Thus, $\widetilde{\Delta} \mathbf{w}_m \widetilde{\Delta} \mathbf{w}_m$ is bounded in $L^2(0, T^*; L^{3/2})$, and

$$\widetilde{\Delta} \mathbf{w}_m \widetilde{\Delta} \mathbf{w}_m \rightharpoonup \widetilde{\Delta} \mathbf{w} \widetilde{\Delta} \mathbf{w} \quad \text{in } L^2(0, T^*; L^{3/2}).$$

This implies that, $\forall \phi \in C^\infty_{per}(Q)$, the following convergence takes place in $L^2(0, T^*)$:

$$\left(\left(\mathbb{I} - \frac{\delta^2}{24} \Delta + \frac{\delta^4}{1152} \Delta^2 \right)^{-1} \left[\widetilde{\Delta} \mathbf{w}_m \widetilde{\Delta} \mathbf{w}_m \right], \nabla \phi \right)$$

$$= \left(\widetilde{\Delta} \mathbf{w}_m \widetilde{\Delta} \mathbf{w}_m, \left(\mathbb{I} - \frac{\delta^2}{24} \Delta + \frac{\delta^4}{1152} \Delta^2 \right)^{-1} \nabla \phi \right)$$

$$\to \left(\widetilde{\Delta} \mathbf{w} \widetilde{\Delta} \mathbf{w}, \left(\mathbb{I} - \frac{\delta^2}{24} \Delta + \frac{\delta^4}{1152} \Delta^2 \right)^{-1} \nabla \phi \right).$$

The convergence of the other nonlinear terms is rather standard and can easily be obtained with the same tools as Sect. 7.2.1. Full details can be found in [33].
□

7.3.3 Numerical Validation and Testing of the HOSFS Model

Single- and Two-mode Analysis

We follow the approach used by Geurts [130] and Katopodes, Street, and Ferziger [183, 182, 184], and compare the HOSFS model with (1) the Rational LES model, (2) the Gradient LES model, and (3) the higher-order gradient model (7.63) for the particular choices $u(x) = e^{iKx}$ and $u(x) = e^{iK_1 x} + e^{iK_2 x}$, where $K_2 = C\,K_1$ and $C = 2, 3, 4, 5$, and 10.

We emphasize that while these tests do not automatically imply the success of an LES model in actual turbulent flow simulations, they give some insight into the way the LES models reconstruct the SFS stress tensor τ (see also Remark 7.34).

To this end, we first present an equivalent of the HOSFS model for the gradient model: the higher-order gradient model. We stress, however, that we consider this last model only to illustrate numerically our theoretical considerations. The higher-order gradient model (HOGR) [183, 182, 184] (that is a model obtained in the same way – see Sect. 7.1.1 – as the Gradient model, but with a Taylor series expansion up to $O(\delta^6)$ of $\widehat{g}_\delta(\mathbf{k})$) reads

$$\boldsymbol{\tau} = \frac{\delta^4}{576}(\Delta\overline{\mathbf{u}}\,\Delta\overline{\mathbf{u}}^T + \overline{\mathbf{u}}\,\Delta^2\overline{\mathbf{u}}^T) - \frac{\delta^2}{12}\overline{\mathbf{u}}\,\Delta\overline{\mathbf{u}}^T + \frac{\delta^2}{24}\Delta(\overline{\mathbf{u}}\,\overline{\mathbf{u}}^T) - \frac{\delta^4}{288}\Delta(\overline{\mathbf{u}}\,\Delta\overline{\mathbf{u}}^T)$$

$$+ \frac{\delta^4}{1152}\Delta^2(\overline{\mathbf{u}}\,\overline{\mathbf{u}}^T). \tag{7.63}$$

This simple scalar and one-dimensional example will give us some insight into the behavior of the SFS stress tensor for the four models considered. We shall compare these results with the exact SFS stress tensor.

Case I. $u(x) = e^{iKx}$.

This case gives us some insight into the stress tensor based on interactions at the same wavenumber. Specifically, we focus on the oscillatory part of e^{iKx}. First, after a simple calculation, we get

$$\overline{u} = e^{-\frac{K^2\delta^2}{24}}u.$$

The exact stress tensor is

$$\tau = \overline{u\,u} - \overline{u}\,\overline{u} = \left(e^{-4\frac{K^2\delta^2}{24}} - e^{-2\frac{K^2\delta^2}{24}}\right)e^{2iKx}.$$

We get the oscillatory part (that is, the term multiplying e^{2iKx}) of the stress tensors corresponding to the gradient model,

$$-\frac{\delta^2 K^2}{12}\mathrm{e}^{-2\frac{\delta^2 K^2}{24}},$$

the Rational LES model ,

$$-\frac{\delta^2 K^2}{12}\frac{\mathrm{e}^{-2\frac{\delta^2 K^2}{24}}}{1+\frac{\delta^2 K^2}{6}},$$

the HOGR model (7.63),

$$\left(-\frac{\delta^2 K^2}{12}-\frac{\delta^4 K^4}{96}\right)\mathrm{e}^{-2\frac{\delta^2 K^2}{24}},$$

and the HOSFS model,

$$\left(-\frac{\delta^2 K^2}{12}-\frac{\delta^4 K^4}{96}\right)\frac{\mathrm{e}^{-2\frac{\delta^2 K^2}{24}}}{1+\frac{\delta^2 K^2}{6}+\frac{\delta^4 K^4}{72}}.$$

The corresponding results are presented in Fig. 7.1. Clearly, the best results correspond to the HOSFS model: its oscillating part in the SFS stress tensor τ is the closest to the *exact* value. Notice also that the Gradient LES model and the HOGR model (7.63) overpredict the correct results, and this is apparent for the higher wave numbers. This is due to the inaccurate approximation to the Fourier transform of the Gaussian filter away from the origin.

Case II. $u(x) = \mathrm{e}^{iK_1 x} + \mathrm{e}^{iK_2 x}$, with $K_2 = CK_1$, $C = 2, 3, 4, 5$, and 10.

This case gives insight into the interaction between large and small scales in the SFS stress tensor. Since we looked at the interaction between the same wavenumbers in Case I, we focus now on the interaction between large (that is, K_2) and small (that is, K_1) wavenumbers. Specifically, we focus on the oscillatory part of $\mathrm{e}^{i(K_1+K_2)x}$.
For the exact SFS stress tensor $\tau = \overline{u\,u} - \overline{u}\,\overline{u}$, this oscillatory part is

$$2\,\mathrm{e}^{-(1+c)^2\,\frac{K_1^2\delta^2}{24}} - 2\,\mathrm{e}^{-(1+c^2)\,\frac{K_1^2\delta^2}{24}}.$$

We get the oscillatory part of $\mathrm{e}^{i(K_1+K_2)x}$ in the stress tensors corresponding to the Gradient LES model,

$$-2\,c\,\frac{\delta^2 K_1^2}{12}\,\mathrm{e}^{-(1+c^2)\,\frac{\delta^2 K_1^2}{24}},$$

the Rational LES model,

$$-2\,c\,\frac{\delta^2 K_1^2}{12}\,\frac{\mathrm{e}^{-(1+c^2)\,\frac{\delta^2 K_1^2}{24}}}{1+(1+c)^2\,\frac{\delta^2 K_1^2}{24}},$$

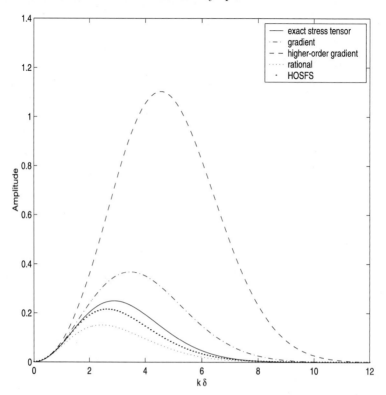

Fig. 7.1. Oscillatory part of the SFS stress tensor τ for Case I (*one wave*): exact stress (*continuous line*), the Gradient LES model (*dash-dotted line*); the higher-order gradient model (7.63) (*dashed line*); the Rational LES model (*thin dotted line*); the HOSFS model (*thick dotted line*)

the HOGR model (7.63),

$$\left[(-2\,(1+c)^2 \;+\; 2\,(1+c^2))\,\left(-\frac{\delta^2 K_1^2}{24}\right) \right.$$

$$\left. +\;\; (2\,(1+c)^4 \;+\; 2\,(1+c^4)\;-\;4\,(1+c^2)\,(1+c)^2\;+\;4\,c^2)\,\left(-\frac{\delta^4 K_1^4}{1152}\right) \right]$$

$$\mathrm{e}^{-(1+c^2)\,\frac{\delta^2 K_1^2}{24}}\;,$$

and the HOSFS model,

$$\left(-2\,c\,\frac{\delta^2 K_1^2}{12}\;-\;2\,(2\,(c+c^3)+2\,c^2)\,\frac{\delta^4 K_1^4}{576}\right)$$

$$\frac{\mathrm{e}^{-(1+c^2)\,\frac{\delta^2 K_1^2}{24}}}{1+(1+c)^2\,\frac{\delta^2 K_1^2}{24}+(1+c)^4\,\frac{\delta^4 K_1^4}{1152}}\cdot$$

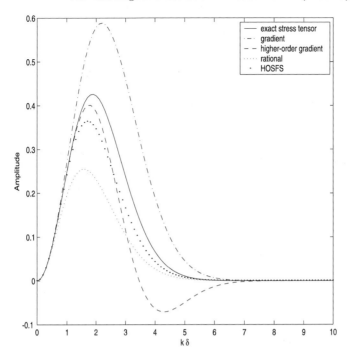

Fig. 7.2. Oscillatory part of the SFS stress tensor τ for Case II (*two wave*), C = 2: exact stress (*continuous line*), the Gradient LES model (*dash-dotted line*); the higher-order gradient model (7.63) (*dashed line*); the Rational LES model (*thin dotted line*); the HOSFS model (*thick dotted line*)

The corresponding results are presented in Figs. 7.2 and 7.3. The best results seem to correspond to the HOSFS model: its SFS stress tensor is the closest to the *exact* SFS stress tensor. The HOGR model (7.63) performs better for low wavenumbers, but it underpredicts drastically the correct stress tensor for large wavenumbers. This behavior is alleviated for larger values for the constant C, when the HOGR model (7.63) performs similarly to the HOSFS model. The pure Gradient LES model performs as in the previous case: it overpredicts the correct value of the oscillating part of the stress tensor. Again, this is due to the inaccurate approximation to the Fourier transform of the Gaussian filter away from the origin.

Remark 7.34. As mentioned at the beginning of the section, this single- and two-mode analysis sheds some light on the ability of the LES models to reconstruct the SFS stress tensor τ.

Of course, this is just a preliminary step in assessing the HOSFS model. We need to use *a priori* and, especially, *a posteriori* tests in actual turbulent flow simulations in order to validate the HOSFS model.

Based on the insight gained from actual simulations, we could improve the performance of the HOSFS models. One such possible improvement could

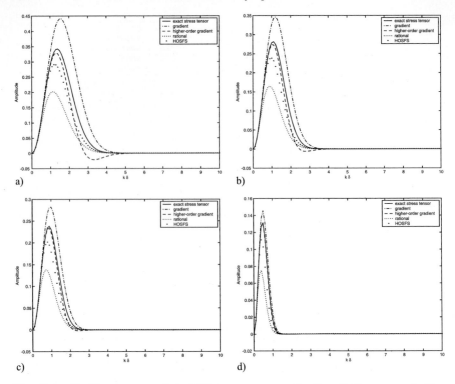

Fig. 7.3. Oscillatory part of the SFS stress tensor τ for Case II (*two waves*), C = 3, 4, 5, 10 (*left-right, top-bottom*): exact stress (*continuous line*), the Gradient LES model (*dash-dotted line*); the higher-order gradient model (7.63) (*dashed line*); the Rational LES model (*thin dotted line*); the HOSFS model (*thick dotted line*)

come from a mixed model obtained by coupling the HOSFS model with an eddy viscosity model accounting for the loss of information in the discretization process, as advocated by Carati *et al.* [55] and used by Winckelmans *et al.* [316]. Then we could compare the HOSFS model with other LES models, such as the dynamic eddy viscosity model in Germano *et al.* [129] or the variational multiscale method of Hughes *et al.* [160, 162].

On Numerical Implementation

Because of the higher-order derivatives appearing in this model, higher-order methods (for example, spectral methods or finite difference methods) seem the most appropriate approaches for the numerical realization of the HOSFS model (7.57). Of course, once the behavior of the higher-order terms is well understood, one may be able to extend this model to finite element methods.

Also, since the inverse operator in (7.57) is just an approximation to convolution with the Gaussian filter g_δ, the most natural numerical approach toward its implementation would probably be through a local operator. Specifically, we need to compute the convolution only locally (just across a few elements in the neighborhood of our grid point) because of the rapidly decaying behavior of the Gaussian filter. This approach could be implemented in an efficient manner. For example, if the finite element method is used, one would need to store at the beginning of the computation the convolution of the finite element basis functions with the spatial filter g_δ at the Gauss points in each element. By using this information, one could then update the SFS stress tensor at each time-step accordingly.

The HOSFS model is very similar in spirit to the approximate deconvolution model of Stolz and Adams [285] described in Chap. 8. Thus, the numerical approach adopted in the implementation of this approximate deconvolution model could guide us in an efficient numerical implementation of the HOSFS model.

7.4 Conclusions

In this chapter we introduced the concept of *SFS-subfilter-scale modeling* in which the goal is to approximate (some of) the information lost in the filtering process (*i.e.* the convolution with the spatial filter g_δ). This approach is also known as *approximate deconvolution*. In this chapter we considered a particular class of approximate deconvolution models – those based on wavenumber asymptotics. Three LES models in this class were presented: the Gradient LES model of Leonard [212] and Clark, Ferziger, and Reynolds [65], the Rational LES model of Galdi and Layton [122], and the Higher-order Subfilter-scale model of Berselli and Iliescu [33]. For all the models a careful derivation and a thorough mathematical analysis was presented. The corresponding numerical results will be presented in Chap. 12.

We want to stress that there is a fundamental difference between the SFS and SGS philosophies: the SFS approach considers the LES modeling process as a sequence of two steps:

Step 1: Approximate the information lost in the spatial filtering process to approximate $\boldsymbol{\tau} = \overline{\mathbf{u}\,\mathbf{u}^T} - \overline{\mathbf{u}}\,\overline{\mathbf{u}}^T$ by terms involving $\overline{\mathbf{u}}$ only (the closure problem).
Step 2: Discretize the resulting SFS model.

Thus, in the SFS modeling approach, the modeling is done sequentially: first at a continuous level (Step 1) and then at a discrete level (Step 2).

In contrast, in the SGS modeling, both the continuum and the discrete approximations in Step 1 and Step 2 are treated unitarily, as a unique source of error. Thus, the modeling is done in one step. A classic example in this class are the pure eddy viscosity models presented in detail in Part II. In these models, it is assumed that the information below the filter-scale δ is irreversibly lost (and thus there is no subfilter-scale information). It is also assumed that the only means of (subgrid-scale) modeling is by using physical insight.

One possible inconsistency of the SGS modeling approach is that although one does not make a distinction between the grid-scale h and the filter-scale δ in the modeling process, in actual implementations δ is a multiple of h. For example, $\delta = 2h$ is a very popular choice in practical LES computations. Moreover, the numerical simulations are very sensitive to the choice of the proportionality constant.

The SFS modeling approach, on the other hand, offers the opportunity of understanding the relationship between the filter-scale δ and the grid-scale h, with the potential of deriving appropriate scalings. This could eventually yield robust and universal LES models able to work in different settings without the tuning necessary for present LES models.

Connecting the two steps in the SFS modeling process (the continuum and the discrete approximation) is a daunting task closely related to the *numerical analysis* of this process. Only the first steps have been made in this direction [168, 177, 164, 99, 100, 102, 153]. Some of these steps have been presented in the exquisite monograph of John [175]. These are the first steps along a tenuous road whose finish line could, however, offer the robustness and level of generality LES is currently needing.

We end by briefly listing some of the contributions to SFS modeling that, due to space limitations, we had to leave out: the inverse modeling of Geurts [130], the velocity estimation model of Domaradzki and collaborators [93, 92], the approximate deconvolution model of Stolz and Adams [285] (a detailed presentation is given in Chap. 8), and the work of Carati and collaborators [55, 315], Vreman and collaborators [308, 307, 310, 194, 309], Winckelmans and collaborators [316, 315], and Borue and Orszag [41], and Katopodes, Street, and Ferziger [183, 182, 184].

A particular class of SFS models are the *scale similarity* models of Bardina, Ferziger, and Reynolds [13, 14]. We give a detailed presentation of these models in Chap. 8.

8

Scale Similarity Models

8.1 Introduction

Scale similarity models were introduced in 1980 by Bardina in his PhD thesis and in a series of papers with Ferziger and Reynolds [13, 14]. The principle behind them can be expressed in various ways. One description is that

> the energy transfer from all unresolved scales to resolved scales is dominated by the transfer from the first, largest unresolved scale to the smallest resolved scale. This transfer across scales is similar to the energy transfer from the smallest resolved scale to the next smallest resolved scale.

One direct realization of this idea is the Bardina model (given by (8.1) below). At another level, scale similarity is an assumption that

> unresolved quantities (ϕ) can be effectively approximated by extrapolating their values from their resolved scales ($\overline{\phi}, \overline{\overline{\phi}}, \overline{\overline{\overline{\phi}}}$, etc.)

They have been tantalizingly close to seeming a universal and accurate model in LES. However, stability problems have also been reported for them, spurring the development of successively more complicated or refined models. In the sequel we shall introduce the most widely known scale similarity models, together with some recent improvements. The reader will find in this chapter the fundamental ideas and will have a chance of understanding the potential and the problems related to designing scale similarity models.

8.1.1 The Bardina Model

To shorten the notation, in this chapter we denote $\mathbf{u}\,\mathbf{u}^T$ simply by $\mathbf{u}\,\mathbf{u}$. We now briefly derive the Bardina model, the first one designed by using scale similarity ideas. In [13] the authors proposed to model terms in the *triple decomposition* (Cross, Reynolds, and Leonard term; see p. 72) by using a second application of the filter, together with a zero-order approximation, *i.e.* $\overline{\phi\,\psi} = \overline{\phi}\,\overline{\psi}$. These lead to

$$R \approx (\overline{\mathbf{u}} - \overline{\overline{\mathbf{u}}})(\overline{\mathbf{u}} - \overline{\overline{\mathbf{u}}})$$
$$C \approx (\overline{\mathbf{u}} - \overline{\overline{\mathbf{u}}})\,\overline{\overline{\mathbf{u}}} + \overline{\overline{\mathbf{u}}}\,(\overline{\mathbf{u}} - \overline{\overline{\mathbf{u}}}).$$

Adding then $L = \overline{\overline{\mathbf{u}}\,\overline{\mathbf{u}}} - \overline{\mathbf{u}}\,\overline{\mathbf{u}}$, the third, and last term in the triple decomposition $\boldsymbol{\tau} = R + C + L$, we get the Bardina model

$$\boldsymbol{\tau}(\mathbf{u}, \mathbf{u}) = \overline{\overline{\mathbf{u}\,\mathbf{u}}} - \overline{\mathbf{u}}\,\overline{\mathbf{u}} \approx (\overline{\overline{\mathbf{u}}\,\overline{\mathbf{u}}} - \overline{\overline{\mathbf{u}}}\,\overline{\overline{\mathbf{u}}}) =: \mathcal{S}_{\text{Bardina}}(\overline{\mathbf{u}}, \overline{\mathbf{u}}). \tag{8.1}$$

Thus, the model (8.1) estimates the effects of the unresolved scales by a simple extrapolation from the smallest resolved scales. *A priori* tests, *i.e.* tests in which a turbulent velocity field $\mathbf{u}$ from a DNS is explicitly filtered and the model's consistency is estimated via

$$\|\boldsymbol{\tau}(\mathbf{u}, \mathbf{u}) - \mathcal{S}_{\text{Bardina}}(\overline{\mathbf{u}}, \overline{\mathbf{u}})\| \tag{8.2}$$

have been consistently positive and, in fact, have been consistently better than analytic studies of the accuracy of the model in the irrotational flow region (specifically, for a smooth $\mathbf{u}$). For smooth $\mathbf{u}$, it is easy to show that the model's consistency error is $O(\delta^2)$. Since $\boldsymbol{\tau}(\mathbf{u}, \mathbf{u})$ is itself $O(\delta^2)$ for smooth $\mathbf{u}$, this only shows that the relative error is $O(1)$ for the Bardina model! Again, it must be emphasized that in *a priori* tests the Bardina model always appears quite accurate.

Proposition 8.1. *Suppose the averaging operator is convolution with a Gaussian. Consider the modeling consistency* (8.2) *of the Bardina model* (8.1). *For the Cauchy problem or the periodic problem, and for smooth functions* $\mathbf{u} \in H^2(\Omega)$

$$\|\boldsymbol{\tau}(\mathbf{u}, \mathbf{u}) - \mathcal{S}_{Bardina}(\overline{\mathbf{u}}, \overline{\mathbf{u}})\| \le C\delta^2 \|\mathbf{u}\|_{L^\infty} \|\mathbf{u}\|_{H^2}.$$

Proof. Adding and subtracting terms and using the triangle inequality gives

$$\|(\overline{\overline{\mathbf{u}\mathbf{u}}} - \overline{\mathbf{u}}\,\overline{\mathbf{u}}) - (\overline{\overline{\mathbf{u}}\,\overline{\mathbf{u}}} - \overline{\overline{\mathbf{u}}}\,\overline{\overline{\mathbf{u}}})\| \le \|\overline{\overline{\mathbf{u}\,(\mathbf{u} - \overline{\mathbf{u}})}}\| + \|\overline{\overline{(\mathbf{u} - \overline{\mathbf{u}})\,\overline{\mathbf{u}}}}\| + \|\overline{\overline{\mathbf{u}}(\mathbf{u} - \overline{\mathbf{u}})}\|$$
$$+ \|\overline{(\mathbf{u} - \overline{\mathbf{u}})\overline{\mathbf{u}}}\|.$$

Simple inequalities (like $\|\mathbf{f}\,\mathbf{g}\| \le \|\mathbf{f}\|_{L^\infty}\|\mathbf{g}\|$) and the basic properties of averaging ($\|\mathbf{u} - \overline{\mathbf{u}}\| \le c\,\delta^2\|\mathbf{u}\|_{H^2}$), give

$$\|\boldsymbol{\tau}(\mathbf{u}, \mathbf{u}) - \mathcal{S}_{\text{Bardina}}(\overline{\mathbf{u}}, \overline{\mathbf{u}})\| \le 4\,\|\mathbf{u}\|_{L^\infty} C\,\delta^2\|\mathbf{u}\|_{H^2}, \tag{8.3}$$

which completes the proof.   □

The Bardina closure model (8.1) gives the following system for the approximation $\mathbf{w}$ to the average velocity $\overline{\mathbf{u}}$ and q to the average pressure $\overline{p}$ in $\Omega \times (0, T)$:

$$\mathbf{w}_t + \nabla \cdot (\mathbf{w}\,\mathbf{w}) - \frac{1}{Re}\Delta\mathbf{w} + \nabla \cdot (\overline{\mathbf{w}\,\mathbf{w}} - \overline{\mathbf{w}}\,\overline{\mathbf{w}}) + \nabla q = \overline{\mathbf{f}}, \tag{8.4}$$

$$\nabla \cdot \mathbf{w} = 0, \tag{8.5}$$

which is supplemented by initial and boundary conditions. In this chapter we follow our usual procedure of studying periodic boundary conditions (2.3) (with the zero mean condition) to isolate the closure problem.

Since the Bardina model (8.4) has been observed to have very small consistency error in various *a priori* tests, its direct usefulness hinges on the analytic criterion of stability. Furthermore, since many *a posteriori* tests (*i.e.* tests using the Bardina model in the actual numerical simulations, as opposed to the *a priori* tests where only DNS data is needed) have raised some questions about the stability of the Bardina model, it is important to try to understand the energy balance of the model (8.4). To that end, we will uncouple the model (8.4) from its boundary conditions by the usual approach of studying (8.4) subject to periodic boundary conditions (2.3). In exploring the energy balance of the model, the next well-known lemma will be essential and we shall make use of the space V of divergence-free functions defined in (7.22).

Lemma 8.2. *Let* $\mathbf{u}, \mathbf{v}, \mathbf{w} \in V$. *Then,*

$$(\nabla \cdot (\mathbf{u}\,\mathbf{v}), \mathbf{w}) = -(\mathbf{u}\,\mathbf{v}, \nabla\mathbf{w}) = -(\mathbf{u} \cdot \nabla\mathbf{w}, \mathbf{v}), \tag{8.6}$$

$$(\mathbf{u} \cdot \nabla\mathbf{v}, \mathbf{w}) = -(\mathbf{u} \cdot \nabla\mathbf{w}, \mathbf{v}), \tag{8.7}$$

and thus,

$$(\mathbf{u} \cdot \nabla\mathbf{v}, \mathbf{v}) = 0. \tag{8.8}$$

Furthermore, in three dimensions (improvable in two dimensions) there is a constant $C = C(\Omega)$ *such that*

$$|(\mathbf{u} \cdot \nabla\mathbf{v}, \mathbf{w})| \leq C \,\|\nabla\mathbf{u}\| \,\|\nabla\mathbf{v}\| \,\sqrt{\|\mathbf{w}\| \,\|\nabla\mathbf{w}\|}, \tag{8.9}$$

and

$$|(\mathbf{u} \cdot \nabla\mathbf{v}, \mathbf{w})| \leq C\sqrt{\|\mathbf{u}\| \,\|\nabla\mathbf{u}\|} \,\|\nabla\mathbf{v}\| \,\|\nabla\mathbf{w}\|. \tag{8.10}$$

Proof. Essentially, we used the above equality and estimates in Chap. 7, but we collect them for simplicity and also because we shall use them several times in the present chapter. The proof of (8.6)–(8.8) follows by applying the divergence theorem. The bounds (8.9) and (8.10) follow from Holder's inequality, the Sobolev inequality and the interpolation inequality.   □

Let us use the lemma to explore the kinetic energy balance of the model (8.4). Assuming $(\mathbf{w}, q)$ to be a smooth enough solution of the Bardina model (for example, a strong solution), we can multiply (8.4) by $\mathbf{w}$ and integrate over the domain Ω. This gives

$$\int_{\Omega} \mathbf{w}_t\,\mathbf{w}\,dx + \int_{\Omega} \nabla \cdot (\mathbf{w}\,\mathbf{w})\,\mathbf{w}\,dx - \frac{1}{Re}\int_{\Omega} \Delta\mathbf{w}\,\mathbf{w}\,dx$$
$$+ \int_{\Omega} \nabla q\,\mathbf{w}\,dx + \int_{\Omega} \nabla \cdot (\overline{\mathbf{w}\,\mathbf{w}} - \overline{\mathbf{w}}\,\overline{\mathbf{w}})\,\mathbf{w}\,dx = \int_{\Omega} \overline{\mathbf{f}}\,\mathbf{w}\,dx. \tag{8.11}$$

Numbering the terms above in (8.11) I, II, ..., VI, we treat most of them exactly as in the case of deriving the energy balance of the Navier–Stokes equations in Chap. 2. Terms II and IV vanish, term III is the energy dissipation rate, and term I is the time derivative of the kinetic energy:

$$\frac{1}{2}\frac{d}{dt}\|\mathbf{w}\|^2 + \frac{1}{Re}\|\nabla\mathbf{w}\|^2 = \int_\Omega \overline{\mathbf{f}}\,\mathbf{w}\,dx - \int_\Omega \nabla\cdot(\overline{\mathbf{w}\mathbf{w}} - \overline{\mathbf{w}}\,\overline{\mathbf{w}})\,\mathbf{w}\,dx.$$

We can apply the estimates of the previous lemma to the last term on the right-hand side as follows: first, note that for a constant filter radius δ, in the absence of boundaries, and for periodic boundary conditions, differentiation and filtering commute and filtering is a self-adjoint operation, meaning

$$(\overline{\mathbf{u}}, \mathbf{v}) = (\mathbf{u}, \overline{\mathbf{v}}).$$

In particular, the above observation can be used to prove the following equalities:

$$\int_\Omega \nabla\cdot(\overline{\mathbf{w}}\,\overline{\mathbf{w}})\,\mathbf{w}\,dx = (\nabla\cdot(\overline{\mathbf{w}}\,\overline{\mathbf{w}}), \mathbf{w}) = (\overline{\nabla\cdot(\mathbf{w}\,\overline{\mathbf{w}})}, \mathbf{w}) = (\nabla\cdot(\mathbf{w}\,\overline{\mathbf{w}}), \overline{\mathbf{w}})$$

$$= (\text{by Lemma 8.2}) = -(\mathbf{w}\cdot\nabla\overline{\mathbf{w}}, \mathbf{w})$$

$$= (\mathbf{w}\cdot\nabla\mathbf{w}, \overline{\mathbf{w}} - \mathbf{w}),$$

since the extra term $(\mathbf{w}\cdot\nabla\mathbf{w}, \mathbf{w})$ is zero. Similarly,

$$\int_\Omega \nabla\cdot(\overline{\mathbf{w}}\,\overline{\mathbf{w}})\cdot\mathbf{w}\,dx = -(\overline{\mathbf{w}}\cdot\nabla\mathbf{w}, \overline{\mathbf{w}}) = (\overline{\mathbf{w}}\cdot\nabla\overline{\mathbf{w}}, \mathbf{w}).$$

Thus, by using (8.8)

$$\int_\Omega \nabla\cdot(\overline{\mathbf{w}\mathbf{w}} - \overline{\mathbf{w}}\,\overline{\mathbf{w}})\cdot\mathbf{w}\,dx = (\mathbf{w}\cdot\mathbf{w}, \overline{\mathbf{w}}) - (\overline{\mathbf{w}}\cdot\nabla\overline{\mathbf{w}}, \mathbf{w})$$

$$= (\mathbf{w}\cdot\nabla\mathbf{w}, \overline{\mathbf{w}} - \mathbf{w}) - (\overline{\mathbf{w}}\cdot\nabla\overline{\mathbf{w}}, (\mathbf{w} - \overline{\mathbf{w}})).$$

We summarize them in the following lemma.

Lemma 8.3. *Provided differentiation and filtering commute and filtering is self-adjoint, then the following identity holds:*

$$\int_\Omega \nabla\cdot(\overline{\mathbf{w}\mathbf{w}} - \overline{\mathbf{w}}\,\overline{\mathbf{w}})\cdot\mathbf{w}\,dx = (\mathbf{w}\cdot\nabla\mathbf{w}, \overline{\mathbf{w}} - \mathbf{w}) + (\overline{\mathbf{w}}\cdot\nabla\overline{\mathbf{w}}, \overline{\mathbf{w}} - \mathbf{w}).$$

Furthermore, there is a constant $C = C(\Omega)$ such that

$$\left|\int_\Omega \nabla\cdot(\overline{\mathbf{w}\mathbf{w}} - \overline{\mathbf{w}}\,\overline{\mathbf{w}})\cdot\mathbf{w}\,dx\right| \le C\,\|\nabla\mathbf{w}\|^2\,\|\mathbf{w} - \overline{\mathbf{w}}\|^{1/2}\,\|\nabla(\mathbf{w} - \overline{\mathbf{w}})\|^{1/2},$$

and

$$\left|\int_\Omega \nabla\cdot(\overline{\mathbf{w}\mathbf{w}} - \overline{\mathbf{w}}\,\overline{\mathbf{w}})\cdot\mathbf{w}\,dx\right| \le C(\Omega)\,\delta^{1/2}\|\nabla\mathbf{w}\|^3$$

$$\le C(\Omega)\,\delta^{1/4}\|\nabla\mathbf{w}\|^{11/4}\|\mathbf{w}\|^{1/4}.$$

Proof. The first identity is just a summary of the manipulations leading up to the lemma. The second one follows by applying Lemma 8.2 and

$$\|\mathbf{w} - \overline{\mathbf{w}}\|^{1/2} \leq C\, \delta^{1/2}\, \|\nabla \mathbf{w}\|^{1/2}.$$

For the third estimate, we note that $\|\mathbf{w} - \overline{\mathbf{w}}\| \leq C\delta\,\|\nabla \mathbf{w}\|$ and $\|\mathbf{w} - \overline{\mathbf{w}}\| \leq C\,\|\mathbf{w}\|$ together imply a group of estimates that interpolate between these two via, for $0 \leq \theta \leq 1$,

$$\|\mathbf{w} - \overline{\mathbf{w}}\| = \|\mathbf{w} - \overline{\mathbf{w}}\|^{\theta}\, \|\mathbf{w} - \overline{\mathbf{w}}\|^{1-\theta} \leq C\,\delta^{1-\theta}\|\nabla \mathbf{w}\|^{1-\theta}\|\mathbf{w}\|^{\theta}.$$

In particular, picking $\theta = 1/2$ and taking square roots gives $\|\mathbf{w} - \overline{\mathbf{w}}\|^{1/2} \leq C\,\delta^{1/4}\,\|\mathbf{w}\|^{1/4}\,\|\nabla \mathbf{w}\|^{1/4}$. The third estimate follows by using this in the second one. □

The two kinetic energy inequalities that follow from Lemma 8.3 are then

$$\frac{d}{dt}\|\mathbf{w}\|^2 + \frac{1}{Re}\|\nabla \mathbf{w}\|^2 \leq \frac{1}{Re}\|\overline{\mathbf{f}}\|^2 + C\,\delta^{1/2}\|\nabla \mathbf{w}\|^3,$$

$$(8.12)$$

$$\frac{d}{dt}\|\mathbf{w}\|^2 + \frac{1}{Re}\|\nabla \mathbf{w}\|^2 \leq \frac{1}{Re}\|\overline{\mathbf{f}}\|^2 + C\,\delta^{1/4}\|\mathbf{w}\|^{1/4}\|\nabla \mathbf{w}\|^{11/4}.$$

While several steps in the above derivation are improvable, the basic difficulty remains: a cubic term on the right-hand side cannot be bounded for large data by the quadratic terms on the left-hand side of either energy inequality. Indeed, this energy inequality predicts stability provided

$$\frac{1}{Re}\|\nabla \mathbf{w}\|^2 \geq C\,\delta^{1/2}\,\|\nabla \mathbf{w}\|^3,$$

that is, while

$$\|\nabla \mathbf{w}\| \leq \frac{C}{\delta^{1/2} Re}.$$

Since turbulence is essentially about high Reynolds numbers and large local changes in velocities (that is, large gradients) these conditions cannot be considered to cover interesting cases of practical turbulent flows, unless δ is small enough that $\delta = O(Re^{-1/2})$, *i.e.* the problem begins to be fully resolved.

Remark 8.4. The Bardina model can be interpreted as an approximate deconvolution method. In fact, if the filter is given by using a second-order differential approximation, then the Bardina model is equivalent to the Gradient model of Sect. 7.1. See Chap. 6.4.1 in [267] for further details. This fact is reflected in the cubic term involving the gradient in the energy estimate, *cfr.* with estimate (7.9).

8.2 Other Scale Similarity Models

The most straightforward attempt at an energy estimate for the Bardina model *just* fails. We could try multiplication of the equation by a linear combination of $\mathbf{w}$ and $\overline{\mathbf{w}}$ or using other inner products. However, the failure of direct assault plus the stability problems reported in simulations suggest that a delicate difficulty might be inherent in the model rather than a failure of analysis. While there is no rigorous mathematical proof that the kinetic energy of the Bardina model can behave catastrophically, this has been sufficient evidence to spur many attempts (see the presentation in Sagaut [267]) at improvements of scale similarity models. It is useful to review a few before proceeding.

8.2.1 Germano Dynamic Model

The most common realization of the dynamic model uses a locally weighted combination of the Smagorinsky model, which is stable but inaccurate, and Bardina's model, which is accurate but with stability problems. In effect, in Germano *et al.* [129] and Lilly [220], the Smagorinsky "constant" C_S is picked to make, in a least squares sense, the Smagorinsky model of $\boldsymbol{\tau}(\mathbf{u}, \mathbf{u})$ as close as possible to the Bardina scale similarity model of $\boldsymbol{\tau}(\mathbf{u}, \mathbf{u})$.

8.2.2 The Filtered Bardina Model

The filtered Bardina model [157] is given by

$$\boldsymbol{\tau}(\mathbf{u}, \mathbf{u}) \approx \overline{\overline{\mathbf{u}\,\overline{\mathbf{u}}}} - \overline{\overline{\mathbf{u}}\,\overline{\mathbf{u}}}.$$

Testing of the model has been performed in [14]. It is still an open problem to understand the energy balance of this model, especially the influence of the extra filtering step, and then develop a mathematical foundation for the filtered Bardina model.

The Bardina model has also been generalized by using two filters (consequently two different cut-off scales) by Liu, Meneveau, and Katz [223], obtaining

$$\boldsymbol{\tau}(\mathbf{u}, \mathbf{u}) \approx C_1 \widetilde{\overline{\mathbf{u}}\,\overline{\mathbf{u}}} - \widetilde{\overline{\mathbf{u}}}\,\widetilde{\overline{\mathbf{u}}}.$$

The cut-off length of the filter, denoted by *tilde*, is larger than that of the filter denoted by *bar*. The constant C_1 is chosen in order to ensure that the average kinetic energy is equal to the exact one. A dynamic version of this model has also been proposed in [223]. In the latter C_1 is not a constant, but it is determined dynamically with the aid of a third filter.

8.2.3 The Mixed-scale Similarity Model

In mixed models an extra p-Laplacian is added to enhance stability of the overall model. For example, for the mixed-Bardina model [14] we have

$$\tau(\mathbf{u}, \mathbf{u}) \approx (\overline{\overline{\mathbf{u}}\,\overline{\mathbf{u}}} - \overline{\overline{\mathbf{u}}}\,\overline{\overline{\mathbf{u}}}) - (C_\alpha \delta)^\alpha |\nabla^s \overline{\mathbf{u}}| \, \nabla^s \overline{\mathbf{u}}. \tag{8.13}$$

The added Smagorinsky term *does* provide stabilization. Mathematically, it is also appealing because it gives a cubic (in $\nabla^s \mathbf{w}$) term in an energy inequality. The "bad" nonlinear interactions in the Bardina model lead to cubic terms as well, and bounding cubic terms by cubic terms has hope. Experimentally, it is also a sensible approach. Difficulties with the kinetic energy evolution of the Bardina model appear to be delicate, and the Smagorinsky term in (8.13) *is* stabilizing. Thus, it is possible that, even in the worst case, the added term causes kinetic energy catastrophes to occur after a much longer time interval – perhaps long enough to be uninteresting. Admittedly, this is all speculation (and an interesting open question). If we seek *sufficient* conditions on C_α and α that ensure the mixed model is stable over $0 \le t < \infty$, then mathematical analysis can contribute some (pessimistic) conditions.

The analysis leading up to (8.12) can easily be adapted to the mixed model, provided $\alpha = 1/2$ and the constant C_α is large enough – certainly pessimistic conditions!

Proposition 8.5. *Let $\alpha = 1/2$ in (8.13) and suppose the constant C_α is large enough. Then, strong solutions of the mixed Bardina model (8.13) are stable and satisfy*

$$\frac{1}{2}\|\mathbf{w}(T)\|^2 + \int_0^T \frac{1}{Re}\|\nabla \mathbf{w}\|^2 + (C - C_\alpha)\,\delta^{1/2}\,\|\nabla^s \mathbf{w}\|_{L^3}^3 \, dt$$

$$= \frac{1}{2}\|\overline{\mathbf{u}}_0\|^2 + \int_0^T \int_\Omega \overline{\mathbf{f}}\,\mathbf{w}\,d\mathbf{x}\,dt.$$

Using this energy inequality, many interesting properties of the mixed model can be developed. Unfortunately, the case $\alpha = 1/2$ is far from the typical choice of $\alpha = 2$. An open problem is to sharpen the analysis of the mixed Bardina model to treat the case $\alpha = 2$ and large data.

8.3 Recent Ideas in Scale Similarity Models

One direction of research in scale similarity models has been to begin with the Bardina model and to develop from it successively extended or refined models with incrementally better stability properties. Each additional modeling step typically has succeeded in increasing stability but the same modeling steps also typically decrease accuracy.

There have been at least three interesting and relatively new ideas in scale similarity models that, so far, seem like good paths to derive models that are both accurate and stable:

1. explicit skew-symmetrization of the nonlinear Bardina term $\int_\Omega (\overline{\mathbf{w}\,\mathbf{w}} - \overline{\mathbf{w}}\,\overline{\mathbf{w}}) : \nabla^s \mathbf{w}\, d\mathbf{x}$ in the energy inequality and other, related skew-symmetric models;
2. a low-order accurate scale similarity model arising by dropping the cross and Reynolds terms;
3. models based on higher-order extrapolations from the resolved to the unresolved scales, such as the Stolz–Adams deconvolution models [285, 289, 290].

Skew-symmetrization

The idea of explicit skew-symmetrization is to mimic in the model the fundamental role the skew-symmetry of the nonlinear interaction terms plays in the mathematics of the NSE. Although there are different ways to develop the model, the quickest is as follows. If we define the new trilinear form

$$B(\mathbf{u}, \mathbf{v}, \mathbf{w}) := \int_\Omega (\overline{\mathbf{u}\,\mathbf{v}} - \overline{\mathbf{u}}\,\overline{\mathbf{v}}) : \nabla^s \mathbf{w}\, d\mathbf{x},$$

then its skew-symmetric part is

$$B_{\text{skew}}(\mathbf{u}, \mathbf{v}, \mathbf{w}) := \frac{1}{2} B(\mathbf{u}, \mathbf{v}, \mathbf{w}) - \frac{1}{2} B(\mathbf{u}, \mathbf{w}, \mathbf{v}).$$

It is easy to check that $B(\cdot, \cdot, \cdot) \neq B_{\text{skew}}(\cdot, \cdot, \cdot)$, so replacing $B(\cdot, \cdot, \cdot)$ by $B_{\text{skew}}(\cdot, \cdot, \cdot)$ commits a possible serious modeling error that needs to be checked. One skew-symmetrized scale similarity model arises by replacing $B(\cdot, \cdot, \cdot)$ by $B_{\text{skew}}(\cdot, \cdot, \cdot)$ in the variational formulation of the Bardina model and then working backwards to find the LES model that gives this new variational formulation. The LES model that results (and which we shall derive next) is

$$\overline{\mathbf{u}\,\mathbf{u}} - \overline{\mathbf{u}}\,\overline{\mathbf{u}} \approx \frac{1}{2} \left(\overline{\overline{\mathbf{u}}\,\overline{\mathbf{u}}} - \overline{\overline{\mathbf{u}}}\,\overline{\overline{\mathbf{u}}} + \overline{\overline{\overline{\mathbf{u}}}\,\overline{\mathbf{u}}} - \overline{\mathbf{u}}\,\overline{\overline{\mathbf{u}}} \right). \tag{8.14}$$

Unfortunately, the modeling question in (8.14) is not that simple: stability is necessary, but so is accuracy and intelligibility.

Let us develop the mathematical properties linked to the model (8.14) that arise from one-step, explicit skew symmetrization of the Bardina model. Calling $\mathbf{v}, \mathbf{w}$, and ϕ filtered quantities, the model is associated with the tensor

$$\boldsymbol{\tau}(\mathbf{v}, \mathbf{w}) := \overline{\mathbf{v}\,\mathbf{w}} - \overline{\mathbf{v}}\,\overline{\mathbf{w}}.$$

(Specifically, it is given by $\overline{\mathbf{u}\,\mathbf{u}} - \overline{\mathbf{u}}\,\overline{\mathbf{u}} \approx \boldsymbol{\tau}(\overline{\mathbf{u}}\,\overline{\mathbf{u}})$.) The variational formulation of the associated Bardina term is

$$B(\mathbf{v}, \mathbf{w}, \phi) := \int_\Omega \nabla \cdot (\overline{\mathbf{v}\,\mathbf{w}} - \overline{\mathbf{v}}\,\overline{\mathbf{w}}) \cdot \phi\, d\mathbf{x}.$$

The skew-symmetric part of B is $B^*(\mathbf{v}, \mathbf{w}, \boldsymbol{\phi}) := \frac{1}{2} B(\mathbf{v}, \mathbf{w}, \boldsymbol{\phi}) - \frac{1}{2} B(\mathbf{v}, \boldsymbol{\phi}, \mathbf{w})$ and is given by

$$B^*(\mathbf{v}, \mathbf{w}, \boldsymbol{\phi}) := \frac{1}{2} \int_\Omega \nabla \cdot (\overline{\mathbf{v}\mathbf{w}} - \overline{\mathbf{v}}\,\overline{\mathbf{w}}) \cdot \boldsymbol{\phi} - \nabla \cdot (\overline{\mathbf{v}\boldsymbol{\phi}} - \overline{\mathbf{v}}\,\overline{\boldsymbol{\phi}}) \cdot \mathbf{w} \, d\mathbf{x}.$$

Integrating by parts, the last term of the RHS repeatedly shows that for smooth enough, divergence-free, periodic functions

$$B^*(\mathbf{v}, \mathbf{w}, \boldsymbol{\phi}) := \frac{1}{2} \int_\Omega \nabla \cdot (\overline{\mathbf{v}\mathbf{w}} - \overline{\mathbf{v}}\,\overline{\mathbf{w}} + \overline{\overline{\mathbf{v}}\,\overline{\mathbf{w}}} - \overline{\mathbf{v}}\,\overline{\mathbf{w}}) \cdot \boldsymbol{\phi} \, d\mathbf{x}.$$

For LES modeling, this would correspond to the closure model (8.14) mentioned earlier.

While the RHS is recognizable as an approximation of the LHS, it is certainly an odd one. For example, the LHS tensor is symmetric, while the RHS tensor, which approximates it, is not. It may well prove that the scale similarity model (8.14) is the best/most accurate/ultimate model. However, it seems like a wrong headed attempt: an ugly model dictated by mathematical formalism.

On the other hand, the basic premise is sound: find a model which is as clear and accurate as possible among the many which lead to a skew-symmetric nonlinear interaction term. We will show a first step in this direction in Sect. 8.4. We will call the resulting model (for obvious reasons) the skew-symmetrized-scale-similarity model, or S^4 model.

The kinetic energy balance of any skew-symmetric LES model is simple and clear. Because of the skew-symmetry of the nonlinear term, it follows exactly as in the Navier–Stokes case.

Lemma 8.6. *Let $\mathbf{w}$ be a strong solution of the skew-symmetrized scale similarity model (8.14). Then, $\mathbf{w}$ satisfies*

$$\frac{1}{2}\|\mathbf{w}(t)\|^2 + \int_0^t \frac{1}{Re}\|\nabla \mathbf{w}(t')\|^2 \, dt' = \frac{1}{2}\|\overline{\mathbf{u}}_0\|^2 + \int_0^t (\overline{\mathbf{f}}, \mathbf{w}) \, dt'.$$

Proof. Multiply by $\mathbf{w}$ and integrate over Ω. Skew-symmetry of the trilinear term implies that it vanishes. The remainder follows exactly like the energy estimate for the Navier–Stokes equations.   □

A related (better) skew-symmetric scale similarity model (the S^4 model) was recently studied in Layton [203, 204]. Briefly, if one seeks a closure model which preserves structures of the true Reynolds stresses (like symmetry) *and* which yields a skew-symmetric nonlinear interaction term, the S^4 model (studied in Sect. 8.4) arises.

The model consists of finding $(\mathbf{w}, q)$ satisfying

$$\mathbf{w}_t + \nabla \cdot (\overline{\mathbf{w}}\,\overline{\mathbf{w}}) + \nabla \cdot (\overline{\mathbf{w}}(\mathbf{w} - \overline{\mathbf{w}}) + (\mathbf{w} - \overline{\mathbf{w}})\overline{\mathbf{w}})$$

$$-\nabla \cdot (\nu_T(\delta, \mathbf{w})\nabla^s \mathbf{w}) - \nabla q - \frac{1}{Re}\Delta \mathbf{w} - A(\delta)\,\mathbf{w} = \overline{\mathbf{f}} \quad \text{in } \Omega \times (0, T),$$

$$\nabla \cdot \mathbf{w} = 0 \quad \text{in } \Omega \times (0, T),$$

subject to the initial $\mathbf{w}(\mathbf{x}, 0) = \overline{\mathbf{u}}_0(\mathbf{x})$ and usual normalization conditions, while $\nu_T(\delta)$ and $A(\delta)$ are described below. The operator $A(\delta)$ takes the general form

$$A(\delta)\,\mathbf{w} = R^* \left[\nabla \cdot \nu_F(R\,[\mathbf{w}]) \right],$$

where R is a restriction defined using its variational representation:

$$(A(\delta)\,\mathbf{w}, \mathbf{v}) = -(\nu_F(\delta)\,\nabla^s(\mathbf{w} - \overline{\mathbf{w}}), \nabla^s(\mathbf{v} - \overline{\mathbf{v}})),$$

where $\nu_F(\delta)$ is the fine-scale fluctuation coefficient. It satisfies minimally the consistency condition

$$\nu_F(\delta) \;\rightarrow\; 0 \qquad \text{as} \qquad \delta \rightarrow 0.$$

There are several possibilities for the "turbulent viscosity" coefficient. The most common ones used in computational practice are $\nu_T = \nu_T(\delta) \rightarrow 0$ as $\delta \rightarrow 0$ and the Smagorinsky model [277]. We shall thus specify either

$$\nu_T = \nu_T(\delta, \mathbf{w}), \quad \text{where } \nu_T \rightarrow 0 \text{ as } \delta \rightarrow 0, \text{ uniformly in } \mathbf{w},$$

or

$$\nu_T(\delta, \mathbf{w}) = C_S \delta^2 \, |\nabla^s \mathbf{w}|.$$

Energy Sponges and Higher Order Models

The second and third new ideas of scale similarity models came about from a search in the other direction. There was a search for models that were provably *more accurate* (to be very precise: of a higher-order consistency error for smooth solutions) than the already accurate Bardina model. There were three important breakthroughs. First, Stolz and Adams [285] developed a family of models of high accuracy that performed well in practical tests. Second, we noticed in [209, 210, 211] that with the right combination of filter plus model, the simplest possible, zero-order model also satisfied a surprising and very strong stability condition: the models acted as a sort of "energy sponge". The third breakthrough was that Dunca and Epshteyn [98] proved a similar stability bound for the entire family.

Thus, the door was opened to *stable* models with *high-order accurate consistency error*.

The first model is actually simpler than the Bardina model (and, although it does not perform nearly as well in *a priori* tests, it has the same order of consistency). We will introduce the stability idea first in Sect. 8.5 for this model, before considering the more accurate models in Sect. 8.6.

These models and their stability properties lead to interesting computational and mathematical developments of LES models in new directions.

8.4 The S^4 = Skew-symmetric Scale Similarity Model

Averaging the NSE with any filter that commutes with differentiation in the absence of boundaries, gives

$$\overline{\mathbf{u}}_t + \nabla \cdot (\overline{\mathbf{u}\,\mathbf{u}}) - \nabla \overline{p} - \frac{1}{Re}\,\Delta \overline{\mathbf{u}} = \overline{\mathbf{f}}. \tag{8.15}$$

The nonlinear term $\overline{\mathbf{u}\,\mathbf{u}}$ can be expanded by using the decomposition of $\mathbf{u}$ into means ($\overline{\mathbf{u}}$) and fluctuations ($\mathbf{u}' = \mathbf{u} - \overline{\mathbf{u}}$)

$$\overline{\mathbf{u}\,\mathbf{u}} = \overline{(\overline{\mathbf{u}} + \mathbf{u}')(\overline{\mathbf{u}} + \mathbf{u}')} = \overline{\overline{\mathbf{u}}\,\overline{\mathbf{u}}} + \overline{\overline{\mathbf{u}}\mathbf{u}'} + \overline{\mathbf{u}'\overline{\mathbf{u}}} + \overline{\mathbf{u}'\mathbf{u}'} \tag{8.16}$$

into the resolved term, cross-terms and turbulent fluctuations. Motivated by the scale similarity ideas, the S^4 model introduced in [203] arises from (8.16) as follows: first, the turbulent fluctuations in (8.16) are modeled by the Boussinesq hypothesis that they are dissipative in the mean, giving

$$\overline{\mathbf{u}'\mathbf{u}'} \approx \nu_T(\delta, \overline{\mathbf{u}})\,\nabla^s \overline{\mathbf{u}}, \tag{8.17}$$

where $\nu_T(\delta, \overline{\mathbf{u}})$ is the turbulent viscosity coefficient. Then, the cross-terms in (8.16) are modeled by scale similarity:

$$\overline{\overline{\mathbf{u}}\mathbf{u}'} + \overline{\mathbf{u}'\overline{\mathbf{u}}} = \overline{\overline{\mathbf{u}}(\mathbf{u} - \overline{\mathbf{u}})} + \overline{(\mathbf{u} - \overline{\mathbf{u}})\overline{\mathbf{u}}} \approx \overline{\overline{\mathbf{u}}}(\overline{\mathbf{u}} - \overline{\overline{\mathbf{u}}}) + (\overline{\mathbf{u}} - \overline{\overline{\mathbf{u}}})\,\overline{\overline{\mathbf{u}}}. \tag{8.18}$$

Finally, the key step which gives a skew-symmetric interaction term is to model the resolved term. We consider an EV model of the resolved term given by

$$\overline{\overline{\mathbf{u}}\,\overline{\mathbf{u}}} \approx \overline{\overline{\mathbf{u}}}\,\overline{\overline{\mathbf{u}}} \; + \; \text{dissipative mechanism on } O(\delta) \text{ scales}.$$

Specifically, we write

$$\nabla \cdot (\overline{\overline{\mathbf{u}}\,\overline{\mathbf{u}}}) \approx \overline{\overline{\mathbf{u}}}\,\overline{\overline{\mathbf{u}}} - A(\delta)\,\overline{\mathbf{u}}. \tag{8.19}$$

The operator $A(\delta) : H_0^1 \to H^{-1}$ has the following variational representation:

$$-(A(\delta)\,\mathbf{w}, \mathbf{v}) = (\nu_F(\delta)\nabla^s(\mathbf{w} - \overline{\mathbf{w}}),\, \nabla^s(\mathbf{v} - \overline{\mathbf{v}})).$$

Using this, we can obtain a more concrete representation for $A(\delta)$. Indeed, integrating by parts and exploiting self-adjointness of the averaging operator, gives

$$-(A(\delta)\,\mathbf{w}, \mathbf{v}) = \left(-\nabla \cdot \left[\nu_F(\delta)\,\nabla^s(\mathbf{w} - \overline{\mathbf{w}}) - \overline{\nu_F(\delta)\,\nabla^s(\mathbf{w} - \overline{\mathbf{w}})}\right], \mathbf{v}\right)$$

$$= (\text{when } \nu_F(\delta) \text{ is independent of } \mathbf{w})$$

$$= (-\nabla \cdot \nu_F(\delta)\,\nabla^s(\mathbf{w} - 2\overline{\mathbf{w}} + \overline{\overline{\mathbf{w}}}), \mathbf{v}).$$

Thus, $A(\delta)$ can be thought of as representing the operator $\nabla \cdot \widehat{A}(\delta)\,\mathbf{w}$, where

$$\widehat{A}(\delta)\,\mathbf{w} \approx (\nu_F(\delta)\,\nabla^s((\mathbf{w} - \overline{\mathbf{w}}) - (\overline{\mathbf{w}} - \overline{\overline{\mathbf{w}}}))). \tag{8.20}$$

To complete the model's specification, insert (8.17), (8.18), (8.19), (8.20) into (8.16), then into (8.15), and call $(\mathbf{w}, q)$ the resulting approximations to $(\overline{\mathbf{u}}, \overline{p})$. This gives

$$\mathbf{w}_t + \nabla \cdot (\overline{\mathbf{w}}\,\overline{\mathbf{w}}) - \nabla q - \tfrac{1}{Re}\,\Delta \mathbf{w} - \widehat{A}(\delta)\,\mathbf{w}$$
$$+ \nabla \cdot (\overline{\mathbf{w}\,(\mathbf{w} - \overline{\mathbf{w}})} + (\mathbf{w} - \overline{\mathbf{w}})\,\overline{\mathbf{w}} - \nu_T \nabla^s \mathbf{w}) = \overline{\mathbf{f}}, \tag{8.21}$$
$$\nabla \cdot \mathbf{w} = 0. \tag{8.22}$$

It will be useful later to rewrite the SFNSE (8.15) in a more convenient form. Adding and subtracting terms gives

$$\overline{\mathbf{u}}_t + \nabla \cdot (\overline{\mathbf{u}}\,\overline{\mathbf{u}}) - \frac{1}{Re}\Delta\overline{\mathbf{u}} + \nabla \cdot \left(\overline{\overline{\mathbf{u}}\,(\mathbf{u} - \overline{\mathbf{u}})} + (\mathbf{u} - \overline{\mathbf{u}})\,\overline{\mathbf{u}} - \nu_T \nabla^s \mathbf{w}\right)$$
$$- \nabla \overline{p} - \nabla \cdot (\widetilde{T} - \overline{\mathbf{u}\,\mathbf{u}}) = \overline{\mathbf{f}} - \nabla \cdot (\widetilde{T}),$$
$$\nabla \cdot \overline{\mathbf{u}} = 0.$$

The tensor $\widetilde{T}$ that approximates $\overline{\mathbf{u}\,\mathbf{u}}$ is given by

$$\widetilde{T}(\mathbf{u}, \mathbf{u}) = \left[\overline{\mathbf{u}}\,\overline{\mathbf{u}} - \widehat{A}(\delta)\,\overline{\mathbf{u}} + \overline{\overline{\mathbf{u}}(\mathbf{u} - \overline{\mathbf{u}})} + (\mathbf{u} - \overline{\mathbf{u}})\overline{\mathbf{u}} - \nu_T \nabla^s \overline{\mathbf{u}}\right].$$

Thus, the magnitude of the tensor difference $\overline{\mathbf{u}\,\mathbf{u}} - \widetilde{T}(\mathbf{u}, \mathbf{u})$ is a measure of the accuracy of the modeling steps employed, *i.e.* the model's consistency error.

We will consider the model (8.21) under periodic boundary conditions for the usual reason of uncoupling the analysis of the interior closure problem from the problems associated with walls.

8.4.1 Analysis of the Model

In an exactly analogous way to the NSE, it is easy to derive the weak formulation of the S^4-model. Indeed, a weak solution $\mathbf{w} : [0, T] \to V$ of the S^4-large eddy model satisfies, $\forall \mathbf{v} \in L^2(0, T; V)$

$$\int_0^t \Big[(\mathbf{w}_t, \mathbf{v}) - (\overline{\mathbf{w}}\,\overline{\mathbf{w}}, \nabla \mathbf{v}) + (\nu_F(\delta)\nabla^s(\mathbf{w} - \overline{\mathbf{w}}), \nabla^s(\mathbf{v} - \overline{\mathbf{v}}))$$
$$- (\overline{\mathbf{w}\,(\mathbf{w} - \overline{\mathbf{w}})} + (\mathbf{w} - \overline{\mathbf{w}})\,\overline{\mathbf{w}}, \nabla\overline{\mathbf{v}}) + (\nu_F(\delta, \mathbf{w})\nabla^s \mathbf{w}, \nabla^s \mathbf{v})$$
$$+ \frac{1}{Re}(\nabla \mathbf{w}, \nabla \mathbf{v})\Big]\,dt' = \int_0^t (\overline{\mathbf{f}}, \mathbf{v})\,dt',$$

where H and V denote respectively the subspaces of L^2 and H^1 of divergence-free, periodic functions, with zero mean; see p. 159 for further details.

Let $b(\mathbf{u}, \mathbf{v}, \mathbf{w})$ denote the (nonstandard) trilinear form

$$b(\mathbf{u}, \mathbf{v}, \mathbf{w}) := -\int_\Omega \overline{\mathbf{u}}\,\overline{\mathbf{v}}\,\nabla\mathbf{w} + \left[\overline{\mathbf{u}}\,(\mathbf{v} - \overline{\mathbf{v}}) + (\mathbf{u} - \overline{\mathbf{u}})\,\overline{\mathbf{v}}\right]\nabla\overline{\mathbf{w}}\,dx. \quad (8.23)$$

It satisfies the following property.

Lemma 8.7. *The trilinear form (8.23) is skew-symmetric:*

$$b(\mathbf{u}, \mathbf{v}, \mathbf{w}) = -b(\mathbf{u}, \mathbf{w}, \mathbf{v}), \quad \text{and thus} \quad b(\mathbf{u}, \mathbf{v}, \mathbf{v}) = 0, \qquad \forall\,\mathbf{u}, \mathbf{v}, \mathbf{w} \in V.$$

Proof. This is a simple calculation given in [203]. First, notice that integrating by parts gives

$$b(\mathbf{u}, \mathbf{v}, \mathbf{w}) = \int_\Omega \overline{\mathbf{u}} \cdot \nabla\mathbf{w} \cdot \overline{\mathbf{v}} + \overline{\mathbf{u}} \cdot \nabla\overline{\mathbf{w}} \cdot (\mathbf{v} - \overline{\mathbf{v}}) + (\mathbf{u} - \overline{\mathbf{u}}) \cdot \nabla\overline{\mathbf{w}} \cdot \overline{\mathbf{v}}\,dx.$$

We will use repeatedly the fact that $(\boldsymbol{\phi} \cdot \nabla\psi, \boldsymbol{\eta}) = -(\boldsymbol{\phi} \cdot \nabla\boldsymbol{\eta}, \psi)$ for all $\boldsymbol{\phi}, \boldsymbol{\eta}, \psi \in V$. Consider $\boldsymbol{\Phi} = b(\mathbf{u}, \mathbf{v}, \mathbf{w}) + b(\mathbf{u}, \mathbf{w}, \mathbf{v})$ which we seek to prove to vanish. We have

$$\boldsymbol{\Phi} = \int_\Omega \overline{\mathbf{u}} \cdot \nabla\mathbf{w} \cdot \overline{\mathbf{v}} + \overline{\mathbf{u}} \cdot \nabla\overline{\mathbf{w}}\,\mathbf{v} - \overline{\mathbf{u}} \cdot \nabla\overline{\mathbf{w}}\,\overline{\mathbf{v}} + \mathbf{u} \cdot \nabla\overline{\mathbf{w}} \cdot \overline{\mathbf{v}} - \overline{\mathbf{u}} \cdot \nabla\overline{\mathbf{w}} \cdot \overline{\mathbf{v}}$$
$$+ \overline{\mathbf{u}} \cdot \nabla\mathbf{v} \cdot \overline{\mathbf{w}} + \overline{\mathbf{u}} \cdot \nabla\overline{\mathbf{v}} \cdot \mathbf{w} - \overline{\mathbf{u}} \cdot \nabla\overline{\mathbf{v}} \cdot \overline{\mathbf{w}} + \mathbf{u} \cdot \nabla\overline{\mathbf{v}}\,\overline{\mathbf{w}} - \overline{\mathbf{u}} \cdot \nabla\overline{\mathbf{v}}\,\overline{\mathbf{w}}\,dx.$$

Canceling the obvious terms gives that $\boldsymbol{\Phi} = 0$ and consequently $b(\cdot, \cdot, \cdot)$ is skew-symmetric. $\square$

Using skew-symmetry of the nonlinear interaction term, it is possible to prove both existence of weak solutions to the model and an energy inequality for the model. (The proof is given in detail in M. Kaya [185] and is an adaptation of the NSE case, so we will omit it here.)

Theorem 8.8. *Consider the S^4-model (8.21) subject to periodic boundary conditions. Let $\nu_T(\delta), \nu_F(\delta)$ be nonnegative constants, and let averaging be with convolution by a Gaussian or a differential filter. Then, for any $\mathbf{u}_0 \in H$ and $\mathbf{f} \in L^2(0, T; H)$ there exists a weak solution to the model (8.21). Any weak solution satisfies the energy inequality*

$$\frac{1}{2}\|\mathbf{w}(t)\|^2 + \int_0^t \frac{1}{Re}\|\nabla\mathbf{w}\|^2 + \nu_T\|\nabla^s\mathbf{w}\|^2 + \nu_F\|\nabla^s(\mathbf{w} - \overline{\mathbf{w}})\|^2\,dt'$$

$$\leq \frac{1}{2}\|\mathbf{u}_0\|^2 + \int_0^t (\overline{\mathbf{f}}(t'), \mathbf{w}(t'))\,dt'.$$

Proof. See [185]. $\square$

The above proof does not take full advantage of the smoothing properties of the mentioned averaging operators. For a smoothing averaging process (hence *not* for the box filter) we expect a better result. Indeed, this existence theorem can be sharpened. (We leave the next proof as a technical exercise.)

Theorem 8.9. *Under the assumptions of the previous theorem, (i) the weak solution of the model (8.19) is a unique solution; (ii) if $\mathbf{u}_0 \in C^\infty(\Omega) \cap H$ and $\mathbf{f} \in C^\infty(\Omega \times (0,T))$, then the unique strong solution is also smooth:*

$$\mathbf{w} \in C^\infty(\Omega \times (0,T));$$

(iii) the energy inequality in the previous theorem is, in fact, an equality.

8.4.2 Limit Consistency and Verifiability of the S^4 Model

We consider herein the S^4 model and the question of limit consistency; supposing $\nu_T(\delta)$ and $\nu_F(\delta)$ vanish as $\delta \to 0$, does $\mathbf{w}(\delta) \to \mathbf{u}_{NSE}$ as $\delta \to 0$? Further, does the error in the model satisfy

$$\|\mathbf{w} - \mathbf{u}_{NSE}\| \le C \|\widetilde{T}(\mathbf{u}, \mathbf{u}) - \overline{\mathbf{u}}\,\overline{\mathbf{u}}\| + o(1) \quad \text{as} \quad \delta \to 0 ?$$

We still restrict our attention to the periodic with zero mean boundary conditions. For this model we show the solution $\mathbf{w}$ to the model for $\overline{\mathbf{u}}$ converges to $\mathbf{u}$ as the averaging radius $\delta \to 0$. We also show that the error $\|\overline{\mathbf{u}} - \mathbf{w}\|$ is bounded by the modeling error (perhaps better termed "modeling residual"), *evaluated on the true solution* $\mathbf{u}$. This last bound suggests one path to validating the model in either computational or physical experiments.

Let us try to first prove limit consistency by direct assault. The first step will be to derive an equation for $\phi = \mathbf{u} - \mathbf{w}$. To this end, rewrite the above equation for $\mathbf{u}$ as

$$\int_0^t \left[(\mathbf{u}_t, \mathbf{v}) + b(\mathbf{u}, \mathbf{u}, \mathbf{v}) + (\nu_F(\delta)\nabla^s(\mathbf{u} - \overline{\mathbf{u}}), \nabla^s(\mathbf{v} - \overline{\mathbf{v}})) + \frac{1}{Re}(\nabla \mathbf{u}, \nabla \mathbf{v}) \right] dt'$$

$$= \int_0^t [(\mathbf{f}, \mathbf{v}) + b(\mathbf{u}, \mathbf{u}, \mathbf{v}) - (\mathbf{u}\,\mathbf{u}, \nabla \mathbf{v}) + (\nu_F(\delta)\nabla^s(\mathbf{u} - \overline{\mathbf{u}}), \nabla^s(\mathbf{v} - \overline{\mathbf{v}}))] \, dt'.$$

Subtracting the equation satisfied by $\mathbf{w}$ from this, gives

$$\int_0^t (\phi_t, \mathbf{v}) + b(\mathbf{u}, \mathbf{u}, \mathbf{v}) - b(\mathbf{w}, \mathbf{w}, \mathbf{v}) + (\nu_F(\delta)\nabla^s(\phi - \overline{\phi}), \nabla^s(\mathbf{v} - \overline{\mathbf{v}}))$$

$$+ \frac{1}{Re}(\nabla \phi, \nabla \mathbf{v}) + (\nu_T \nabla^s \mathbf{w}, \nabla^s \mathbf{v}) \, dt' \tag{8.24}$$

$$= \int_0^t (\mathbf{f} - \overline{\mathbf{f}}, \mathbf{v}) + [b(\mathbf{u}, \mathbf{u}, \mathbf{v}) - (\mathbf{u}\,\mathbf{u}, \nabla \mathbf{v})] + (\nu_F(\delta)\nabla^s(\mathbf{w} - \overline{\mathbf{w}}), \nabla^s(\mathbf{v} - \overline{\mathbf{v}})) \, dt'.$$

Next, we need the following result on the trilinear form $b(\cdot, \cdot, \cdot)$.

Lemma 8.10. *The trilinear form $b(\cdot, \cdot, \cdot)$ is skew-symmetric. It satisfies the following bound in two or three dimensions:*

$$|b(\mathbf{u}, \mathbf{v}, \mathbf{w})| \le C \left[\|\overline{\mathbf{u}}\|^{1/2}\|\nabla\overline{\mathbf{u}}\|^{1/2}\|\nabla \mathbf{w}\| \, \|\nabla\overline{\mathbf{v}}\| \right.$$

$$+ \|\overline{\mathbf{u}}\|^{1/2}\|\nabla\overline{\mathbf{u}}\|^{1/2}\|\nabla(\mathbf{v} - \overline{\mathbf{v}})\| \, \|\nabla\overline{\mathbf{w}}\|$$

$$\left. + \|\overline{\mathbf{v}}\|^{1/2}\|\nabla\overline{\mathbf{v}}\|^{1/2}\|\nabla(\mathbf{u} - \overline{\mathbf{u}})\| \, \|\nabla\overline{\mathbf{w}}\| \right], \quad \forall \mathbf{u}, \mathbf{v}, \mathbf{w} \in V.$$

Proof. Skew-symmetry was proven in [203]. The above bound follows directly from analogous estimates on the $(\mathbf{u} \cdot \nabla \mathbf{v}, \mathbf{w})$ term, occurring in the usual Navier–Stokes case (see [121] and calculations in Chap. 7).   □

Next, we show that the solution of the S^4 model satisfies

$$\mathbf{w}(\mathbf{x},t) \rightarrow \mathbf{u}(\mathbf{x},t) \quad \text{as} \quad \delta \rightarrow 0,$$

provided $\mathbf{u} \in L^r(0,t;L^s(\Omega))$ for some r and s satisfying Serrin's uniqueness criteria (2.31) for solutions of the NSE.

Theorem 8.11. *Let* $\mathbf{u}, \mathbf{w}$ *be strong solutions of the NSE and the model* (8.21) *respectively,* $\mathbf{u}_0 \in H$, *and* $\mathbf{f} \in L^2(0,T;H)$. *Let*

$$\mathbf{u} \in L^r(0,t;L^s(\Omega)) \text{ for some } r, s \text{ satisfying } \frac{3}{s} + \frac{2}{r} = 1 \text{ and } s \in [3,\infty[.$$

Then, for $0 < T < \infty$

$$\mathbf{w} \rightarrow \mathbf{u}, \text{ as } \delta \rightarrow 0, \quad \text{in } L^\infty(0,T;L^2(\Omega)) \cap L^2(0,T;H^1(\Omega)).$$

Remark 8.12. This "consistency in the limit" result is a fundamental mathematical requirement for model consistency, yet (to the authors' knowledge) there are few models for which it has been rigorously proven; see Foiaş, Holm, and Titi [110] and Berselli and Grisanti [31]. The condition that $\mathbf{u} \in L^r(0,T;L^s(\Omega))$ for these r and s is a central open question in three dimensions. This assumption implies uniqueness of weak solutions in $\mathbb{R}^3$ (see Chap. 2 in this book, or the presentations in Ladyžhenskaya [196], Galdi [121], and Serrin [275]).

Proof. Setting $\mathbf{v} = \boldsymbol{\phi}$ in (8.24) gives

$$\frac{1}{2}\|\boldsymbol{\phi}(t)\|^2 + \int_0^t \frac{1}{Re}\|\nabla\boldsymbol{\phi}\|^2 + \nu_F(\delta)\|\nabla^s(\boldsymbol{\phi} - \overline{\boldsymbol{\phi}})\|^2 + (\nu_F \nabla^s \mathbf{w}, \nabla^s \boldsymbol{\phi})\, dt'$$

$$= \frac{1}{2}\|\boldsymbol{\phi}(0)\|^2 + \int_0^t (\mathbf{f} - \overline{\mathbf{f}}, \boldsymbol{\phi}) + \left[b(\mathbf{w},\mathbf{w},\boldsymbol{\phi}) - b(\mathbf{u},\mathbf{u},\boldsymbol{\phi})\right] \qquad (8.25)$$

$$+ \left(\nu_F(\delta)\nabla^s(\mathbf{u} - \overline{\mathbf{u}}), \nabla^s(\boldsymbol{\phi} - \overline{\boldsymbol{\phi}})\right) + \left[b(\mathbf{u},\mathbf{u},\boldsymbol{\phi}) - (\mathbf{u}\,\mathbf{u}, \nabla\boldsymbol{\phi})\right] dt'.$$

The first bracketed term on the RHS simplifies to

$$\int_0^t b(\mathbf{w},\mathbf{w},\boldsymbol{\phi}) - b(\mathbf{u},\mathbf{u},\boldsymbol{\phi})\, dt' = \int_0^t b(\boldsymbol{\phi},\mathbf{u},\boldsymbol{\phi})\, dt',$$

by skew-symmetry of $b(\cdot,\cdot,\cdot)$. The other terms on the RHS can be bounded as follows:

$$\int_0^t (\mathbf{f} - \overline{\mathbf{f}}, \phi)\, dt' \le \int_0^t \epsilon \, \|\nabla\phi\|^2 + \frac{C}{\epsilon}\, \|\mathbf{f} - \overline{\mathbf{f}}\|_{-1}^2\, dt',$$

$$\int_0^t (\nu_F(\delta)\nabla^s(\mathbf{u} - \overline{\mathbf{u}}), \nabla^s(\phi - \overline{\phi}))\, dt' \le \frac{1}{2}\nu_F(\delta) \int_0^t \|\nabla^s(\mathbf{u} - \overline{\mathbf{u}})\|^2\, dt'$$

$$+ \frac{1}{2}\nu_F(\delta) \int_0^t \|\nabla^s(\phi - \overline{\phi})\|^2\, dt'.$$

Inserting these bounds into (8.25) gives

$$\frac{1}{2}\|\phi(t)\|^2 + \int_0^t \left(\frac{1}{Re} - \epsilon\right) \|\nabla\phi\|^2 + \frac{1}{2}\nu_F(\delta) \|\nabla^s(\phi - \overline{\phi})\|^2\, dt'$$

$$\le \frac{C}{\epsilon} \int_0^t \|\mathbf{f} - \overline{\mathbf{f}}\|_{-1}^2\, dt' + \frac{1}{2}\nu_F(\delta) \int_0^t \|\nabla^s(\mathbf{u} - \overline{\mathbf{u}})\|^2\, dt' \qquad (8.26)$$

$$- \int_0^t (\nu_T\nabla^s\mathbf{w}, \nabla^s\phi)\, dt' + \int_0^t b(\phi, \mathbf{u}, \phi)\, dt' + \int_0^t \left[b(\mathbf{u}, \mathbf{u}, \phi) - (\mathbf{u}\,\mathbf{u}, \nabla\phi)\right] dt'.$$

Note that

$$\left| \int_0^t (\nu_T\nabla^s\mathbf{w}, \nabla^s\phi)\, dt' \right| \le \epsilon \int_0^t \|\nabla\phi\|^2 dt' + \frac{C(\epsilon)\,\delta^2}{Re} \int_0^t \|\nabla\mathbf{w}\|^2 dt',$$

where $\nu_T(\delta) \to 0$ as $\delta \to 0$. Note further that, due to the *a priori* bounds in the energy estimates for solutions of the S^4 model,

$$\frac{\delta}{Re} \int_0^t \|\nabla\mathbf{w}\|^2 dt' \to 0 \quad \text{as} \quad \delta \to 0.$$

Thus, (8.26) becomes

$$\frac{1}{2}\|\phi(t)\|^2 + \int_0^t \left(\frac{1}{Re} - 2\epsilon\right) \|\nabla\phi\|^2 + \frac{1}{2}\nu_F(\delta) \|\nabla^s(\phi - \overline{\phi})\|^2 dt'$$

$$\le \int_0^t \frac{C}{\epsilon} \|\mathbf{f} - \overline{\mathbf{f}}\|_{-1}^2 + \frac{1}{2}\nu_F(\delta) \|\nabla^s(\mathbf{u} - \overline{\mathbf{u}})\|^2 + C(\epsilon)\,\nu_{\max}(\delta) \|\nabla\mathbf{w}\|^2 dt'$$

$$+ \int_0^t b(\phi, \mathbf{u}, \phi))dt' - \int_0^t ((\mathbf{u} - \overline{\mathbf{u}}) \cdot \nabla\phi, \mathbf{u} - \overline{\mathbf{u}})\, dt', \qquad (8.27)$$

where the last term was simplified using the identity

$$\int_0^t b(\mathbf{u}, \mathbf{u}, \phi) - (\mathbf{u}\,\mathbf{u}, \nabla\phi)\, dt' = - \int_0^t ((\mathbf{u} - \overline{\mathbf{u}}) \cdot \nabla\phi, \mathbf{u} - \overline{\mathbf{u}})\, dt'.$$

Consider this last term. We wish to show that, modulo a term which can be dominated by the $\int_0^t \|\nabla\phi\|^2 dt'$ term on the LHS, it approaches zero as $\delta \to 0$. To this end, we will use an inequality originally due to Serrin in 1963 [275] in the form presented in Galdi ([121]; Lemma 4.1, page 30). Specifically, in 3D (improvable in 2D), for any r and s satisfying $3/s + 2/r = 1$, $s \in [3, \infty[$,

$$\left| \int_0^t ((\mathbf{u} - \overline{\mathbf{u}}) \cdot \nabla \phi, \mathbf{u} - \overline{\mathbf{u}}) \, dt' \right| \leq$$

$$C \left[\int_0^t \|\nabla \phi\|^2 \, dt' \right]^{1/2} \left[\int_0^t \|\nabla (\mathbf{u} - \overline{\mathbf{u}})\|^2 \, dt' \right]^{3/2s} \left[\int_0^t \|\mathbf{u} - \overline{\mathbf{u}}\|_{L^s}^r \|\mathbf{u} - \overline{\mathbf{u}}\|^2 dt' \right]^{1/r}.$$

Elementary inequalities then imply that, for any $\epsilon > 0$,

$$\left| \int_0^t ((\mathbf{u} - \overline{\mathbf{u}}) \cdot \nabla \phi, \mathbf{u} - \overline{\mathbf{u}}) dt' \right| \leq \epsilon \int_0^t \|\nabla \phi\|^2 dt' +$$

$$C(\epsilon) \left[\int_0^t \|\nabla (\mathbf{u} - \overline{\mathbf{u}})\|^2 dt' \right]^{3/s} \left(\sup_{0 \leq t \leq t'} \|\mathbf{u} - \overline{\mathbf{u}}\|^{4/r} \right) \left[\int_0^t \|\mathbf{u} - \overline{\mathbf{u}}\|_{L^s}^r dt' \right]^{2/r}.$$

By the first *a priori* estimate for $\mathbf{u}$ and elementary properties of mollifiers, we obtain that $\int_0^t \|\nabla (\mathbf{u} - \overline{\mathbf{u}})\|^2 \, dt' \to 0$, as $\delta \to 0$, and $\sup_{0 \leq t \leq t'} \|\mathbf{u} - \overline{\mathbf{u}}\| \to 0$, as $\delta \to 0$. Furthermore, by assumption $\mathbf{u} \in L^r(0, t; L^s(\Omega))$. Thus, the term $\int_0^t \|\mathbf{u} - \overline{\mathbf{u}}\|_{L^s}^r dt' \to 0$ as $\delta \to 0$.

Inserting these into (8.27) gives

$$\frac{1}{2} \|\phi(t)\|^2 + \int_0^t \left[\left(\frac{1}{Re} - 3\epsilon \right) \|\nabla \phi\|^2 + \frac{1}{2} \nu_F(\delta) \|\nabla^s (\phi - \overline{\phi})\|^2 \right] dt'$$

$$\leq \int_0^t b(\phi, \mathbf{u}, \phi) \, dt' + Z(\delta),$$

where $Z(\delta)$ denotes all the remaining terms, which vanish as $\delta \to 0$.

Consider the remaining term $\int_0^t b(\phi, \mathbf{u}, \phi) \, dt'$. First, note that

$$\int_0^t b(\phi, \mathbf{u}, \phi) \, dt' = -\int_0^t (\phi \cdot \nabla \overline{\phi}, \mathbf{u}) + (\overline{\phi} \cdot \nabla \overline{\phi}, \mathbf{u} - \overline{\mathbf{u}}) + ((\phi - \overline{\phi}) \cdot \nabla \overline{\phi}, \overline{\mathbf{u}}) \, dt'.$$

Applying Serrin's inequality term by term, gives

$$\left| \int_0^t b(\phi, \mathbf{u}, \phi) \, dt' \right| \leq C \left(\int_0^t \|\nabla \phi\|^2 \, dt' \right)^{1 - \frac{1}{r}} \cdot \left(\int_0^t \|\mathbf{u}\|_{L^s}^r \|\phi\|^2 \, dt' \right)^{1/r}.$$

(This bound is, of course, improvable, but this form suffices for our purposes here.) Thus,

$$\left| \int_0^t b(\phi, \mathbf{u}, \phi) \, dt' \right| \leq \epsilon \int_0^t \|\nabla \phi\|^2 \, dt' + C(\epsilon) \int_0^t \|\mathbf{u}\|_{L^s}^r \|\phi\|^2 \, dt'.$$

Picking $\epsilon = 1/(8 \, Re)$ and inserting this in (8.28), gives

$$\frac{1}{2} \|\phi(t)\|^2 + \int_0^t \left[\frac{1}{2} \frac{1}{Re} \|\nabla \phi\|^2 + \frac{1}{2} \nu_F(\delta) \|\nabla^s (\phi - \overline{\phi})\|^2 \right] dt'$$

$$\leq C(Re) \int_0^t \|\mathbf{u}\|_{L^s}^r \|\phi\|^2 dt' + Z(\delta).$$

The first result of the theorem now follows from Gronwall's lemma.

For the second result, subtract the equation for $\overline{\mathbf{u}}$ and $\mathbf{w}$, multiply by $\phi = \overline{\mathbf{u}} - \mathbf{w}$, and integrate over Ω. This gives, as before,

$$
\frac{1}{2}\|\phi(t)\|^2 + \int_0^t \left(\frac{1}{Re} + \nu_T\right)\|\nabla\phi\|^2 dt'
$$
$$
= \frac{1}{2}\|\phi(0)\|^2 + \int_0^t \left[b(\mathbf{w}, \mathbf{w}, \phi) - b(\mathbf{u}, \mathbf{u}, \phi) + (\widetilde{T} - \overline{\mathbf{u}\,\mathbf{u}}, \nabla^s\phi)\right] dt'.
$$

The difference $\int_0^t b(\mathbf{w}, \mathbf{w}, \phi) - b(\mathbf{u}, \mathbf{u}, \phi)\, dt'$ is treated exactly as in the last proof, while the last term is bounded by

$$
|(\widetilde{T} - \overline{\mathbf{u}\,\mathbf{u}}, \nabla^s\phi)| \le \frac{\epsilon}{2}\|\nabla\phi\|^2 + \frac{1}{2\epsilon}\|\widetilde{T} - \overline{\mathbf{u}\,\mathbf{u}}\|^2.
$$

Inserting this and applying Gronwall's lemma yields the result.   □

To conclude this section, we prove the verifiability.

Theorem 8.13. *Let $\mathbf{u}, \mathbf{w}$ be strong solutions of the NSE and the S^4 model, respectively. Under the assumptions of the previous theorem, for any $t \in (0, T]$,*

$$
\frac{1}{2}\|\overline{\mathbf{u}}(t) - \mathbf{w}(t)\|^2 + \left(\frac{1}{Re} + \nu_T(\delta)\right)\int_0^t \|\nabla(\overline{\mathbf{u}} - \mathbf{w})\|^2\, dt'
$$
$$
\le C^* \int_0^t \|\overline{\mathbf{u}\,\mathbf{u}} - \widetilde{T}(\mathbf{u}, \mathbf{u})\|^2\, dt',
$$

where $C^* = C^*(Re, \|\mathbf{u}\|_{L^r(0,T;L^s)})$.

Proof. We just sketch out this results, whose proof follows the same path of the previous (for full details, see [185]). Subtracting and multiplying by ϕ, gives

$$
\frac{1}{2}\|\phi(t)\|^2 + \int_0^t \left[(\nu_T(\delta)\, 2\,\nabla^s\phi, \nabla^s\phi) + \frac{1}{Re}\|\nabla\phi\|^2\right] dt'
$$
$$
= \frac{1}{2}\|\phi(0)\|^2 + \int_0^t \left[b(\mathbf{w}, \mathbf{w}, \phi) - b(\mathbf{u}, \mathbf{u}, \phi) - (\widetilde{T} - \overline{\mathbf{u}\,\mathbf{u}}, \nabla^s\phi)\right] dt'.
$$

Using this result and proceeding exactly as in the previous proof, completes the proof of verifiability.   □

Conclusions on the S^4 Model

The ultimate test of an LES model is naturally how close its predicted velocity $\mathbf{w}$ matches $\overline{\mathbf{u}}$. Such studies are difficult and scarce, as noted by Jimenez [172]. Thus, it is also interesting to seek qualitative (analytical) tests for reasonableness. Since the kinetic energy in $\overline{\mathbf{u}}$ is provably finite for all time, one such test is that the kinetic energy in the model is provably bounded, which the S^4 model satisfies.

Since $\overline{\mathbf{u}} \to \mathbf{u}$ as $\delta \to 0$, another test is that $\mathbf{w} \to \mathbf{u}$ as $\delta \to 0$. This condition was also established for the S^4 model.

Further, we also show that the difference for this model between $\mathbf{w}$ and $\overline{\mathbf{u}}$ is bounded by a residual type modeling error term evaluated on the solution. Thus, the quantitative accuracy of the model $\mathbf{w}$ as an approximation for $\overline{\mathbf{u}}$ can be evaluated by estimating the L^2 norm of this residual in either a (moderate Re) direct numerical simulation of $\mathbf{u}$, or by data from observations of real flows. Thus, the analytic information suggests that the S^4 model is a reasonable attempt. Its accuracy in practical settings remains an open question for which computational studies are needed.

8.5 The First Energy-sponge Scale Similarity Model

Recall that the Bardina models approximation to $\boldsymbol{\tau}(\mathbf{u}, \mathbf{u})$ is given by

$$(\overline{\mathbf{u}\,\mathbf{u}} - \overline{\mathbf{u}}\,\overline{\mathbf{u}}) \approx \mathcal{S}_{\text{Bardina}}(\overline{\mathbf{u}}, \overline{\mathbf{u}}) := \overline{\overline{\mathbf{u}}\,\overline{\mathbf{u}}} - \overline{\overline{\mathbf{u}}}\,\overline{\overline{\mathbf{u}}}.$$

The Bardina model can be thought of as using the simplest possible approximation to both terms in $\boldsymbol{\tau}(\mathbf{u}, \mathbf{u})$. The only simpler model possible is to approximate only one term in $\boldsymbol{\tau}(\mathbf{u}, \mathbf{u})$ – the term depending on $\mathbf{u}$ and not $\overline{\mathbf{u}}$. This yields the model

$$(\overline{\mathbf{u}\,\mathbf{u}} - \overline{\mathbf{u}}\,\overline{\mathbf{u}}) \approx \mathcal{S}(\overline{\mathbf{u}}, \overline{\mathbf{u}}) := \overline{\overline{\mathbf{u}}\,\overline{\mathbf{u}}} - \overline{\mathbf{u}}\,\overline{\mathbf{u}}.$$

This new model is equivalent to the simple approximation to $\overline{\mathbf{u}\,\mathbf{u}}$ given by $\overline{\mathbf{u}\,\mathbf{u}} \approx \overline{\overline{\mathbf{u}}\,\overline{\mathbf{u}}}$ which is $O(\delta^2)$ consistent. Thus, in the usual expansion into resolved, cross, and subgrid-scale terms,

$$\overline{\mathbf{u}\,\mathbf{u}} = \overline{(\overline{\mathbf{u}} + \mathbf{u}')(\overline{\mathbf{u}} + \mathbf{u}')} = \overline{\overline{\mathbf{u}}\,\overline{\mathbf{u}}} + \overline{\overline{\mathbf{u}}\,\mathbf{u}'} + \overline{\mathbf{u}'\,\overline{\mathbf{u}}} + \overline{\mathbf{u}'\,\mathbf{u}'}, \tag{8.28}$$

$\mathcal{S}(\overline{\mathbf{u}}, \overline{\mathbf{u}})$ is equivalent to simply dropping the last two terms which are of formal order of $O(\delta^2)$ and $O(\delta^4)$ on the right-hand side.

The cross-terms $\overline{\overline{\mathbf{u}}\,\mathbf{u}'} + \overline{\mathbf{u}'\,\overline{\mathbf{u}}}$ and $\overline{\mathbf{u}'\,\mathbf{u}'}$ on the right-hand side of (8.28) might be small in laminar regions but they can be dominant in turbulent regions. Thus, simply dropping them cannot be the ultimate answer: models are needed (and will be given!) which are $O(\delta^4)$ and $O(\delta^6)$ accurate, and thus include the effects of these terms.

With that said (and accepting for the moment that we are considering a first step model whose accuracy will be increased), calling $(\mathbf{w}, q)$ (as usual) the resulting approximation to $(\overline{\mathbf{u}}, \overline{p})$, we arrive at the problem in $\Omega \times (0, T)$:

$$\mathbf{w}_t + \nabla \cdot (\overline{\mathbf{w}\,\mathbf{w}}) + \nabla q - \frac{1}{Re}\Delta \mathbf{w} = \overline{\mathbf{f}} \tag{8.29}$$

$$\nabla \cdot \mathbf{w} = 0 . \tag{8.30}$$

Since we are studying the interior closure problem, we consider (8.29), (8.30) on $\Omega = (0, 2\pi)^3$ and subject to periodic boundary conditions, the initial condition $\mathbf{w}(\mathbf{x}, 0) = \overline{\mathbf{u}}_0(\mathbf{x})$ in Ω, and under the usual zero mean condition on all data. In many ways, differential filters are very promising and the development of this model supports this as well. In the development of the model (8.29), the energy balance is also clearest with a simple differential filter.

Definition 8.14. *Given a function $\phi \in L^2(\Omega)$, the differential filter of ϕ, $\overline{\phi}$, is the solution of the boundary value problem*

$$\text{find } \ \overline{\phi} \in L^2(\Omega) : \qquad -\delta^2 \Delta \overline{\phi} + \overline{\phi} = \phi \quad \text{in } \Omega,$$

subject to periodic boundary conditions.

For suitable functions ϕ, for example, $\phi \in L^2(\Omega)$,

$$\overline{\phi} = (-\delta^2 \Delta + \mathbb{I})^{-1} \phi.$$

Thus, this averaging has the following properties:

$$(-\delta^2 \Delta + \mathbb{I}) \, \overline{\phi} = \phi \quad \text{and} \quad \overline{(-\delta^2 \Delta \phi + \phi)} = \phi.$$

The model (8.29) is a zeroth-order model in a very precise sense. The extrapolation $\mathbf{u} \simeq \overline{\mathbf{u}}$ is exact on constant velocities (degree 0) flows. Further, if we expand $\mathbf{u} = \overline{\mathbf{u}} + \mathbf{u}'$, we see that

$$\overline{\mathbf{u}\,\mathbf{u}} = \overline{\overline{\mathbf{u}}\,\overline{\mathbf{u}}} + \overline{\overline{\mathbf{u}}\,\mathbf{u}'} + \overline{\mathbf{u}'\,\overline{\mathbf{u}}} + \overline{\mathbf{u}'\,\mathbf{u}'}$$

and thus (8.29) is equivalent to simply dropping the cross-terms and the fluctuation terms and keeping only the Leonard term $\overline{\overline{\mathbf{u}}\,\overline{\mathbf{u}}}$.

If this term is further approximated by asymptotic approximation, such as a Taylor series in δ, we obtain exactly the model studied by Leonard [212] in one of the pioneering papers in LES. However, we wish to study the above model, with no further approximation. In some sense, variations on (8.29) should arise as a sort of primitive model in every family of LES models of different orders. Thus, it could be called the primitive model, a Leonard model, a zeroth-order model, and so on. Mathematically, in one case it has interesting energy balance (derived in [210, 209]), which we develop next. This energy balance leads us to think of it as an "energy sponge" model.

Before writing equations, recall that the LES closure model is thought of as having two functions. The first is to accurately represent the unresolved scales by the resolved scales. This first function is essential in having a model which has high accuracy in smooth and transitional flow regions. The second function is to subtract energy from the system to represent, in a statistical sense, the energy lost to the resolved scales by breakdown of eddies from resolved scales to unresolved ones. This lost energy must go somewhere. It can be dissipated (that is, lost down an energy drain), or converted from one type of system energy to another, conserving the total kinetic energy of the model. In this latter case it is acting as a sort of "energy sponge." (The authors learned this evocative description of energy drains versus energy sponges from Scott Collis.)

The mathematical development of the model (8.29) is based upon *a priori* bounds for weak solutions. These are proven using a limiting argument from the kinetic energy equality for strong solutions. Thus, the next proposition is the key to both the stability of the model and its mathematical development.

Proposition 8.15. *Let* $\mathbf{u}_0 \in H$ *and* $\mathbf{f} \in L^2(0, T; H)$. *For* $\delta > 0$, *let the averaging be defined through the application of* $(-\delta^2 \Delta + \mathbb{I})^{-1}$. *If* $\mathbf{w}$ *is a strong solution of the model* (8.29), *then* $\mathbf{w}$ *satisfies*

$$\frac{1}{2} \left[\|\mathbf{w}(t)\|^2 + \delta^2 \|\nabla \mathbf{w}(t)\|^2 \right] + \int_0^t \frac{1}{Re} \|\nabla \mathbf{w}(t')\|^2 + \frac{\delta^2}{Re} \|\Delta \mathbf{w}(t')\|^2 \, dt'$$
$$= \frac{1}{2} \left[\|\overline{\mathbf{u}}_0(t)\|^2 + \delta^2 \|\nabla \overline{\mathbf{u}}_0(t)\|^2 \right] + \int_0^t (\mathbf{f}(t'), \mathbf{w}(t')) dt'. \tag{8.31}$$

Remark 8.16. The kinetic energy balance of the model in the previous proposition has two terms which reflect extraction of energy from resolved scales. The energy dissipation in the model

$$\varepsilon_{model}(t) := \frac{1}{Re} \|\nabla \mathbf{w}(t)\|^2 + \frac{\delta^2}{Re} \|\Delta \mathbf{w}(t)\|^2$$

is enhanced by the extra term $Re^{-1} \delta^2 \|\Delta \mathbf{w}(t)\|^2$. This term acts as an irreversible energy drain localized at large local fluctuations. The second term, $\delta^2 \|\nabla \mathbf{w}(t)\|^2$, occurs in the model's kinetic energy

$$k_{model}(t) := \frac{1}{2} \left[\|\mathbf{w}(t)\|^2 + \delta^2 \|\nabla \mathbf{w}(t)\|^2 \right].$$

The true kinetic energy, $\frac{1}{2} \|\mathbf{w}(t)\|^2$, in regions of large deformations is thus extracted, conserved and stored in the kinetic energy penalty term $\delta^2 \|\nabla \mathbf{w}(t)\|^2$. Thus, this reversible term acts as a kinetic "energy sponge." Both terms have an obvious regularizing effect.

Remark 8.17. The key idea in the proof of the energy equality is worth noting and emphasizing. The Navier–Stokes equations are well posed[1] primarily because the nonlinear term $\nabla \cdot (\mathbf{u}\,\mathbf{u})$ is a mixing term which redistributes kinetic energy rather than increasing it. Mathematically this is because of the skew-symmetry property $(\nabla \cdot (\mathbf{u}\,\mathbf{u}), \mathbf{u}) = 0$. The main idea in the proof is to lift this property of the NSE by deconvolution, to understand the energy balance in the model. This is done as follows: noting that all operations are self-adjoint

---

[1] In the sense described in Chap. 2: it is possible to prove existence of weak solutions (due to the energy inequality) but the global existence of smooth solutions is still an open problem!

and, by definition of the averaging used, $(-\delta^2\Delta + \mathbb{I})\,\overline{\phi} = \overline{(-\delta^2\Delta + \mathbb{I})\,\phi} = \phi$, we have

$$
\begin{aligned}
(\nabla \cdot (\overline{\mathbf{w}}\,\overline{\mathbf{w}}), (-\delta^2\Delta + \mathbb{I})\,\mathbf{w}) &= (\overline{\nabla \cdot (\mathbf{w}\,\mathbf{w})}, (-\delta^2\Delta + \mathbb{I})\,\mathbf{w}) \\
&= (\nabla \cdot (\mathbf{w}\,\mathbf{w}), \overline{(-\delta^2\Delta + \mathbb{I})\,\mathbf{w}}) \\
&= (\nabla \cdot (\mathbf{w}\,\mathbf{w}), \mathbf{w}) \\
&= 0.
\end{aligned}
\tag{8.32}
$$

This idea was first used, to our knowledge, in the analysis of the Rational LES model in [29].

Proof. Motivated by (8.32), multiply the model by $(-\delta^2\Delta + \mathbb{I})\,\mathbf{w}$, and integrate over the domain Ω. This gives

$$
(\mathbf{w}_t, (-\delta^2\Delta + \mathbb{I})\,\mathbf{w}) + (\nabla \cdot (\overline{\mathbf{w}}\,\overline{\mathbf{w}}), (-\delta^2\Delta + \mathbb{I})\,\mathbf{w}) + (\nabla q, (-\delta^2\Delta + \mathbb{I})\,\mathbf{w})
$$
$$
- \left(\frac{1}{Re}\Delta\mathbf{w}, (-\delta^2\Delta + \mathbb{I})\,\mathbf{w} \right) = (\overline{\mathbf{f}}, (-\delta^2\Delta + \mathbb{I})\,\mathbf{w}).
$$

The second term vanishes by (8.32). The third term vanishes because $\nabla \cdot \mathbf{w} = 0$, and the last term equals $(\mathbf{f}, \mathbf{w})$. Integrating by parts the first and fourth terms, gives the differential equality

$$
\frac{1}{2}\frac{d}{dt}\left[\|\mathbf{w}(t')\|^2 + \delta^2 \|\nabla\mathbf{w}(t')\|^2 \right] + \frac{1}{Re}\|\nabla\mathbf{w}(t')\|^2 + \frac{\delta^2}{Re}\|\Delta\mathbf{w}(t')\|^2
$$
$$
= (\mathbf{f}(t'), \mathbf{w}(t')).
$$

Then, the result follows by integrating this equality from 0 to t. □

The stability bound in Proposition 8.15 is very strong. Using Galerkin approximations and this stability bound to extract a limit, it is straightforward to prove existence for the model (following Layton and Lewandowski [209, 210]).

Theorem 8.18. *Let the averaging operator be given by $(-\delta^2\Delta + \mathbb{I})^{-1}$, $\delta > 0$ be fixed, and suppose $\mathbf{u}_0 \in H$ and $\mathbf{f} \in L^2(0, T; H)$. Then, there exists a unique strong solution to the model (8.29). Furthermore, that solution satisfies the energy equality (8.31) and thus, for $\delta > 0$,*

$$
\mathbf{w} \in L^\infty(0, T; H^1) \cap L^2(0, T; H^2).
\tag{8.33}
$$

Proof. The proof is very easy once the energy balance of Proposition 8.15 is identified. Indeed, let ψ_r be the orthogonal basis for V of eigenfunctions of the Stokes operator under periodic with zero mean boundary conditions. These are also the eigenfunctions of $(-\delta^2\Delta + \mathbb{I})$ in the same setting. Thus, let $-\Delta\psi_r = \lambda_r\psi_r$. Let $V_k := \mathrm{span}\,\{\psi_r : r = 1, \dots, k\}$. The Galerkin approximation $\mathbf{w}_k : [0, T] \to V_k$ satisfies, for all $\psi \in V_k$,

$$
(\partial_t\mathbf{w}_k, \psi) + (\nabla \cdot (\overline{\mathbf{w}_k\mathbf{w}_k}), \psi) - \frac{1}{Re}(\Delta\mathbf{w}_k, \psi) = (\overline{\mathbf{f}}, \psi).
\tag{8.34}
$$

As usual, the Galerkin approximation (8.34) reduces to a system of ordinary differential equations for the undetermined coefficients $C_{k,r}(t)$. Existence for (8.34) will follow from an *a priori* bound on its solution. Since $\mathbf{w}_k(t) \in V_k$, it follows that $(-\delta^2 \Delta + \mathbb{I}) \mathbf{w}_k = \sum_{r=1}^{k} (\delta^2 \lambda_r + 1) C_{k,r} \psi_r(\mathbf{x}) \in V_k$. Thus it is permissible to set $\psi = (-\delta^2 \Delta + \mathbb{I}) \mathbf{w}_k$ in (8.34). By exactly the same proof as in Proposition 8.15, we have

$$\frac{1}{2} \left[\|\mathbf{w}_k(t)\|^2 + \delta^2 \|\nabla \mathbf{w}_k(t)\|^2 \right] + \int_0^t \frac{1}{Re} \|\nabla \mathbf{w}_k(t')\|^2 + \frac{\delta^2}{Re} \|\Delta \mathbf{w}_k(t')\|^2 \, dt'$$
$$= \frac{1}{2} \left[\|\overline{\mathbf{u}}_0(t)\|^2 + \delta^2 \|\nabla \overline{\mathbf{u}}_0(t)\|^2 \right] + \int_0^t (\mathbf{f}(t'), \mathbf{w}_k(t')) \, dt'.$$

The Cauchy–Schwartz inequality then immediately implies

$$\|\mathbf{w}_k\|_{L^\infty(0,T;H^1)} \leq M_1 = M_1(\mathbf{f}, \mathbf{u}_0, \delta) < \infty;$$
$$\|\mathbf{w}_k\|_{L^\infty(0,T;L^2)} \leq M_2 = M_2(\mathbf{f}, \mathbf{u}_0) < \infty;$$
$$\|\mathbf{w}_k\|_{L^2(0,T;H^2)} \leq M_3 = M_3(\mathbf{f}, \mathbf{u}_0, \delta, Re) < \infty;$$

and (8.34) thus has a unique solution.

From the above *a priori* bounds and using exactly the same approach as in the NSE case (following the beautiful and clear presentation of Galdi [121]) letting $k \to \infty$, we recover a limit, $\mathbf{w}$, which is (using the above stronger *a priori* bounds) a unique strong solution of the model satisfying the energy equality and belonging to (8.33), for $\delta > 0$. □

8.5.1 "More Accurate" Models

The regularity proven for the solution of this first model is very strong and many more mathematical properties can be developed for it. However, it is also important not to forget that it is a "mathematical toy" and not of sufficient accuracy. Thus, it is important to find models with similar strong mathematical properties which are more accurate. The critical condition "more accurate" is presently evaluated in two ways: analytical studies in smooth flow regions, and experimental studies in turbulent flow regions.

Next, we turn to the modeling error in the simple model. Our goal is to give an analytical study of the modeling error, in other words, to give an *a priori* bound upon norm $\|\overline{\mathbf{u}} - \mathbf{w}\|$. To do this, we need a strong enough regularity condition upon $\mathbf{u}$ to apply Gronwall's equality uniformly in δ.

A sufficient condition for this is $\|\nabla \mathbf{u}\| \in L^4(0,T)$. This can obviously be weakened in many ways. Next, we need a strong enough regularity condition on $\mathbf{u}$ to extract a bound on the models *consistency error evaluated at the true solution* in the norm $L^2(\Omega \times (0,T))$, *i.e.* on

$$\|\mathbf{u}\,\mathbf{u} - \overline{\mathbf{u}}\,\overline{\mathbf{u}}\|_{L^2(\Omega \times (0,T))}.$$

It will turn out that a sufficient condition for this to be $O(\delta^2)$, is

$$\mathbf{u} \in L^4(0, T; H^2).$$

To proceed, filtering the Navier–Stokes equations, shows that $\overline{\mathbf{u}}$ satisfies, after rearrangement,

$$\overline{\mathbf{u}}_t + \nabla \cdot (\overline{\mathbf{u}}\,\overline{\mathbf{u}}) + \nabla \overline{p} - \frac{1}{Re}\Delta\overline{\mathbf{u}} = \overline{\mathbf{f}} - \nabla \cdot \overline{\boldsymbol{\rho}}$$
$$\nabla \cdot \overline{\mathbf{u}} = 0,$$

where $\boldsymbol{\rho}$ is the consistency error term, in this case given by

$$\boldsymbol{\rho} := \mathbf{u}\,\mathbf{u} - \overline{\mathbf{u}}\,\overline{\mathbf{u}}.$$

Theorem 8.19. *Let the filtering be* $\overline{\boldsymbol{\phi}} = (-\delta^2\Delta + \mathbb{I})^{-1}\boldsymbol{\phi}$*; let* $\mathbf{u}$ *be a unique strong solution of the NSE satisfying the regularity condition* $\mathbf{u} \in L^4(0, T; H^1)$ *or Serrin's uniqueness condition (2.31). Then, there exists a positive constant* $C^* = C^*(Re, T, \|\mathbf{u}\|_{L^4(0,T;H^1)})$ *such that the modeling error* $\boldsymbol{\phi} := \overline{\mathbf{u}} - \mathbf{w}$ *satisfies*

$$\|\boldsymbol{\phi}\|_{L^\infty(0,T;L^2)}^2 + \delta^2 \|\nabla\boldsymbol{\phi}\|_{L^\infty(0,T;L^2)}^2 + \frac{1}{Re}\|\nabla\boldsymbol{\phi}\|_{L^2(0,T;L^2)}^2$$
$$+ \frac{\delta^2}{Re}\|\Delta\boldsymbol{\phi}\|_{L^2(0,T;L^2)}^2 \le C^* \|\boldsymbol{\rho}\|_{L^2(\Omega\times(0,T))}^2.$$

If additionally $\mathbf{u} \in L^4(0, T; H^2)$*, then the consistency error satisfies*

$$\|\boldsymbol{\rho}\|_{L^2(\Omega\times(0,T))}^2 \le C\,\delta^4 \|\mathbf{u}\|_{L^4(0,T;H^2(\Omega))}^2.$$

Proof. First, we note that by the definition of $\boldsymbol{\rho}$ and the Sobolev inequality:

$$\|\boldsymbol{\rho}\|_{L^2(\Omega\times(0,T))} = \|\overline{\mathbf{u}}\,\overline{\mathbf{u}} - \overline{\mathbf{u}}\,\mathbf{u} + \overline{\mathbf{u}}\,\mathbf{u} - \mathbf{u}\,\mathbf{u}\| \le 2\|\mathbf{u}\|_{L^\infty(\Omega)}\|\mathbf{u} - \overline{\mathbf{u}}\|$$
$$\le 2\,\delta^2\|\Delta\mathbf{u}\|^2.$$

Thus, by squaring and integrating, we have, as claimed: $\|\boldsymbol{\rho}\|_{L^2(\Omega\times(0,T))}^2 \le C\,\delta^4\|\mathbf{u}\|_{L^4(0,T;H^2(\Omega))}^2$.

For the proof that the modeling error is bounded by the model's consistency error $\boldsymbol{\rho}$, we subtract $\mathbf{w}$ from $\overline{\mathbf{u}}$ and mimic the proof of the model's energy estimate. Indeed, the modeling error $\boldsymbol{\phi}$ satisfies $\boldsymbol{\phi}(0) = \mathbf{0}$, $\nabla \cdot \boldsymbol{\phi} = 0$, and

$$\boldsymbol{\phi}_t + \nabla \cdot (\overline{\mathbf{u}}\,\overline{\mathbf{u}} - \mathbf{w}\,\mathbf{w}) + \nabla(\overline{p} - q) - \frac{1}{Re}\Delta\boldsymbol{\phi} = -\nabla \cdot \overline{\boldsymbol{\rho}}, \quad \text{in } \Omega \times (0, T).$$

Under the above regularity assumptions, $\mathbf{u}$ is a strong solution of the Navier–Stokes equations and $\boldsymbol{\phi}$ also satisfies the above equation strongly. Thus, only two paths are reasonable to bound $\boldsymbol{\phi}$ by $\boldsymbol{\rho}$:

(i) multiply by ϕ and integrate,

(ii) multiply by $(-\delta^2 \Delta + \mathbb{I})\phi$ and integrate.

Following the proof of the energy estimate, we use (ii). This gives, after steps which follow exactly those in the energy estimate,

$$\frac{1}{2}\frac{d}{dt}\left[\|\phi\|^2 + \delta^2 \|\phi\|^2\right] + \frac{1}{Re}\left[\|\nabla\phi\|^2 + \|\Delta\phi\|^2\right] - \int_\Omega (\overline{\mathbf{u}}\,\overline{\mathbf{u}} - \mathbf{w}\,\mathbf{w})\nabla\phi\,dx$$
$$= \int_\Omega \boldsymbol{\rho}\,\nabla\phi\,dx.$$

The third term is handled in the standard way: adding and subtracting $\mathbf{w}\,\overline{\mathbf{u}}$. This gives

$$\int_\Omega (\overline{\mathbf{u}}\,\overline{\mathbf{u}} - \mathbf{w}\,\mathbf{w}) : \nabla\phi\,dx = \int_\Omega \phi \cdot \nabla\phi \cdot \overline{\mathbf{u}}\,dx.$$

Next, use the following inequalities, which are valid in two and three dimensions (and improvable in two dimensions),

$$\left|\int_\Omega \phi \cdot \nabla\phi \cdot \overline{\mathbf{u}}\,dx\right| \le \frac{1}{4\,Re}\|\nabla\phi\|^2 + C(Re)\|\overline{\mathbf{u}}\|^4 \|\phi\|^2,$$

$$\left|\int_\Omega \boldsymbol{\rho} : \nabla\phi\,dx\right| \le \frac{1}{4\,Re}\|\boldsymbol{\rho}\|^2 + C(Re)\|\nabla\phi\|^2.$$

These give

$$\frac{1}{2}\frac{d}{dt}\left[\|\phi\|^2 + \delta^2 \|\phi\|^2\right] + \frac{1}{Re}\left[\|\nabla\phi\|^2 + \|\Delta\phi\|^2\right]$$
$$\le C(Re)\|\boldsymbol{\rho}\|^2 + C(Re)\|\overline{\mathbf{u}}\|^4 \|\phi\|^2.$$

The theorem then follows by Gronwall's inequality. $\quad\square$

8.6 The Higher Order, Stolz–Adams Deconvolution Models

The first model of Sect. 8.5 is based on an extrapolation from resolved to unresolved scales which is exact on constants: $\mathbf{u} \simeq \overline{\mathbf{u}} + O(\delta^2)$. It is immediately clear how to generate more accurate models by higher order extrapolations in δ. For example, with the differential filter $\overline{\phi} = (-\delta^2\Delta + \mathbb{I})^{-1}\phi$, we can approximate $\mathbf{u} \simeq 2\overline{\mathbf{u}} - \overline{\overline{\mathbf{u}}}$, which is an exactly linear extrapolation in δ. This gives the closure model

$$\overline{\mathbf{u}}\,\overline{\mathbf{u}} \simeq \overline{(2\overline{\mathbf{u}} - \overline{\overline{\mathbf{u}}})(2\overline{\mathbf{u}} - \overline{\overline{\mathbf{u}}})} + O(\delta^4).$$

Quadratic extrapolation reads $\phi \simeq 3\overline{\phi} - 3\overline{\overline{\phi}} + \overline{\overline{\overline{\phi}}}$, so the corresponding closure model is

$$\overline{\mathbf{u}}\,\overline{\mathbf{u}} \simeq \overline{(3\overline{\phi} - 3\overline{\overline{\phi}} + \overline{\overline{\overline{\phi}}})(3\overline{\phi} - 3\overline{\overline{\phi}} + \overline{\overline{\overline{\phi}}})} + O(\delta^6).$$

In this way, by using successively higher order extrapolations, we can generate closure models of any formal asymptotic accuracy we desire. For other filters, such as sharp Fourier cut off, $\overline{\mathbf{u}} = \overline{\overline{\mathbf{u}}}$, so these must be modified in the obvious way, replacing $\overline{\mathbf{u}}$ by $g_\delta * \mathbf{u}$ and $\overline{\overline{\mathbf{u}}}$ by $g_{\sqrt{2}\delta} * \mathbf{u}$ and so on.

Using successively higher order models we can then investigate analytically and experimentally the right balance between accuracy and cost. The first step is obviously to test the models and analyze their stability.

The family of models, so generated, coincide with the family of models developed by Stolz and Adams [285, 289, 290, 3], by adapting the van Cittert [36] deconvolution method from image processing to the closure problem in LES. We thus turn to considering the very interesting Stolz–Adams deconvolution/scale similarity models.

8.6.1 The van Cittert Approximations

Let $G\phi = \overline{\phi}$ denote the filtering operator, either by convolution with a smooth kernel or by the differential filter $(-\delta^2\Delta + \mathbb{I})^{-1}$.

Since $G = \mathbb{I} - (\mathbb{I} - G)$, an inverse to G can be written formally as the nonconvergent Neumann series

$$G^{-1} \sim \sum_{n=0}^{\infty} (\mathbb{I} - G)^n.$$

Truncating the series gives the van Cittert Approximate Deconvolution operators [36],

$$G_N := \sum_{n=0}^{N} (\mathbb{I} - G)^n.$$

The approximations G_N are not convergent as N goes to infinity, but rather are asymptotic as δ approaches zero, as the next lemma shows.

Lemma 8.20. *For smooth $\mathbf{u}$, the approximate deconvolution G_N has error*

$$\mathbf{u} - G_N\overline{\mathbf{u}} = (-1)^{N+1} \delta^{2N+2} \Delta^{N+1}\, \mathbf{u} \quad \text{pointwise and}$$
$$\|\mathbf{u} - G_N\overline{\mathbf{u}}\| \leq \delta^{2N+2} \|\mathbf{u}\|_{H^{2N+2}(\Omega)}, \quad \text{globally.}$$

Proof. This is a simple algebraic argument. Let $A := (\mathbb{I} - G)$ and note that $A\phi = \phi - \overline{\phi} = -\delta^2\Delta\overline{\phi}$. Then, with $\mathbf{e} = \mathbf{u} - G_N\overline{\mathbf{u}}$, we have, by definition of G_N, $\mathbf{u} = \overline{\mathbf{u}} + A\overline{\mathbf{u}} + \cdots + A^N\overline{\mathbf{u}} + \mathbf{e}$. Applying to both sides the operator A and subtracting, gives, since $\mathbb{I} - A = G$,

$$G\mathbf{u} = \overline{\mathbf{u}} - A^{N+1}\overline{\mathbf{u}} + G\mathbf{e}.$$

Or, as $G\mathbf{u} = \overline{\mathbf{u}}$, $G\mathbf{e} = \overline{\mathbf{e}}$, applying $(-\delta^2\Delta + \mathbb{I})$ to both sides, implies $\mathbf{e} = A^{N+1}\mathbf{u}$, which, after rearrangement proves the lemma. $\square$

Lemma 8.20 shows that $G_N \overline{\mathbf{u}}$ gives an approximation to $\mathbf{u}$ to accuracy $O(\delta^{2N+2})$ in the smooth flow regions. Thus, it is justified to use it as a closure approximation. Doing so, results in the family of Stoltz–Adams deconvolution models

$$\overline{\mathbf{u}\,\mathbf{u}} \simeq \overline{G_N\overline{\mathbf{u}}\,G_N\overline{\mathbf{u}}} + O(\delta^{2N+2}).$$

If $\boldsymbol{\tau}$ denotes the usual subfilter-scale stress tensor $\boldsymbol{\tau}(\mathbf{u},\mathbf{u}) := \overline{\mathbf{u}\,\mathbf{u}} - \overline{\mathbf{u}}\,\overline{\mathbf{u}}$, then this closure approximation is equivalent to the closure model

$$\boldsymbol{\tau}(\mathbf{u},\mathbf{u}) \approx \boldsymbol{\tau}_N(\overline{\mathbf{u}},\overline{\mathbf{u}}) := \overline{G_N\overline{\mathbf{u}}\,G_N\overline{\mathbf{u}}} - \overline{\mathbf{u}}\,\overline{\mathbf{u}}. \tag{8.35}$$

We recall (see Chap. 6 for further details) that a tensor function $\boldsymbol{\tau}(\mathbf{u},\mathbf{v})$ of two vector variables is reversible if $\boldsymbol{\tau}(-\mathbf{u},-\mathbf{v}) = \boldsymbol{\tau}(\mathbf{u},\mathbf{v})$. In addition a tensor $\boldsymbol{\tau}$ is Galilean invariant if, for any divergence-free periodic vector field $\mathbf{w}$ and any constant vector $\mathbf{U}$, $\nabla \cdot \boldsymbol{\tau}(\mathbf{w} + \mathbf{U}, \mathbf{w} + \mathbf{U}) = \nabla \cdot \boldsymbol{\tau}(\mathbf{w},\mathbf{w})$. These are requirements for any satisfactory closure model; see p. 136. The interest in reversibility and Galilean invariance is that the true subfilter-scale stress tensor $\boldsymbol{\tau}$ is both reversible and invariant. Thus, many feel that appropriate closure models should (at least to leading-order effects) share these two properties. (For a more detailed discussion on the properties that the subfilter-scale stress tensor $\boldsymbol{\tau}$ should satisfy, the reader is referred to Chap. 6.) We next show that the model (8.35) is both reversible and Galilean invariant.

Lemma 8.21. *For each $N = 0,1,2,\ldots$, the Stolz–Adams closure model $\boldsymbol{\tau}_N$ is both reversible and Galilean invariant.*

Proof. Reversibility is immediate. Galilean invariance also follows easily once it is noted that $\overline{\mathbf{U}\,\mathbf{w}} = \mathbf{U}\,\overline{\mathbf{w}}$, $\overline{\mathbf{U}\,\mathbf{U}} = \mathbf{U}\,\mathbf{U}$, $G_N\mathbf{U}\,\overline{\mathbf{w}} = \mathbf{U}\,G_N\overline{\mathbf{w}}$, and $\nabla \cdot \overline{\mathbf{u}} = \nabla \cdot G_N\overline{\mathbf{u}} = \cdots = 0$.   □

The Stolz–Adams models are thus highly accurate, in the sense that their consistency error is asymptotically small as δ approaches zero; they are reversible and Galilean invariant. Their usefulness thus hinges on their stability properties. These were established by Dunca and Epshteyn [98] by an argument similar to the one we give now.

Consider the model

$$\mathbf{w}_t + \nabla \cdot (\overline{G_N\mathbf{w}\,G_N\mathbf{w}}) + \nabla q - \frac{1}{Re}\Delta\mathbf{w} = \overline{\mathbf{f}}$$
$$\nabla \cdot \mathbf{w} = 0$$

under periodic boundary conditions and zero spatial mean on all data and on $\mathbf{w}$. If the filter chosen is the differential filter $\overline{\phi} = (-\delta^2\Delta + \mathbb{I})^{-1}\phi$, then the natural lifting to the model (8.35) of the skew-symmetry property of the nonlinearity in the Navier–Stokes equations is

$$(\nabla \cdot (\overline{G_N\mathbf{w}\,G_N\mathbf{w}}), (-\delta^2\Delta + \mathbb{I})\,G_N\mathbf{w}) = 0.$$

Thus, the natural idea is to multiply (8.35) by $(-\delta^2 \Delta + \mathbb{I}) G_N \mathbf{w}$ and integrate over Ω. This gives, after obvious simplification which follows the work in the previous section,

$$(\mathbf{w}_t, (-\delta^2 \Delta + \mathbb{I}) G_N \mathbf{w}) - \frac{1}{Re}(\Delta \mathbf{w}, (-\delta^2 \Delta + \mathbb{I}) G_N \mathbf{w})$$

$$= (\overline{\mathbf{f}}, (-\delta^2 \Delta + \mathbb{I}) G_N \mathbf{w}) = (\mathbf{f}, G_N \mathbf{w}).$$

By the choice of filter, it follows that Δ, $(-\delta^2 \Delta + \mathbb{I})$, G_N, and G all commute. If G_N is a positive operator, then it has a positive square root A, *i.e.* an operator such that $A^2 = G_N$. If this A exists, then we can integrate this last equation by parts to get

$$(A \mathbf{w}_t, (-\delta^2 \Delta + \mathbb{I}) A \mathbf{w}) + \frac{1}{Re}(\nabla A \mathbf{w}, \nabla A \mathbf{w}) + \frac{\delta^2}{Re}(\Delta A \mathbf{w}, \Delta A \mathbf{w}) = (\mathbf{f}, G_N \mathbf{w}),$$

or

$$\frac{1}{2}\frac{d}{dt}\left[\|A\mathbf{w}(t)\|^2 + \delta^2 \|\nabla A \mathbf{w}(t)\|^2 \right] + \frac{1}{Re}\|\nabla A \mathbf{w}(t)\|^2 + \frac{\delta^2}{Re}\|\Delta A \mathbf{w}(t)\|^2$$
$$= (\mathbf{f}(t), G_N \mathbf{w}(t)).$$

From this, flows an energy inequality which is the key turning the lock, opening the mathematical foundation of existence, uniqueness, and regularity for the model. Thus, the essential point is to verify that G_N is a positive operator.

Lemma 8.22. *Let the averaging be defined by* $(-\delta^2 \Delta + \mathbb{I})^{-1}$. *Then* $G_N :$ $L_0^2(\Omega)^d \to L_0^2(\Omega)^d$ *is a bounded, symmetric positive-definite operator.*

Proof. First, since $G\phi = (-\delta^2 \Delta + \mathbb{I})^{-1}\phi$, is symmetric and bounded and G_N is a function of G it follows that G_N is itself symmetric and bounded. To show G_N is positive definite we use Fourier series. Namely, we expand ϕ in terms of its Fourier coefficients

$$\phi = \sum_{\mathbf{j} \in \mathbb{Z}^d, \mathbf{j} \neq \mathbf{0}} a_{\mathbf{j}} e^{i \mathbf{j} \cdot \mathbf{x}},$$

where $\mathbf{j} \neq \mathbf{0}$, derives from the fact that $\int_\Omega \phi \, d\mathbf{x} = 0$. In wavenumber space the operator G and G^{-1} act in a very simple manner and this implies

$$\overline{\phi} = \sum_{\mathbf{j} \in \mathbb{Z}^d, \mathbf{j} \neq \mathbf{0}} (\delta^2 |\mathbf{j}|^2 + 1)^{-1} a_{\mathbf{j}} e^{i \mathbf{j} \cdot \mathbf{x}}.$$

Then, it follows that

$$(\mathbb{I} - G)^k \phi = \sum_{\mathbf{j} \in \mathbb{Z}^d, \mathbf{j} \neq \mathbf{0}} a_{\mathbf{j}} \left[1 - (\delta^2 |\mathbf{j}|^2 + 1)^{-1} \right]^k e^{i \mathbf{j} \cdot \mathbf{x}}.$$

or, after simplification,

$$(\mathbb{I} - G)^k \phi = \sum_{\mathbf{j} \in \mathbb{Z}^d, \mathbf{j} \neq \mathbf{0}} \left(\frac{\delta^2 |\mathbf{j}|^2}{\delta^2 |\mathbf{j}|^2 + 1} \right)^k a_\mathbf{j} \, e^{i\mathbf{j} \cdot \mathbf{x}}.$$

Taking the L^2-scalar product with ϕ we finally obtain

$$(\phi, (\mathbb{I} - G)^k \phi) = \sum_{\mathbf{j} \in \mathbb{Z}^d, \mathbf{j} \neq \mathbf{0}} \left(\frac{\delta^2 |\mathbf{j}|^2}{\delta^2 |\mathbf{j}|^2 + 1} \right)^k |a_\mathbf{j}|^2.$$

Thus, $(\mathbb{I} - G)$ is positive-definite provided the multiplier on the right-hand side is positive. That is, provided $\frac{\delta^2 |\mathbf{j}|^2}{\delta^2 |\mathbf{j}|^2 + 1} > 0$ for $\mathbf{j} \neq \mathbf{0}$. By direct inspection, this is true, and $(\mathbb{I} - G)$ is positive. Since $(\mathbb{I} - G)$ is positive, G_N is then a sum of symmetric positive definite operators, and hence symmetric positive definite itself. The lemma is proven. $\square$

Since existence of a positive square root of G_N, $A = G_N^{\frac{1}{2}}$, is proven, existence follows together with uniqueness, regularity, and an energy equality. We will omit the proof here since it follows the pattern of the proof in Sect. 8.5 with only added technical complexities.

Theorem 8.23. *Let the averaging be defined by $(-\delta^2 \Delta + \mathbb{I})^{-1}$. Suppose $\mathbf{u}_0 \in V$, and $\mathbf{f} \in L^2(0, T; H)$. Let $\delta > 0$. Then, the Stolz–Adams deconvolution model has a unique strong solution. That solution satisfies the energy equality below:*

$$\frac{1}{2} \left[\|A\mathbf{w}(t)\|^2 + \delta^2 \|\nabla A\mathbf{w}(t)\|^2 \right] + \int_0^t \frac{1}{Re} \|\nabla A\mathbf{w}(t')\|^2 + \frac{\delta^2}{Re} \|\Delta A\mathbf{w}(t')\|^2 \, dt'$$

$$= \frac{1}{2} \left[\|A\overline{\mathbf{u}}_0(t)\|^2 + \delta^2 \|\nabla A\overline{\mathbf{u}}_0(t)\|^2 \right] + \int_0^t (\mathbf{f}(t'), G_N \mathbf{w}(t')) \, dt',$$

where $A = G_N^{\frac{1}{2}}$.

8.7 Conclusions

Scale similarity models in general, and the Stolz–Adams approximate deconvolution models in particular, represent an extremely promising path for the future development of LES. In particular, the Stolz–Adams approximate deconvolution approach gives a family of models that are both highly accurate and have excellent stability properties. It seems appropriate to call all these model predictive models and to lump many other models based on phenomenology into the category of descriptive models.

The Stolz–Adams approximate deconvolution models are very recent and many open questions remain for their development. One important topic we have not discussed is the development of good *algorithms* for these models. The filtering in the operator G_N almost forces these terms to be treated *explicitly*. While quite easy in typical compressible flow problems in gas dynamics, explicit treatment requires more algorithmic finesse in incompressible flows.

The van Cittert is only the simplest (and possibly least effective) deconvolution procedure from image processing. Thus, an important open path for the development of LES is to incorporate and test more accurate deconvolution methods in LES models.

Part IV

Boundary Conditions

Filtering on Bounded Domains

One basic problem which is reported in many experimental assessments of LES is:

> LES continues to have difficulties predicting near wall turbulence and to have still more difficulties predicting turbulence driven by flow/boundary interactions.

These reports clearly provide strong motivations to explore carefully the closure errors related to filtering on a bounded domain. To emphasize further the importance of boundaries and wall treatments, recall the result of classical mathematical fluid mechanics (e.g. in Serrin's 1959 article [274] or Poincaré [257]) that in problems with irrotational initial conditions and potential body forces, *all vorticity comes from the boundary*. Indeed, in this case the vorticity $\boldsymbol{\omega} = \nabla \times \mathbf{u}$ satisfies

$$\boldsymbol{\omega}_t + \mathbf{u} \cdot \nabla \boldsymbol{\omega} - \frac{1}{Re} \Delta \boldsymbol{\omega} = \boldsymbol{\omega} \cdot \nabla \mathbf{u} \quad \text{in } \Omega \times (0, T) \tag{9.1}$$

and if $\boldsymbol{\omega}(\mathbf{x}, 0) = \mathbf{0}$ and $\boldsymbol{\omega}|_{\partial\Omega} = \mathbf{0}$ (no vorticity is generated at the boundary) then all problem data is zero and it is easy to show that thereafter $\boldsymbol{\omega} \equiv \mathbf{0}$ in $\Omega \times (0, T)$.[1]

One popular treatment of boundary conditions in LES is to let the radius of the filter, δ, approach 0 at the solid surface,

$$\delta = \delta(\mathbf{x}) \to 0 \qquad \text{as} \qquad \mathbf{x} \to \partial\Omega.$$

---

[1] In practical flows, $\boldsymbol{\omega} = \mathbf{0}$ on the boundary is not an appropriate boundary condition and this is a strong limitation in the use of the vorticity equation in the presence of boundaries. In several cases there is a lack of knowledge of the values of the vorticity on the boundary. From the mathematical point of view the main problem is due to the fact that the boundary integrals arising in the integration by parts needed in the derivation of energy estimates for $\boldsymbol{\omega}$ do not vanish.

Then, the boundary conditions at the wall are clear: $\overline{\mathbf{u}} = \mathbf{0}$ on $\partial\Omega$. Thus, loosely speaking, the LES model is reduced to a DNS at the wall. There are, however, serious mathematical and algorithmic open questions associated with this approach.

From the theoretical point of view it is clear that by considering a variable filter width $\delta(\mathbf{x})$, commutation errors $\mathcal{E}_i[u](\mathbf{x})$ (depending on the function u, on the point $\mathbf{x}$, and on the direction $\mathbf{e}_i$) are introduced in LES models since, for variable δ,

$$\mathcal{E}_i[u](\mathbf{x}) := \frac{\partial \overline{u}(\mathbf{x})}{\partial x_i} - \overline{\frac{\partial u(\mathbf{x})}{\partial x_i}} \neq 0 \qquad \text{for } i = 1, \ldots, d,$$

even for a very smooth scalar function u. Some progress has been made on the numerical analysis and the estimation of the size of the error that is committed, see Fureby and Tabor [119], Ghosal and Moin [136], and Vasilyev *et al.* [304], but many questions remain unanswered.

In the above references, the proofs are mainly based on one-dimensional Taylor series expansion for very smooth functions. It has been argued that these commutation errors can be neglected in applications, provided special filters with vanishing moments are used over smooth enough functions. There are, however, interesting and relevant mathematical challenges associated with this approach.

There is also an intense study of the associated commutation errors. For important recent advances, see the work of van der Bos and Geurts [299], Iovieno and Tordella [171], and Berselli, Grisanti, and John [32], where the commutation error is estimated in the presence of functions with low regularity properties.

The second drawback is more practical in nature: since the filter radius $\delta(\mathbf{x})$ is decreased near the wall, the numerical resolution needed is greatly increased, because the mesh size must be accordingly reduced to resolve at least the inner layers near the wall. By using the approach introduced by Chapman [58], recently Piomelli and Balaras [253] presented an estimate for the computational cost for LES of turbulent channel flows. The flow is divided into an *inner layer* in which the effects of viscosity are important, and an *outer layer* in which the direct effects of viscosity on the mean flow are negligible. First, grid-resolution requirements are presented for the inner and outer layers separately. Then, the computational cost associated with the time integration is derived, based on the need to resolve the life of the smallest eddy. Based on these estimates, the total computational cost scales like $Re^{0.5}$ for the outer layer, and $Re^{2.4}$ for the inner layer. In a wide range of flows in the geophysical sciences (meteorology and oceanography) and engineering (ship hydrodynamics and aircraft aerodynamics) the Reynolds number is very high, of the order of tens or hundreds of millions. Based on the above estimate, the computational cost for the LES approaches that aim at resolving the inner layer is unfeasible for these applications.

A common attempt to overcome the presence of commutation errors (and related problems) is through *near wall models*. We review some of these models in Chap. 10. In the next sections we sketch out the main ideas and results in the use of nonuniform (namely, with nonconstant radius) filters. If the computational complexity of LES is to be truly independent of the Reynolds number, some sort of filtering through the boundary must be performed in the modeling step. We also present an analysis of one such approach: a constant filter width δ is used and the filtering goes through the boundaries.

Plan of the Chapter

Part of this chapter is rather technical and could seem difficult at first reading. This is due to the fact that Chap. 9 presents some delicate topics that represent, to some extent, the state-of-the-art of current research in LES. In fact, many of the results we report are going to appear or are recent findings in the mathematical study of turbulent flows. A strong interest in the commutation error is rather new in the mathematical LES community and our intent is, at least, to get the reader interested in this new and challenging topic.

In particular, at some points in this chapter a deeper knowledge of analysis is needed. Understanding filtering in the presence of boundaries is necessarily technical, even if some hand-waving arguments are allowed. Our aim is to give the main ideas of this important topic. We will also emphasize what could happen if the flow variable to be filtered is not smooth, a common situation in computational fluid dynamics.

First, we will derive the basic equations involving filtering with nonconstant radius, together with their error estimates that involve delicate issues (and ugly formulas) regarding Taylor series expansion.

Then, we will briefly consider the problems arising in the filtering after a zero extension outside the physical domain. In the last section we will use advanced tools in distribution theory (some knowledge of geometric measure theory will be necessary for a better understanding) to properly write the filtered equations and to derive suitable estimates for the commutation error.

9.1 Filters with Nonconstant Radius

In this section we summarize the approach of Ghosal and Moin [136] and we show how it is possible to properly define filtering also in complex geometries, and hence in the presence of boundaries. We start by considering the one-dimensional case. The filtering is defined, as usual, by

$$\overline{u}(x,t) = (g_\delta * u)(x,t) := \frac{1}{\delta} \int_{\mathbb{R}} g\left(\frac{x-y}{\delta}\right) u(y,t)\, dy,$$

where the function g is generally smooth, even, and fast decaying at infinity, see Chap. 1. In situations where the domain is finite (or at least semi-infinite) the above definition may have several generalizations. In the case of the box filter a generalization might be, if $u : (a, b) \to \mathbb{R}$ is given,

$$\bar{u}(x, t) = \frac{1}{\delta_+(x) - \delta_-(x)} \int_{x - \delta_-(x)}^{x + \delta_+(x)} u(y, t)\, dy,$$

where $\delta_+(x)$ and $\delta_-(x)$ are nonnegative functions and $\delta_+(x) - \delta_-(x)$ is the "effective filter width" at location x.

In this case (others can be treated similarly), both $\delta_+(x)$ and $\delta_-(x)$ must go to zero sufficiently fast at the boundaries, so that

$$(x - \delta_-(x), x + \delta_+(x)) \subseteq (a, b),$$

i.e. the window of values used in the filtering must remain always in the domain of definition of the function u. In this special case it is well known that the commutation error does not vanish. To stress the importance of this source of error, we cite Ghosal and Moin [136]:

> One would like to believe that the commutation error would be small for some reasonable class of non-uniform filters, but this has never been conclusively demonstrated . . .

Some analysis of this topic will be given later. Therefore, a new closure problem arises not only for the nonlinear term, as we extensively analyzed in the previous chapters, but also for the linear terms.

9.1.1 Definition of the Filtering

In order to extend the above "generalized box filter" to other filters, a possible approach is that of mapping the interval (a, b) onto the whole real line, by means of a *mapping function* $f : (a, b) \to \mathbb{R}$ that is monotonically increasing and smooth, such that

$$\lim_{x \to a} f(x) = -\infty \quad \text{and} \quad \lim_{x \to b} f(x) = +\infty.$$

A nonuniform radius $\delta(x)$ is then defined by

$$\delta(x) = \frac{\delta}{f'(x)}.$$

This implies that both $f'(a)$ and $f'(b)$ must be infinite, thus the filtering kernel becomes the "Dirac's delta-function" (see p. 243) at the finite boundaries. A classical choice for the function f is

$$f(x) = \tanh^{-1}\left(\frac{2x}{b - a} - \frac{a + b}{b - a}\right) \quad \text{for } a \leq x \leq b,$$

which is related to the "tan-hyperbolic grid," used for instance in channel flow computations [240].

Given an arbitrary scalar function $u(x)$, we first make a change of variables, to obtain the new function

$$\phi(\xi) = u(f^{-1}(\xi)) \qquad \forall \xi \in \mathbb{R}.$$

The function $\phi : \mathbb{R} \to \mathbb{R}$ is then filtered according to the usual definition by means of a convolution. Finally, we transform back to the variable x. Thus,

$$\overline{\phi}(\xi) = \frac{1}{\delta} \int_{\mathbb{R}} g\left(\frac{\xi - \eta}{\delta}\right) \phi(\eta)\, d\eta \tag{9.2}$$

or, by using the mapping function f,

$$\overline{u}(x) = \frac{1}{\delta} \int_a^b g\left(\frac{f(x) - f(y)}{\delta}\right) u(y) f'(y)\, dy. \tag{9.3}$$

The above equivalent expressions (9.2) and (9.3) are called second-order commuting filter. This definition is motivated by the fact that the commutation error satisfies (for smooth u and non uniformly in Re)

$$\mathcal{E}[u] = O(\delta^2).$$

The proof of this result uses in an essential manner the fact that the kernel g is symmetric and also that the function u and f can be expanded as a Taylor series, up to a certain order; see [136].

In the three-dimensional case the filtering may be defined in a similar way through a kernel that is the product of three one-dimensional kernels. If

$$\mathbf{X} = H(\mathbf{x}) \tag{9.4}$$

defines the change of variables from the physical domain Ω to $\mathbb{R}^3$, we transform the field $\mathbf{u}(\mathbf{x})$ (as well as a scalar or a tensor valued function) to be filtered into $\phi(\mathbf{X}) = \mathbf{u}(H^{-1}(\mathbf{X}))$. Then, the function $\phi(\mathbf{X})$ is filtered in the usual way:

$$\overline{\phi}(\mathbf{X}) = \frac{1}{\delta^3} \int_{\mathbb{R}^3} \prod_{i=1}^3 g\left(\frac{X_i - Y_i}{\delta}\right) \mathbf{u}(H^{-1}(\mathbf{Y}))\, d\mathbf{Y}$$

and, coming back to the physical space,

$$\overline{\mathbf{u}}(\mathbf{x}) = \frac{1}{\delta^3} \int_{\mathbb{R}^3} \prod_{i=1}^3 g\left(\frac{H_i(\mathbf{x}) - H_i(\mathbf{y})}{\delta}\right) \mathbf{u}(\mathbf{y}) J(\mathbf{y})\, d\mathbf{y},$$

where $J(\mathbf{x})$ is the Jacobian of the transformation (9.4).

Remark 9.1. The mapping technique from the computational space to a computational domain is not very easily applicable in the case of unstructured meshes and when using finite-volume or finite-element methods. Alternative methods have been proposed, see for instance Fureby and Tabor [119] and the references in the rest of the chapter.

A way to "reduce" the error is proposed in Vasilyev, Lund, and Moin [304] and Marsden, Vasilyev, and Moin [231]. They consider (for simplicity we reduce again to the 1D case) a general filtering (studied also by van der Ven [301]) defined by

$$\overline{u}(x) = \frac{1}{\delta(x)} \int_a^b g\left(\frac{x-y}{\delta(x)}, x\right) u(y)\, dy = \int_{\frac{x-b}{\delta(x)}}^{\frac{x-a}{\delta(x)}} g(\eta, x) u(x - \delta(x)\eta)\, d\eta, \quad (9.5)$$

where $g(x, y)$ is a "location-dependent" filter function. By using a Taylor series expansion it is possible to deduce better estimates on the commutation error. Namely, by defining the "moments" of g as

$$M^l(x) = \int_{\frac{x-b}{\delta(x)}}^{\frac{x-a}{\delta(x)}} \eta^l\, g(\eta, x)\, d\eta$$

and by taking the Taylor series expansion[2] of $u(x - \delta(x)\eta)$ in powers of δ gives

$$u(x - \delta(x)\eta) = \sum_{l=0}^{\infty} \frac{(-1)^l}{l!} \delta^l(x) \eta^l \frac{d^l}{dx^l} u(x).$$

Substituting the expansion in (9.5) we obtain

$$\overline{u}(x) = \sum_{l=0}^{\infty} \frac{(-1)^l}{l!} \delta^l(x) M^l(x) \frac{d^l}{dx^l} u(x)$$

and if we suppose that

$$M^l(x) = \begin{cases} 1, & l = 0 \\ 0, & l = 1, \ldots, N-1, \end{cases} \quad (9.6)$$

it is possible to show that the commutation error has the following expression:

$$\mathcal{E}[u](x) = \sum_{l=N}^{\infty} \frac{(-1)^l}{l!} \frac{d^l}{dx^l} u(x) \frac{d}{dx} \left[\delta^l(x) M^l(x)\right].$$

---

[2] In particular, this series is convergent in the case of uniform δ, by assuming that the Fourier spectrum of u does not include wavenumbers higher than some finite cut-off wavenumber. To some extent, this is the real critical point when filtering is applied to nonsmooth functions.

This shows that if (9.6) holds[3], then

$$\mathcal{E}[u](x) = O(\delta^N(x)), \tag{9.7}$$

provided $\delta'(x) = O(\delta)$.

Remark 9.2. Other generalizations are possible. In particular, since generally the function g is even, the first moment vanishes. A particular important case of skewed filter is the box filter if $\delta^+ \neq \delta^-$. In this case the first moment is nonvanishing and some properties are dramatically different.

A recent analysis of skewed (or a-symmetric) filters is also performed in van der Bos and Geurts [300, 299]. In these papers it is shown how the commutation error may be relatively big in comparison to the SFS stress tensor. By using the same assumption (9.6) on the moments it is shown that the commutation error satisfies (9.7), while

$$\tau[u] = \overline{u\,u} - \overline{u}\,\overline{u} = O(\delta^N(x)) \quad \text{for} \quad N \geq 2.$$

Consequently, the relevant $\frac{d\tau[u]}{dx}$ scales with terms of $O(\delta^N)$ as well as with terms of $O(\delta'\delta^{N-1})$. This result is confirmed through *a priori* tests on a turbulent mixing layer set of data. The numerical experiments show (especially if the filter is skewed) that the contribution from the commutation error is not negligible if compared to the subfilter-scale term and so it is necessary to take this observation into account in the design of advanced LES models. Note that the tests described in [299] do not allow the filter width to degenerate, *i.e.* $\delta(x) \geq \delta_0 > 0$. To continue with the theoretical analysis, in the recent report [34] we compared explicitly (and also asymptotically) these two terms in the case of a couple of near wall models. We used the box filter and a nonuniform filter of radius $\delta(x)$, which do vanish at the boundaries. In these simple, but significant, cases we found that the commutation error may have the same (or even worse) asymptotic behavior as the divergence of the SFS stress tensor but, in the boundary layer,

$$\mathcal{E}[u](x) \geq \frac{d\tau[u](x)}{dx} \quad \text{if } x \text{ is "near" } \partial\Omega.$$

Details on the possible implementation of high order commuting filters are also given in [304, 231], where a discrete version of the filter is defined, achieving both commutation (up to some given order) and an acceptable filter shape in wavenumber space. The main idea, in the case of a 1D grid with points x_i, is to define the value of the filtered variable at the grid point x_i by the relation

$$\overline{u}_i = \sum_{l=-N}^{l=N} a_l u_{i+l},$$

---

[3] Clearly a hidden hypothesis in this application of the Taylor series expansion is that the all derivatives of u, up to the order N, exist and are bounded.

where N is now the radius if the discrete filter stencil. The discrete filter is symmetric if $a_l = a_{-l}$; the constant preservation is represented by $\sum_{l=-N}^{N} a_l = 1$. A connection between discrete filters and convolution kernel or between discrete filters and continuous differential operators is explained in [267], Chap. 10. The main point in defining these filters is to properly choose the coefficients a_l. Common choices are the following three-point symmetric filters:

Table 9.1. Coefficients of some discrete symmetric filters

a_{-1}	a_0	a_1
1/4	1/2	1/4
1/6	2/3	1/6

Full details and the implementation in the multi-dimensional case, can be found in Sect. 3 of [231].

9.1.2 Some Estimates of the Commutation Error

In the previous section we showed some properties of the commutation error arising from filters with variable width and also some possible strategies to reduce it. The estimates proposed in the cited references show some good asymptotic behavior of the commutation error, but unfortunately they are based on the assumption that sufficiently precise Taylor series expansions are known for the functions to be filtered. This is very unlikely in the case of turbulent flows, since the fields to be filtered are generally nonsmooth. In this respect, we derived some estimates by requiring less restrictive constraints on the functions. In particular, in [32] we derived, in some cases, estimates on the commutation error that require just Hölder continuity of the functions to be filtered.

We now specialize to a particular class of filters. We start by showing the effect of a nonuniform filter width for a filter similar to the Gaussian filter over the whole space, i.e. without the presence of boundaries. This section "sets the stage" for the next one. The type of filtering is essentially that of van der Ven [301].

Let $u \in C^1(\mathbb{R}^d) \cap C_b(\mathbb{R}^d)$ (space of continuously differentiable and bounded functions) be a given function and let $\delta_k(\mathbf{x}) \in C_b^1(\mathbb{R}^d)$ denote the width of the filter in the direction of x_k. The average of u is then defined by a tensor product of 1D filters:

$$\overline{u}(\mathbf{y}) = \prod_{k=1}^{d} \frac{1}{\delta_k(\mathbf{y})} \int_{\mathbb{R}^d} \prod_{l=1}^{d} g\left(\frac{x_l}{\delta_l(\mathbf{y})}\right) u(\mathbf{y} - \mathbf{x}) \, d\mathbf{x}.$$

For the moment let us suppose that g is a filter without compact support, but decaying fast enough at infinity in such a way that (possibly after "normalization") g is constant preserving, the first moment of the filter kernel vanishes, and the second one is bounded, *i.e.*

$$\int_{-\infty}^{\infty} g(x)\,dx = 1, \qquad \int_{-\infty}^{\infty} g(x)x\,dx = 0, \qquad \int_{-\infty}^{\infty} g(x)x^2\,dx = M_2 < +\infty.$$

The most popular example for such a filter kernel is the Gaussian (1.17) that we have encountered several times and for which it is well known that

$$\int_{-\infty}^{\infty} g(x)\,x^k dx = \begin{cases} 0 & \text{if } k \text{ is odd} \\ \dfrac{1}{2^k}\dfrac{1}{3^{k/2}} 3\,5\,\cdots\,(k-1) & \text{if } k \text{ is even.} \end{cases}$$

To keep the notation concise, we define the abbreviations $\mathcal{A}(\mathbf{y}) = \prod_{k=1}^{d} \delta_k(\mathbf{y})$,

$$\mathcal{G}(\mathbf{x},\mathbf{y}) = \prod_{k=1}^{d} g\left(\frac{x_k}{\delta_k(\mathbf{y})}\right) \quad \text{and} \quad \mathcal{G}_l(\mathbf{x},\mathbf{y}) = \prod_{k=1,k\neq l}^{d} g\left(\frac{x_k}{\delta_k(\mathbf{y})}\right),$$

such that

$$\overline{u}(\mathbf{y}) = \frac{1}{\mathcal{A}(\mathbf{y})} \int_{\mathbb{R}^d} \mathcal{G}(\mathbf{x},\mathbf{y})u(\mathbf{y}-\mathbf{x})\,d\mathbf{x}.$$

A direct and explicit calculation shows that

$$\begin{aligned}
\partial_i \overline{u}(\mathbf{y}) = -\frac{1}{\mathcal{A}(\mathbf{y})} &\left[\left(\sum_{k=1}^{d} \frac{\partial_i \delta_k(\mathbf{y})}{\delta_k(\mathbf{y})}\right) \int_{\mathbb{R}^d} \mathcal{G}(\mathbf{x},\mathbf{y})u(\mathbf{y}-\mathbf{x})\,d\mathbf{x} \right. \\
&- \int_{\mathbb{R}^d} \left(\sum_{l=1}^{d} \mathcal{G}_l(\mathbf{x},\mathbf{y})g'\left(\frac{x_l}{\delta_l(\mathbf{y})}\right)\frac{x_l\partial_i\delta_l(\mathbf{y})}{\delta_l(\mathbf{y})^2}\right) u(\mathbf{y}-\mathbf{x})\,d\mathbf{x} \qquad (9.8) \\
&\left. + \int_{\mathbb{R}^d} \mathcal{G}(\mathbf{x},\mathbf{y})\partial_i u(\mathbf{y}-\mathbf{x})\,d\mathbf{x} \right],
\end{aligned}$$

where $\partial_i = \partial/\partial x_i$. The last term in (9.8) is just $\overline{\partial_i u}$; consequently the ith component of the commutation error is the sum of the other two terms. The commutation error is now transformed, by using the following integration by parts:

$$\frac{1}{\delta_l(\mathbf{y})} \int_{\mathbb{R}^d} \mathcal{G}_l(\mathbf{x},\mathbf{y})\,g'\left(\frac{x_l}{\delta_l(\mathbf{y})}\right)u(\mathbf{y}-\mathbf{x})\,d\mathbf{x} = \int_{\mathbb{R}^d} \mathcal{G}(\mathbf{x},\mathbf{y})\partial_l u(\mathbf{y}-\mathbf{x})\,d\mathbf{x},$$

$$\begin{aligned}
\frac{1}{\delta_l(\mathbf{y})} \int_{\mathbb{R}^d} \mathcal{G}_l(\mathbf{x},\mathbf{y})\,g'\left(\frac{x_l}{\delta_l(\mathbf{y})}\right)x_l u(\mathbf{y}-\mathbf{x})\,d\mathbf{x} = &-\int_{\mathbb{R}^d} \mathcal{G}(\mathbf{x},\mathbf{y})u(\mathbf{y}-\mathbf{x})\,d\mathbf{x} \\
&+ \int_{\mathbb{R}^d} \mathcal{G}(\mathbf{x},\mathbf{y})x_l\partial_l u(\mathbf{y}-\mathbf{x})\,d\mathbf{x}.
\end{aligned}$$

The vanishing of these terms at infinity follows from the assumption on the fast decay of g as $|\mathbf{x}| \to +\infty$. Inserting these expressions into (9.8) shows that the first term of (9.8) cancels out and yields the following lemma:

Lemma 9.3. *Let $u \in C^1(\mathbb{R}^d) \cap C_b(\mathbb{R}^d)$, and $\delta_l \in C_b^1(\mathbb{R}^d)$, for $l = 1, \ldots, d$. Then, the i-th component of the commutation error can be written in the following special form:*

$$\mathcal{E}_i[u](\mathbf{y}) = \sum_{l=1}^{d} \left[\frac{\partial_i \delta_l(\mathbf{y})}{\delta_l(\mathbf{y})} \left(x_l \overline{\partial_l u} - y_l \overline{\partial_l u} \right) \right].$$

The above representation formula, derived and studied in [32], can be used to obtain a pointwise estimate for the commutation error, as stated in the following proposition:

Proposition 9.4. *Let $u \in C_b^2(\mathbb{R}^d)$, the first moment of the filter kernel vanish and the second moment simply exist, and $\delta_l \in C_b^1(\mathbb{R}^d)$, for $l = 1, \ldots, d$. Then,*

$$|\mathcal{E}_i[u](\mathbf{y})| \leq \|u\|_{C^2(\mathbb{R}^d)} |M_2| \left(\sum_{l=1}^{d} |\partial_i \delta_l(\mathbf{y}) \delta_l(\mathbf{y})| \right).$$

The proof is elementary but it is rather long and not necessary at a first reading, so we prefer not to reproduce it here. We refer to [32] for the proof also in the presence of a more general filter that allows "translation" of the center of the domain of integration.

Remark 9.5. The result in the above proposition shows, in particular, that the value of the commutation error associated with the derivative with respect to x_i depends not only on the derivative with respect to x_i of $\delta_i(\mathbf{y})$, but also on the x_i-derivative of all the filter widths. The commutation error has contributions from all the different directions: the value of $\mathcal{E}_i$ will depend on the variations of all δ_l with respect to the direction x_i.

An Example of a Filter with Compact Support: The Box Filter

We now study filters with compact kernels which are applied to functions u defined on a bounded domain Ω. An essential feature is that the application of the filter must lead to integrals whose domain of integration is a subset of $\overline{\Omega}$, *i.e.* in any direction the filter width at a point $\mathbf{y}$ is not allowed to be larger than the distance of $\mathbf{y}$ to the boundary, in that direction. This situation has the appealing property that an extension of u outside Ω is not necessary. As we have seen in the 1D case, this requirement implies that the filter width has to tend to zero (at least in one direction) as the point $\mathbf{y}$ in which u is filtered tends to the boundary $\partial\Omega$. Thus, necessarily, the filter width is a function of $\mathbf{y}$. We also study the case in which the center of the (asymmetric or skewed filter) filter kernel is not in $\mathbf{y}$.

Let g be a filter kernel with support in $[-1/2, 1/2]$ (without loss of generality) which is normalized. Moreover, we assume again that

$$\int_{-1/2}^{1/2} g(x)\,dx = 1\;, \qquad \int_{-1/2}^{1/2} g(x)x\,dx = 0\;, \qquad \int_{-1/2}^{1/2} g(x)x^2\,dx = M_2,$$

and the most popular filter which fits into this framework is the box or the top-hat filter (1.15).

Let $\Omega \subset \mathbb{R}^d$ be a bounded domain, $u \in C^1(\overline{\Omega})$, $\delta_l(\mathbf{y}) \in C^1(\overline{\Omega})$ be the scalar filter widths with $\delta_l(\mathbf{y}) \geq 0$ for all $\mathbf{y} \in \overline{\Omega}$ and $\delta_l(\mathbf{y}) > 0$ for all $\mathbf{y} \in \Omega$, $l = 1, \ldots, d$. We denote by $\mathcal{B}(\mathbf{y}) = [-\delta_1(\mathbf{y}), \delta_1(\mathbf{y})] \times \cdots \times [-\delta_d(\mathbf{y}), \delta_d(\mathbf{y})]$ and we assume that

$$\mathbf{y} + \mathcal{B}(\mathbf{y}) := [y_1 - \delta_1(\mathbf{y}), y_1 + \delta_1(\mathbf{y})] \times \cdots \times [y_d - \delta_d(\mathbf{y}), y_d + \delta_d(\mathbf{y})] \subset \overline{\Omega}$$

for all $\mathbf{y} = (y_1, \ldots, y_d) \in \overline{\Omega}$. Denoting $\mathcal{A}(\mathbf{y}) = 1/\prod_{l=1}^{d}(2\delta_l(\mathbf{y}))$, the average of u is defined by

$$\overline{u}(\mathbf{y}) = \frac{1}{\mathcal{A}(\mathbf{y})} \int_{\mathbf{y}+\mathcal{B}(\mathbf{y})} \prod_{l=1}^{d} g\left(\frac{y_l - x_l}{2\delta_l(\mathbf{y})}\right) u(\mathbf{x})\,d\mathbf{x}$$

$$= \frac{1}{\mathcal{A}(\mathbf{y})} \int_{\mathcal{B}(\mathbf{y})} \prod_{l=1}^{d} g\left(\frac{x_l}{2\delta_l(\mathbf{y})}\right) u(\mathbf{y} - \mathbf{x})\,d\mathbf{x}.$$

By using the same elementary tools as integration by parts and direct calculations it is possible to derive the following representation formula.

Lemma 9.6. *Let* $u \in C^1(\overline{U(\mathbf{y})})$, *where* $U(\mathbf{y})$ *is a neighborhood of* $\mathbf{y}$ *such that* $\mathbf{y} + \mathcal{B}(\mathbf{y}) \subset U(\mathbf{y})$, $\delta_l^+(\mathbf{y}) \in C^1(\overline{U(\mathbf{y})})$ *and* $\delta_l^-(\mathbf{y}) \in C^1(\overline{U(\mathbf{y})})$, $l = 1, \ldots, d$. *Then, the* i-*th component of the commutation error has the form*

$$\mathcal{E}_i[u](\mathbf{y}) = \sum_{l=1}^{d} \left[\frac{\partial_i \delta_l^+(\mathbf{y}) + \partial_i \delta_l^-(\mathbf{y})}{\delta_l^+(\mathbf{y}) + \delta_l^-(\mathbf{y})} \left(x_l \overline{\partial_l u} - y_l \overline{\partial_l u}\right)(\mathbf{y}) \right.$$

$$\left. + \frac{\partial_i \delta_l^+(\mathbf{y})\delta_l^-(\mathbf{y}) - \partial_i \delta_l^-(\mathbf{y})\delta_l^+(\mathbf{y})}{\delta_l^+(\mathbf{y}) + \delta_l^-(\mathbf{y})} \overline{\partial_l u}(\mathbf{y}) \right].$$

From this formula it is possible to prove an estimate similar to the previous one.

Proposition 9.7. *Let* $u \in C^2(\overline{U(\mathbf{y})})$, *where* $U(\mathbf{y})$ *is defined in Lemma 9.6. Assume that the first moment of the filter kernel vanishes, the second moment exists,* $\delta_l^+(\mathbf{y}) \in C^1(\overline{U(\mathbf{y})})$, *and* $\delta_l^-(\mathbf{y}) \in C^1(\overline{U(\mathbf{y})})$, $l = 1, \ldots, d$. *Then*

$$|\mathcal{E}_i[u](\mathbf{y})| \leq \left| \sum_{l=1}^{d} \frac{\partial_i \delta_l^+(\mathbf{y}) - \partial_i \delta_l^-(\mathbf{y})}{2} \partial_l u(\mathbf{y}) \right|$$

$$+ \|u\|_{C^2(\overline{U(\mathbf{y})})} \left[\sum_{k,l=1}^{d} \frac{|\delta_k^+(\mathbf{y}) - \delta_k^-(\mathbf{y})| \, |\partial_i \delta_l^+(\mathbf{y}) - \partial_i \delta_l^-(\mathbf{y})|}{4} \right.$$

$$\left. + \sum_{l=1}^{d} |M_2| \, |\partial_i \delta_l^+(\mathbf{y}) + \partial_i \delta_l^-(\mathbf{y})| \, \left(\delta_l^+(\mathbf{y}) + \delta_l^-(\mathbf{y}) \right) \right].$$

In the case of a symmetric filter, *i.e.* $\delta^+ = \delta^-$, the estimate becomes much shorter and it is really important to note that the norm in C^2 must be evaluated only in a small neighborhood $\overline{U(\mathbf{y})}$ of the point $\mathbf{y}$. This reflects the "local" nature of the process of filtering.

The Case of Non-very-smooth Functions

In [32], we also studied the problem of the commutation error arising in the filtering of nonsmooth functions. This is motivated by the fact that velocity in weak solutions to the Navier–Stokes equations is found in $W^{1,2}(\Omega)$ and, in both 2D and 3D, $W^{1,2}(\Omega) \not\subset L^\infty(\Omega)$. In special cases (for instance small data or for small time intervals), we have seen that it is possible to prove that the solutions to the Navier–Stokes equations are "strong" and that $\mathbf{u}$ belongs for instance to $W^{2,2}(\Omega) \subset C^{0,1/2}(\Omega)$, see Chap. 2. This means that the study of Hölder-continuous functions may be a first step toward the analysis of functions with the regularity of a weak solution. Furthermore, the corresponding weak pressure solution is even less regular than the velocity.

It is not possible to prove the same results when the domain $\Omega \subset \mathbb{R}^3$ is a polyhedral domain. A regularity result for such a problem has been found in [88]:

$$\mathbf{u} \in W^{3/2-\epsilon,2}(\Omega) \quad \text{and} \quad p \in W^{1/2-\epsilon,2}(\Omega), \qquad \forall \epsilon > 0.$$

An "interior regularity" result still holds and $\mathbf{u} \in W^{2,2}(\Omega')$, with $p \in W^{1,2}(\Omega')$ for each Ω' such that $\overline{\Omega'} \subset \Omega$. We observe that the Sobolev embedding theorem (see [4]) implies, for instance, that $p \in L^\gamma(\Omega)$ for $2 \leq \gamma < 3$. This means that near a possibly singular point $\mathbf{x}_0$ the pressure may have a behavior of the form

$$|p| \sim \frac{1}{\|\mathbf{x} - \mathbf{x}_0\|^\alpha} \quad \text{for} \quad \alpha < 1.$$

In the presence of re-entrant corners, like the backward-facing step, the behavior could be even worse.

Having in mind this motivation, we proved (together with some numerical illustrations) in [32] the following result:

Proposition 9.8. *Let $u \in C^{0,\alpha}(\Omega)$, $\alpha \in (0,1]$ and $\delta_l(\mathbf{y}) \in C^1(\overline{U(\mathbf{y})})$, for $l = 1, \dots, d$ Then*

$$\mathcal{E}_i[u](\mathbf{y}) \leq M_\alpha \left| \sum_{l=1}^d \frac{\partial_i \delta_l(\mathbf{y})}{\delta_l(\mathbf{y})} \right| \left(\sum_{l=1}^d 4\delta_l(\mathbf{y})^2 \right)^{\alpha/2}.$$

This result can be obtained through explicit use of the expression for the commutation error and needs just the Hölder continuity of the function u.

Remark 9.9. As pointed out in [133] (together with an estimate of the commutation error and a comparison with the SFS stress tensor), it is also necessary to take into account more subtle properties. In fact, the estimates used to derive the leading term of the commutation error do not give any information about the spectral content of the analyzed signal. Due to the presence of significant energy in the high frequency portion of the LES spectrum, the commutation error could be large even if it is smaller or comparable with the SFS stress tensor. In this respect see the recent advances on the use of the "local spectrum" analysis in Vasilyev and Goldstein [303]. For details, we refer the reader to the bibliography. Here we just present the main ideas used in [136] to study the *spectral distribution* of the commutation error. Consider a function $u(x) = \sum_k \widehat{u}_k e^{ikx}$, i being the imaginary unit. The two main operations become

$$\overline{\frac{du}{dx}} = \overline{ik\,u} \qquad \text{and} \qquad \frac{d\overline{u}}{dx} = ik\,\overline{u}.$$

A possible way to measure the commutation error is to compare the wavenumber k with the "modified wavenumber" k', the latter being chosen such that

$$\overline{ik\,u} = ik'\,\overline{u}.$$

Then, the departure of k' from k is a measure of the commutation error, which clearly vanishes if $k = k'$. Some manipulations lead to the following expression for the ratio between k' and k:

$$\frac{k'}{k} = 1 - i\delta \frac{f''}{f'^2} \frac{\int_{-\infty}^{+\infty} \eta\, g(\eta) \sin(k\delta\eta/f')\, d\eta}{\int_{-\infty}^{+\infty} \eta\, g(\eta) \cos(k\delta\eta/f')\, d\eta}.$$

On Differential Filters

The presence of a commutation error has been also observed by Germano [126, 127] who introduced differential filters in the study of LES. In particular, he considered the linear differential operator

$$L\,u = u + \alpha_i(\mathbf{x}) \frac{\partial u}{\partial x_i} - \alpha_{ij}(\mathbf{x}) \frac{\partial^2 u}{\partial x_i \partial x_j} \qquad \text{with } \alpha_{ij} = \alpha_{ji},$$

where α_i, and α_{ij} are given smooth functions of $\mathbf{x}$ such that:

$$\exists \, \alpha > 0 : \quad \sum_{ij=1}^{3} \alpha_{ij}(\mathbf{x}) x_i x_j \geq \alpha |\mathbf{x}|^2 \qquad \text{a.e. } \mathbf{x} \in \mathbb{R}^d.$$

The "principal fundamental solution" $G(\mathbf{x}, \mathbf{x}')$ is a "solution" of $L\,u = 0$, defined in the whole space, such that

$$G = O(1/r) \text{ as } r \to 0 \qquad \text{and} \qquad G = O(\mathrm{e}^{-r}) \text{ as } r \to \infty,$$

where $r = |\mathbf{x} - \mathbf{x}'|$. It is well known from the theory of elliptic equations (see for instance Miranda [238]) that for each regular function f, increasing to infinity at most polynomially as $|\mathbf{x}| \to \infty$, we have the following representation formula:

$$\overline{f}(\mathbf{x}) = \int_{\mathbb{R}^3} G(\mathbf{x}, \mathbf{x}') f(\mathbf{x}') \, d\mathbf{x}'$$

for the solution of

$$\overline{f} + \alpha_i(\mathbf{x}) \frac{\partial \overline{f}}{\partial x_i} - \alpha_{ij}(\mathbf{x}) \frac{\partial^2 \overline{f}}{\partial x_i \partial x_j} = f.$$

An important fact is that, if α_i, α_{ij} are not constant, then the function G is not simply a function of $\mathbf{x} - \mathbf{x}'$. This implies that the process of derivation and filtering do not commute.

However, by explicit computation it follows that

$$\mathcal{E}_k[f] = \frac{\overline{\partial f}}{\partial x_k} - \frac{\partial \overline{f}}{\partial x_k} = \overline{\sum_{i=1}^{3} \frac{\partial \alpha_i}{\partial x_k} \frac{\partial \overline{f}}{\partial x_i} - \sum_{ij=1}^{3} \frac{\partial \alpha_{ij}}{\partial x_k} \frac{\partial^2 \overline{f}}{\partial x_i \partial x_j}}.$$

The commutation error can be expressed exactly on the resolvable scale, like the well-known Leonard stresses. This peculiar property of linear differential filters can be utilized in numerical computations.

Other differential filters may be introduced and they can be classified as elliptic, parabolic or hyperbolic, according to the type of linear differential operator involved. In this section, in the spirit of the book, we focused on the mathematical properties of some filters. Other important physical constraints must be taken into account in the choice of the filter. This is itself a broad and complex problem, *e.g.* Sagaut [267] and Germano [128].

9.2 Filters with Constant Radius

In this section, we analyze an approach that is, to some extent, *dual* to that of allowing for variations of the filter radius. Specifically, we analyze in more detail the mathematical consequences of using a constant filter, in the presence of boundaries. The main advantages of using a constant radius filter are:

(i) the commutation error $\mathcal{E}_i[u](\mathbf{x})$ disappears; and (ii) the prohibitive computational cost (scaling as $Re^{2.4}$ to resolve the inner layer, as mentioned in the introduction of this chapter) could be dramatically reduced. Obviously, as we will see, this different approach yields different challenges. The results of this section are essentially those recently proved by Dunca, John, and Layton [101].

9.2.1 Derivation of the Boundary Commutation Error (BCE)

The starting point of our considerations is the NSE in a bounded domain. In order to apply a convolution operator, first one has to extend all functions outside the domain. These functions will fulfill the NSE in a suitable "distributional sense." Then, the convolution operator can be applied, filtering and differentiation commute, and the space averaged Navier–Stokes equations are obtained.

As usual, Ω is a bounded domain in $\mathbb{R}^d$, $d = 2, 3$, with Lipschitz boundary $\partial\Omega$ having the $(d-1)$-dimensional measure $|\partial\Omega| < \infty$. We consider the incompressible NSE (2.1), (2.2) with homogeneous Dirichlet boundary conditions. We assume that the initial boundary value problem associated with the NSE has a unique strong solution $(\mathbf{u}, p)$ in $[0, T]$, hence satisfying

$$\begin{cases} \mathbf{u} \in \left[H^2(\Omega) \cap H_0^1(\Omega) \right]^d & \text{if } t \in [0, T] \\ \mathbf{u} \in \left[H^1(0, T) \right]^d & \text{if } \mathbf{x} \in \overline{\Omega} \\ p \in H^1(\Omega) \cap L_0^2(\Omega) & \text{if } t \in (0, T]. \end{cases} \tag{9.9}$$

We have to extend now $\mathbf{u}$, p, $\mathbf{f}$, and $\mathbf{u}_0$ outside Ω (the first three functions for all times t). Because of the homogeneous Dirichlet boundary conditions, it is natural to extend $\mathbf{u}$ and $\mathbf{u}_0$ by $\mathbf{0}$; from the physical point of view there is no particular reason to extend p and $\mathbf{f}$ in a different way. Thus, we have

$$\mathbf{u}^* = \begin{cases} \mathbf{u} \text{ for } \mathbf{x} \in \Omega \\ \mathbf{0} \text{ for } \mathbf{x} \notin \overline{\Omega} \end{cases} \qquad \mathbf{u}_0^* = \begin{cases} \mathbf{u}_0 \text{ for } \mathbf{x} \in \Omega \\ \mathbf{0} \text{ for } \mathbf{x} \notin \overline{\Omega} \end{cases}$$

$$p^* = \begin{cases} p \text{ for } \mathbf{x} \in \Omega \\ 0 \text{ for } \mathbf{x} \notin \overline{\Omega} \end{cases} \qquad \mathbf{f}^* = \begin{cases} \mathbf{f} \text{ for } \mathbf{x} \in \Omega \\ \mathbf{0} \text{ for } \mathbf{x} \notin \overline{\Omega}. \end{cases}$$

The extended functions satisfy the following:

$$\begin{cases} \mathbf{u}^* \in \left[H_0^1(\mathbb{R}^d) \right]^d & \text{if } t \in [0, T] \\ \mathbf{u}^* \in \left[H^1(0, T) \right]^d & \text{if } \mathbf{x} \in \mathbb{R}^d \\ p^* \in L_0^2(\mathbb{R}^d) & \text{if } t \in (0, T]. \end{cases} \tag{9.10}$$

From (9.9) and (9.10) it follows that $\mathbf{u}_t^*$, $\nabla\mathbf{u}^*$, $\nabla\cdot\mathbf{u}^*$, and $\nabla\cdot(\mathbf{u}^*\mathbf{u}^{*T})$ are well defined for $\mathbf{x}\in\mathbb{R}^d$:

$$\mathbf{u}_t^* = \begin{cases} \mathbf{u}_t \text{ if } \mathbf{x}\in\Omega \\ 0 \quad \text{otherwise} \end{cases} \qquad \nabla\cdot(\mathbf{u}^*\mathbf{u}^{*T}) = \begin{cases} \nabla\cdot(\mathbf{u}\mathbf{u}^T) \text{ if } \mathbf{x}\in\Omega \\ 0 \qquad\qquad \text{otherwise} \end{cases}$$

$$\nabla\mathbf{u}^* = \begin{cases} \nabla\mathbf{u} \text{ if } \mathbf{x}\in\Omega \\ 0 \quad \text{otherwise} \end{cases} \qquad \nabla\cdot\mathbf{u}^* = 0 \quad \text{if } \mathbf{x}\in\mathbb{R}^d.$$

$$(9.11)$$

On the Notion of Distribution

We briefly recall some basic facts about distributions, needed in this section. For more details we refer the reader to Kolmogorov and Fomīn [192]. A reference text, with rigorous results, but also with many interesting applications to mathematical physics, is the textbook by Schwartz [273].

We first define (using standard notation) the linear space

$$\mathcal{D} = \{\phi \in C_0^\infty(\mathbb{R}^d)\},$$

i.e. the space of infinitely differentiable functions, whose support is contained in a bounded set. The space $\mathcal{D}$ is endowed with the following notion of convergence[4]: the sequence $\{\phi_j\}_{j\geq 1} \subset \mathcal{D}$ converges to $\phi \in \mathcal{D}$ if

(a) the support of the functions ϕ_j is contained in the same bounded closed set K, for each $j\in\mathbb{N}$;
(b) there is uniform convergence of all the derivatives of ϕ_j, toward the corresponding derivative of ϕ, *i.e.* for each multi-index $\alpha = (\alpha_1,\ldots,\alpha_d)$

$$\lim_{j\to+\infty}\sup_{\mathbf{x}\in\mathbb{R}^d}\left|\frac{\partial^{|\alpha|}\phi_j}{\partial x_1^{\alpha_1}\ldots\partial x_d^{\alpha_d}} - \frac{\partial^{|\alpha|}\phi}{\partial x_1^{\alpha_1}\ldots\partial x_d^{\alpha_d}}\right| = 0.$$

We can define now the notion of distribution.

Definition 9.10. *We say that T is a distribution if T is a linear and continuous functional over the space $\mathcal{D}$.*

This means that T associates to each $\phi\in\mathcal{D}$ a real number denoted by

$$\langle T, \phi\rangle,$$

and the following properties are satisfied:

$$\begin{cases} \langle T, \phi_1 + \phi_2\rangle = \langle T, \phi_1\rangle + \langle T, \phi_2\rangle & \forall\,\phi_1,\,\phi_2\in\mathcal{D}; \\ \langle T, \lambda\,\phi\rangle = \lambda\langle T, \phi\rangle & \forall\,\lambda\in\mathbb{R}, \quad \forall\,\phi\in\mathcal{D}; \\ \langle T, \phi_j\rangle \to \langle T, \phi\rangle & \text{if } \phi_j \to \phi \text{ in } \mathcal{D}. \end{cases}$$

---

[4] We specify the notion of convergence since the space $\mathcal{D}$ is *not* a Banach space; so it is not possible to find a norm describing this notion of convergence.

We now give a couple of simple examples of distributions. The first is the generalization of the usual concept of function: if $f \in L^1_{loc}(\mathbb{R}^d)$ (that means the integral of $|f|$ is bounded if performed over closed bounded sets of $\mathbb{R}^d$) we can associate to it the distribution T_f defined by

$$\langle T_f, \phi \rangle = \int_{\mathbb{R}^d} f(\mathbf{x}) \phi(\mathbf{x}) \, d\mathbf{x}.$$

The other relevant example is the "Dirac delta function" (note that is not a function but a distribution)

$$\langle \delta, \phi \rangle = \phi(\mathbf{0})$$

that is used to interpret distribution of electric charges, material masses, and so on.

The notion of "support" of a distribution is also of importance. We say that a distribution T is vanishing in an open set $\Omega \subset \mathbb{R}^d$ if $\langle T, \phi \rangle = 0$ for each function $\phi \in \mathcal{D}$ whose support is contained in Ω. We now define the notion of derivative of a distribution, that generalizes the usual notion of derivative.

Definition 9.11. *The partial derivative $\partial T / \partial x_i$ of the distribution T is again a distribution and is defined through the formula*

$$\left\langle \frac{\partial T}{\partial x_i}, \phi \right\rangle = - \left\langle T, \frac{\partial \phi}{\partial x_i} \right\rangle \qquad \forall \phi \in \mathcal{D}.$$

Roughly speaking, this definition uses the integration by parts formula to give meaning to the derivative of a distribution since in the case of a smooth (say C^1) function it implies that

$$T_{\partial f / \partial x_i} = \frac{\partial T_f}{\partial x_i}.$$

Furthermore, since $\partial T / \partial x_i$ is itself a distribution, it may be derived (in the sense of distributions) again. Hence distributions possess infinite-order derivatives, defined by

$$\left\langle \frac{\partial^{|\alpha|} T}{\partial x_1^{\alpha_1} \dots \partial x_d^{\alpha_d}}, \phi \right\rangle = (-1)^{|\alpha|} \left\langle T, \frac{\partial^{|\alpha|} \phi}{\partial x_1^{\alpha_1} \dots \partial x_d^{\alpha_d}} \right\rangle.$$

The last point in our summary is the definition of $\overline{T} = T * g_\delta$, the convolution between a distribution T and a smooth function g_δ.

Definition 9.12. *Let T be a distribution with compact support which has the form*

$$\langle T, \phi \rangle = - \int_{\mathbb{R}^d} f(\mathbf{x}) \partial_\alpha \varphi(\mathbf{x}) \, d\mathbf{x},$$

where ∂_α is the derivative of ϕ with respect to the multi-index α and $f \in L^1_{loc}$. Then

$$\overline{T}(\mathbf{x}) = \langle T, g_\delta(\mathbf{x} - \cdot) \rangle = - \int_{\mathbb{R}^d} f(\mathbf{y}) \partial_\alpha g_\delta(\mathbf{x} - \mathbf{y}) \, d\mathbf{y}. \qquad (9.12)$$

It turns out that $\overline{T} = T * g_\delta \in C^\infty(\mathbb{R}^d)$. For precise definitions and further properties of the convolution of a distribution with a function, see for instance Rudin [266], Chap. 6.

We can now check that all the "starred variables" in this section are well defined in the sense of distributions, too.

Remark 9.13. With a small abuse of notation, we remove from now on the star superscript from all variables. Each function we will study in the sequel is the null extension to $\mathbb{R}^d$ of the corresponding function previously defined on Ω.

For instance, we check the second term in (9.11). If $\varphi \in [\mathcal{D}]^d$ and we define $\nabla \cdot (\mathbf{u}\mathbf{u}^T)$ in the sense of distributions, we obtain

$$\langle \nabla \cdot (\mathbf{u}\mathbf{u}^T)(\mathbf{x}), \varphi \rangle := -\int_{\mathbb{R}^d} (\mathbf{u}\mathbf{u}^T)(\mathbf{x}) \cdot \nabla \varphi(\mathbf{x}) \, d\mathbf{x}$$

$$= -\int_\Omega (\mathbf{u}\mathbf{u}^T)(\mathbf{x}) \cdot \nabla \varphi(\mathbf{x}) \, d\mathbf{x}$$

$$= \int_\Omega \nabla \cdot (\mathbf{u}\mathbf{u}^T)(\mathbf{x}) \, \varphi(\mathbf{x}) \, d\mathbf{x},$$

due to the fact that the boundary integral vanishes ($\mathbf{u} = \mathbf{0}$ on $\partial\Omega$).

The terms $Re^{-1}\triangle\mathbf{u}$ and ∇p require some care since their definitions in the sense of distributions are not trivial at all. Let $\varphi \in [\mathcal{D}]^d$ and, as usual, let $\mathbf{n}$ be the outward normal vector on $\partial\Omega$. Then, from the definition of p on $\mathbb{R}^d$ it follows that

$$\langle \nabla p, \varphi \rangle := -\int_{\mathbb{R}^d} p(\mathbf{x}) \nabla \cdot \varphi(\mathbf{x}) \, d\mathbf{x} = -\int_\Omega p(\mathbf{x}) \nabla \cdot \varphi(\mathbf{x}) \, d\mathbf{x}$$

$$= \int_\Omega \nabla p(\mathbf{x}) \cdot \varphi(\mathbf{x}) \, d\mathbf{x} - \int_{\partial\Omega} p(\mathbf{s})\varphi(\mathbf{s}) \cdot \mathbf{n}(\mathbf{s}) \, dS(\mathbf{s}),$$

where $dS(\mathbf{s})$ is the surface (line in 2D) element.

In the same way, one obtains

$$\langle \triangle\mathbf{u}, \varphi \rangle = \langle \nabla \cdot \nabla\mathbf{u}, \varphi \rangle := -\int_{\mathbb{R}^d} \nabla\mathbf{u}(\mathbf{x})\nabla\varphi(\mathbf{x}) \, d\mathbf{x}$$

$$= -\int_\Omega \nabla\mathbf{u}(\mathbf{x})\nabla\varphi(\mathbf{x}) \, d\mathbf{x}$$

$$= \int_\Omega \triangle\mathbf{u}(\mathbf{x})\varphi(\mathbf{x}) \, d\mathbf{x} - \int_{\partial\Omega} \varphi(\mathbf{s})\nabla\mathbf{u}(\mathbf{s})\mathbf{n}(\mathbf{s}) \, dS(\mathbf{s}).$$

Both distributions have compact support.

The extended functions fulfill the following distributional form of the momentum equation

$$\mathbf{u}_t - \frac{1}{Re}\triangle\mathbf{u} + \nabla\cdot(\mathbf{u}\mathbf{u}^T) + \nabla p = \mathbf{f} + \int_{\partial\Omega} \left(\frac{1}{Re}\nabla\mathbf{u}\cdot\mathbf{n} - p\,\mathbf{n}\right)(\mathbf{s}) \, \varphi(\mathbf{s}) \, dS(\mathbf{s}). \quad (9.13)$$

The space-averaged Navier–Stokes equations are now derived by convolving (9.13) with a filter function $g_\delta(\mathbf{x}) \in C^\infty(\mathbb{R}^d)$.

By filtering (9.13) via convolution with g_δ, by using the fact that convolution and differentiation commute, and by convolving the extra term on the right-hand side according to (9.12), we obtain the space-averaged momentum equation:

$$\overline{\mathbf{u}}_t - \frac{1}{Re}\Delta\overline{\mathbf{u}} + \nabla \cdot (\overline{\mathbf{uu}^T}) + \nabla\overline{p} = \overline{\mathbf{f}}$$
$$+ \int_{\partial\Omega} g_\delta(\mathbf{x} - \mathbf{s}) \left[\frac{1}{Re}\nabla\mathbf{u}(\mathbf{s})\mathbf{n}(\mathbf{s}) - p(\mathbf{s})\mathbf{n}(\mathbf{s})\right] dS(\mathbf{s}) \quad \text{in } (0,T) \times \mathbb{R}^d.$$

$$\text{(9.14)}$$

Remark 9.14. Very often, the deformation-tensor-formulation of the momentum equation of the NSE

$$\mathbf{u}_t - \frac{2}{Re}\nabla \cdot (\nabla^s\mathbf{u}) + \nabla \cdot (\mathbf{uu}^T) + \nabla p = \mathbf{f} \quad \text{in } (0,T) \times \Omega$$

is used as starting point in LES where, as usual, $\nabla^s\mathbf{u}$ is the symmetric part of the gradient. The same considerations as for the gradient formulation of the momentum equation lead to the following space-averaged deformation-tensor-formulation:

$$\overline{\mathbf{u}}_t - \frac{2}{Re}\nabla \cdot (\nabla^s\overline{\mathbf{u}}) + \nabla \cdot (\overline{\mathbf{uu}^T}) + \nabla\overline{p} = \mathbf{f}$$
$$+ \int_{\partial\Omega} g(\mathbf{x} - \mathbf{s}) \left[\frac{2}{Re}\nabla^s\mathbf{u}(\mathbf{s}) \cdot \mathbf{n}(\mathbf{s}) - p(\mathbf{s})\mathbf{n}(\mathbf{s})\right] dS(\mathbf{s}) \quad \text{in } (0,T) \times \mathbb{R}^d.$$

$$\text{(9.15)}$$

Thus, the space-averaged NSE arising from the NSE on a bounded domain possess an extra boundary integral. Omitting this integral results in the so-called *Boundary Commutation Error* (BCE). This integral poses a new modeling question since the BCE depends on $(\mathbf{u}, p)$ and not on the space-averaged quantities $(\overline{\mathbf{u}}, \overline{p})$.

Remark 9.15. The term

$$\frac{2}{Re}\nabla^s\mathbf{u} \cdot \mathbf{n} - p\,\mathbf{n} \tag{9.16}$$

is the *Cauchy stress (or traction) vector*. It is naturally closely related to the vector with the full gradient

$$\frac{1}{Re}\nabla\mathbf{u} \cdot \mathbf{n} - p\,\mathbf{n}. \tag{9.17}$$

In general, the Cauchy stress vector does not vanish on the whole boundary. From the regularity hypotheses (9.9) and (9.10) it follows that both terms in (9.16) belong to $\left[H^{1/2}(\partial\Omega)\right]^d$. Thus, in particular they belong to

$[L^2(\partial\Omega)]^d$. From Galdi [120] Chap. 2, and $|\partial\Omega| < \infty$ it follows that the terms in (9.16) are in $[L^q(\partial\Omega)]^d$ with $1 \leq q < \infty$ if $d = 2$ and $1 \leq q \leq 4$ if $d = 3$. The Sobolev embeddings imply that:

$$\left\| \frac{1}{Re}\nabla\mathbf{u}\cdot\mathbf{n} - p\,\mathbf{n} \right\|_{[L^q(\partial\Omega)]^d} \leq C\left(\frac{1}{Re}\|\mathbf{u}\|_{[H^2(\Omega)]^d} + \|p\|_{H^1(\Omega)} \right),$$

$$\left\| \frac{2}{Re}\nabla^s\mathbf{u}\cdot\mathbf{n} - p\,\mathbf{n} \right\|_{[L^q(\partial\Omega)]^d} \leq C\left(\frac{1}{Re}\|\mathbf{u}\|_{[H^2(\Omega)]^d} + \|p\|_{H^1(\Omega)} \right).$$

9.2.2 Estimates of the Boundary Commutation Error Term

In this section we give some estimates for the BCE. In particular, we show that the BCE, belongs to $[L^p(\mathbb{R}^d)]^d$. We derive a sufficient and necessary condition for the convergence to zero of the commutation error in the norm of $[L^p(\mathbb{R}^d)]^d$, as the filter width δ tends to zero. It turns out that, in general, this condition will not be satisfied in practice.

In view of Remark 9.15, it is necessary to study terms defined on the whole space, of the following special form:

$$F(\mathbf{x}) := \int_{\partial\Omega} g_\delta(\mathbf{x} - \mathbf{s})\psi(\mathbf{s})\,dS(\mathbf{s}) \qquad \mathbf{x} \in \mathbb{R}^d, \tag{9.18}$$

with $\psi(\mathbf{s}) \in L^q(\partial\Omega)$, for $1 \leq q \leq \infty$. We will first show that (9.18) belongs to any Lebesgue space, if g_δ is the Gaussian kernel.

Proposition 9.16. *Let $\psi(\mathbf{s}) \in L^q(\partial\Omega)$, $1 \leq q \leq \infty$, and let g_δ be defined by (1.17). Then (9.18) belongs to $L^p(\mathbb{R}^d)$, for $1 \leq p \leq \infty$.*

Proof. By Hölder's inequality with $r^{-1} + q^{-1} = 1$, $r < \infty$, one obtains

$$\left| \int_{\partial\Omega} g_\delta(\mathbf{x} - \mathbf{s})\psi(\mathbf{s})\,dS(\mathbf{s}) \right| \leq \left(\int_{\partial\Omega} g_\delta^r(\mathbf{x} - \mathbf{s})\,dS(\mathbf{s}) \right)^{1/r} \|\psi\|_{L^q(\partial\Omega)}$$

$$= \left(\int_{\partial\Omega} \left(\frac{6}{\delta^2\pi} \right)^{rd/2} \exp\left(-\frac{6r}{\delta^2}\|\mathbf{x} - \mathbf{s}\|_2^2 \right) \right)^{1/r} \|\psi\|_{L^q(\partial\Omega)}.$$

By the triangle inequality and Young's inequality, it follows that

$$2\|\mathbf{x} - \mathbf{s}\|_2^2 \geq \|\mathbf{x}\|_2^2 - 2\|\mathbf{s}\|_2^2 \qquad \forall\,\mathbf{x}, \mathbf{s} \in \mathbb{R}^d,$$

which implies

$$\exp\left(-\frac{6r\|\mathbf{x} - \mathbf{s}\|_2^2}{\delta^2} \right) \leq \exp\left(3r\frac{-\|\mathbf{x}\|_2^2 + 2\|\mathbf{s}\|_2^2}{\delta^2} \right).$$

It follows that

$$\left| \int_{\partial\Omega} g_\delta(\mathbf{x} - \mathbf{s})\psi(\mathbf{s}) \, dS(\mathbf{s}) \right|$$

$$\leq \left(\frac{6}{\delta^2 \pi} \right)^{d/2} \|\psi\|_{L^q(\partial\Omega)} \left[\int_{\partial\Omega} \exp\left(\frac{6r\|\mathbf{s}\|_2^2}{\delta^2} \right) dS(\mathbf{s}) \right]^{1/r} \exp\left(-\frac{3\|\mathbf{x}\|_2^2}{\delta^2} \right) < \infty,$$

since $\partial\Omega$ is compact and the exponential is a bounded function. This proves the statement for $L^\infty(\mathbb{R}^d)$. The proof for $p \in [1, \infty)$ is obtained by raising both sides of the latter equation to the power p, integrating on $\mathbb{R}^d$, and using

$$\int_{\mathbb{R}^d} \exp\left(-\frac{3p\|\mathbf{x}\|_2^2}{\delta^2} \right) d\mathbf{x} < \infty.$$

If $q = 1$, we have for $1 \leq p < \infty$

$$\int_{\mathbb{R}^d} \left| \int_{\partial\Omega} g_\delta(\mathbf{x} - \mathbf{s})\psi(\mathbf{s}) \, dS(\mathbf{s}) \right|^p d\mathbf{x} \leq \|\psi\|_{L^1(\partial\Omega)}^p \int_{\mathbb{R}^d} \sup_{\mathbf{s} \in \partial\Omega} g_\delta^p(\mathbf{x} - \mathbf{s}) \, d\mathbf{x}$$

$$= \|\psi\|_{L^1(\partial\Omega)}^p \int_{\mathbb{R}^d} g_\delta^p(d(\mathbf{x}, \partial\Omega)) \, d\mathbf{x}.$$

We choose a ball $B(\mathbf{0}, R)$ with radius R such that $d(\mathbf{x}, \partial\Omega) > \|\mathbf{x}\|_2/2$ for all $\mathbf{x} \notin B(\mathbf{0}, R)$. Then, the integral on $\mathbb{R}^d$ is split into a sum of two integrals. The first integral is computed on $B(\mathbf{0}, R)$. This is finite since the integrand is a continuous function on $\overline{B(\mathbf{0}, R)}$. The second integral on $\mathbb{R}^d \setminus B(\mathbf{0}, R)$ is also finite because

$$\int_{\mathbb{R}^d \setminus B(\mathbf{0}, R)} g_\delta^p(d(\mathbf{x}, \partial\Omega)) \, d\mathbf{x} \leq \int_{\mathbb{R}^d} g_\delta^p\left(\frac{\|\mathbf{x}\|_2}{2} \right) d\mathbf{x},$$

due to the integrability of the Gaussian filter. This concludes the proof for $p < \infty$. For $p = \infty$, we have

$$\sup_{\mathbf{x} \in \mathbb{R}^d} \left| \int_{\partial\Omega} g_\delta(\mathbf{x} - \mathbf{s})\psi(\mathbf{s}) \, dS(\mathbf{s}) \right| \leq \sup_{\mathbf{x} \in \mathbb{R}^d} \sup_{\mathbf{s} \in \partial\Omega} g_\delta(\mathbf{x} - \mathbf{s}) \|\psi\|_{L^1(\partial\Omega)}$$

$$\leq g_\delta(\mathbf{0}) \|\psi\|_{L^1(\partial\Omega)} < \infty.$$

$\square$

In the next proposition, we study the behavior of the $L^p(\mathbb{R}^d)$-norm of the function F defined in (9.18), as $\delta \to 0$.

Proposition 9.17. Let $\psi(\mathbf{s}) \in L^p(\partial\Omega)$, $1 \leq p \leq \infty$. A necessary and sufficient condition for

$$\lim_{\delta \to 0} \left\| \int_{\partial\Omega} g_\delta(\mathbf{x} - \mathbf{s})\psi(\mathbf{s}) \, dS(\mathbf{s}) \right\|_{L^p(\mathbb{R}^d)} = 0 \qquad \forall p \in [1, \infty], \tag{9.19}$$

is that $\psi(\mathbf{s})$ vanishes almost everywhere on $\partial\Omega$.

Proof. It is obvious that the condition is sufficient.

Conversely, let (9.19) hold. From Hölder's inequality, we obtain for an arbitrary function $\varphi \in \mathcal{D}$

$$\lim_{\delta \to 0} \left| \int_{\mathbb{R}^d} \varphi(\mathbf{x}) \left(\int_{\partial \Omega} g_\delta(\mathbf{x} - \mathbf{s}) \psi(\mathbf{s}) \, dS(\mathbf{s}) \right) d\mathbf{x} \right| \tag{9.20}$$
$$\leq \lim_{\delta \to 0} \|\varphi\|_{L^q(\mathbb{R}^d)} \left\| \int_{\partial \Omega} g_\delta(\mathbf{x} - \mathbf{s}) \psi(\mathbf{s}) \, dS(\mathbf{s}) \right\|_{L^p(\mathbb{R}^d)} = 0,$$

where $p^{-1} + q^{-1} = 1$. By Fubini's theorem and the symmetry of the Gaussian filter, we have

$$\lim_{\delta \to 0} \int_{\mathbb{R}^d} \varphi(\mathbf{x}) \left(\int_{\partial \Omega} g_\delta(\mathbf{x} - \mathbf{s}) \psi(\mathbf{s}) \, dS(\mathbf{s}) \right) d\mathbf{x}$$
$$= \lim_{\delta \to 0} \int_{\partial \Omega} \psi(\mathbf{s}) \left(\int_{\mathbb{R}^d} g_\delta(\mathbf{x} - \mathbf{s}) \varphi(\mathbf{x}) \, d\mathbf{x} \right) dS(\mathbf{s})$$
$$= \int_{\partial \Omega} \psi(\mathbf{s}) \lim_{\delta \to 0} (g_\delta * \varphi)(\mathbf{s}) \, dS(\mathbf{s}).$$

Since $g_\delta * \varphi$ converges to φ as $\delta \to 0$ (see Proposition 2.32) it follows by the trace theorem that $g_\delta * \varphi \to \varphi$ as $\delta \to 0$ in $L^p(\partial \Omega)$. Thus, from (9.20) it follows that

$$\left| \int_{\partial \Omega} \psi(\mathbf{s}) \varphi(\mathbf{s}) \, dS(\mathbf{s}) \right| = 0 \qquad \forall \varphi \in C_0^\infty(\mathbb{R}^d).$$

This is true if and only if $\psi(\mathbf{s})$ vanishes almost everywhere on $\partial \Omega$. □

Remark 9.18. Proposition 9.17 implies that the commutation error terms in (9.14) and (9.15) vanish in $[L^p(\mathbb{R}^d)]^d$ if and only if the Cauchy stress vectors (9.16) vanish almost everywhere. However, this property is in general not satisfied, since it implies that **there is no interaction between the fluid and the boundary**.

We will now bound the $L^p(\mathbb{R})$-norm of (9.18) in terms of δ.

Proposition 9.19. *Let Ω be a bounded domain in $\mathbb{R}^d$ with Lipschitz boundary $\partial \Omega$, $\psi \in L^p(\partial \Omega)$ for some $p > 1$, and $p^{-1} + q^{-1} = 1$. Then, for every $\alpha \in (0, 1)$ and $k \in (0, \infty)$ there exist constants $C > 0$ and $\epsilon > 0$ such that*

$$\int_{\mathbb{R}^d} \left| \int_{\partial \Omega} g_\delta(\mathbf{x} - \mathbf{s}) \psi(\mathbf{s}) \, dS(\mathbf{s}) \right|^k d\mathbf{x} \leq C \delta^{1 + k\left(\frac{(d-1)\alpha}{q} - d \right)} \|\psi\|_{L^p(\partial \Omega)}^k$$

for every $\delta \in (0, \epsilon)$, where C and ϵ depend on α, k, and $|\partial \Omega|$.

The proof of this proposition is technical and relies on some geometrical properties of the domain, together with a complicated construction of an appropriate mesh on $\partial \Omega$ on which calculations are performed. We have only stated the final result and we refer the reader to [101, 175] for more details.

9.2.3 Error Estimates for a Weak Form of the Boundary Commutation Error Term

In this section, we consider a weak form of the boundary commutation error term, *i.e.* the BCE term (9.18) is multiplied by a test function φ and integrated on $\mathbb{R}^d$. This is very interesting since it can be found in the weak formulation of the space-averaged NSE and in the numerical studies using a discretization based on a variational formulation. In addition, if we consider a weak formulation, we can hope to have better convergence, as $\delta \to 0$, for the BCE.

The following proposition shows how the weak form converges to zero as δ tends to zero with some estimates on its rate. For $d = 2$, Proposition 9.21 shows that the convergence is almost of order one if $\psi(\mathbf{s})$ is sufficiently smooth.

Proposition 9.20. *Let $v \in H^1(\mathbb{R}^d)$ such that $v|_\Omega \in H_0^1(\Omega) \cap H^2(\Omega)$ and $v(\mathbf{x}) = 0$ if $\mathbf{x} \notin \overline{\Omega}$ and let $\psi \in L^p(\partial\Omega)$, $1 \leq p \leq \infty$. Then*

$$\lim_{\delta \to 0} \int_{\mathbb{R}^d} \overline{v}(\mathbf{x}) \left(\int_{\partial\Omega} g_\delta(\mathbf{x} - \mathbf{s}) \psi(\mathbf{s}) \, dS(\mathbf{s}) \right) d\mathbf{x} = 0,$$

where $\overline{v}(\mathbf{x}) = (g_\delta * v)(\mathbf{x})$.

Proof. By Fubini's theorem and the symmetry of g_δ, we obtain

$$\lim_{\delta \to 0} \int_{\mathbb{R}^d} \overline{v}(\mathbf{x}) \left(\int_{\partial\Omega} g_\delta(\mathbf{x} - \mathbf{s}) \psi(\mathbf{s}) \, dS(\mathbf{s}) \right) d\mathbf{x}$$

$$= \lim_{\delta \to 0} \int_{\partial\Omega} \psi(\mathbf{s}) \left(\int_{\mathbb{R}^d} g_\delta(\mathbf{s} - \mathbf{x}) \overline{v}(\mathbf{x}) \, d\mathbf{x} \right) dS(\mathbf{s}).$$

By a Sobolev embedding theorem it follows that $v \in L^\infty(\mathbb{R}^d)$. In addition, by using twice the results of convergence of the Gaussian filter for $\delta \to 0$ and the fact that $\overline{v}$ is uniformly continuous on $\partial\Omega$ (see Proposition 2.32) it follows that

$$\lim_{\delta \to 0} \int_{\mathbb{R}^d} g_\delta(\mathbf{s} - \mathbf{x}) \overline{v}(\mathbf{x}) \, d\mathbf{x} = v(\mathbf{s}).$$

By using the fact that v vanishes on $\partial\Omega$, it follows that

$$\lim_{\delta \to 0} \int_{\mathbb{R}^d} \overline{v}(\mathbf{x}) \left(\int_{\partial\Omega} g_\delta(\mathbf{x} - \mathbf{s}) \psi(\mathbf{s}) \, ds \right) d\mathbf{x} = \int_{\partial\Omega} \psi(\mathbf{s}) v(\mathbf{s}) \, ds = 0.$$

$\square$

With the result of Proposition 9.19, it is possible (again we only state the result, without proofs) to study the order of convergence with respect to δ of the weak form of the BCE term.

Proposition 9.21. *Let v and ψ be defined as in Proposition 9.20 and let the assumption of Proposition 9.19 be fulfilled. Then, there exists an $\epsilon > 0$ such that for $\delta \in (0, \epsilon)$,*

$$\int_{\mathbb{R}^d} \left| \overline{v}(\mathbf{x}) \int_{\partial\Omega} g_\delta(\mathbf{x} - \mathbf{s}) \psi(\mathbf{s}) \, dS(\mathbf{s}) \right|^k dx$$

$$\leq C\delta^{1+(-d+\frac{(d-1)\alpha}{q}+\beta\alpha)k} \|\psi\|_{L^p(\partial\Omega)}^k \|v\|_{H^2(\Omega)}^k,$$

where $k \in [1, \infty)$, $\beta \in (0, 1)$ if $d = 2$ and $\beta = 1/2$ if $d = 3$, $p^{-1} + q^{-1} = 1$, $p > 1$, and C and ϵ depend on α, k, and $|\partial\Omega|$.

An easy consequence of Proposition 9.21 is the following:

Corollary 9.22. *Let the assumptions of Proposition 9.21 be fulfilled. Then, for the weak form of the BCE term, the following inequality holds:*

$$\left| \int_{\mathbb{R}^d} \overline{v}(\mathbf{x}) \int_{\partial\Omega} g_\delta(\mathbf{x} - \mathbf{s}) \psi(\mathbf{s}) \, dS(\mathbf{s}) \right| dx \leq C\delta^{1-d+\frac{(d-1)\alpha}{q}+\beta\alpha} \|\psi\|_{L^p(\partial\Omega)} \|v\|_{H^2(\Omega)}.$$

$$(9.21)$$

Remark 9.23. Let $d = 2$ and $p < \infty$ arbitrarily large. Then q is arbitrarily close to one. Choosing α and β also arbitrarily close to one leads to the following power of δ in (9.21):

$$1 + (-2 + (1 - \epsilon_1) + (1 - \epsilon_2)) = 1 - (\epsilon_1 + \epsilon_2) = 1 - \epsilon_3$$

for arbitrarily small $\epsilon_1, \epsilon_2, \epsilon_3 > 0$. In this case, the convergence is almost of first order.

The result of Proposition 9.21 does not provide an order of convergence for $d = 3$. Following Remark 9.15, let us choose $p = 4$, *i.e.* $q = 4/3$. Then, the power of δ in (9.21) becomes $2(\alpha - 1)$, which is negative for $\alpha < 1$.

9.2.4 Numerical Approximation of the Boundary Commutation Error

Recently, there have been some interesting developments in the numerical approximation of the boundary commutation error.

Das and Moser proposed in [83] the following approach to approximate the boundary commutation error $A_\delta(\sigma)$: to estimate the shear stresses, the authors included in the computational domain a buffer region outside the wall. In this region, the velocities are set to zero, and the wall stresses are determined to minimize the kinetic energy in the buffer region. The resulting system can be thought of as an LES version of embedded boundary techniques. The approach has been tested on several model problems, including the heat equation, Burgers equation, and turbulent channel flow, with good results.

A different approach has recently been proposed by Borggaard and Iliescu [39]. The authors used an *approximate deconvolution* (AD) approach to approximate the boundary commutation error $A_\delta(\sigma)$. The AD was presented at length in Chap. 7. It is based on the following idea: by using the mathematical properties of the particular spatial filter g_δ and the numerical approximation of $\overline{\mathbf{u}}$, one can obtain an *approximation* of (some of) the subfilter-scale information contained in $\mathbf{u} - \overline{\mathbf{u}}$. AD was combined with physical insight and was successfully used in challenging test problems, such as compressible flows and shock-turbulent-boundary-layer interaction [290, 3].

Thus, AD appears as a natural approach in developing NWMs. The applications that would probably benefit most from this approach would be those in which the boundary conditions are time dependent (such as in a flow control setting).

In [39], the authors modeled the commutation error $A_\delta(\sigma)$ using an AD approach. As a first step, they illustrated their *Approximate Deconvolution Boundary Conditions* (ADBC) algorithm for the heat equation. This linear problem was chosen to decouple the boundary treatment from the closure problem. The numerical tests indicated that the commutation error should be included in the numerical model. The ADBC algorithm yielded appropriate numerical approximations for the boundary commutation error.

These first tests were encouraging. Obviously, the algorithm should be tested on realistic turbulent flows (at the time of writing, the ADBC algorithm is being tested on channel flows with time-dependent boundary conditions).

9.3 Conclusions

The twin problems of correctly adapting a filter radius near the wall and of modeling the boundary commutation error when filtering through a wall are *central* problems in the traditional approach to LES. At the moment, these problems are complex and technically intricate – a clear sign that the right approach has not yet been found.

In this chapter we tried to give a general presentation of the accomplishments and, more importantly, the critical challenges in filtering on bounded domains. We also tried to introduce the necessary mathematical background for an inherently technical topic.

Admittedly, this chapter ends with more open questions (and thus, research opportunities for fresh minds!) than answers. Much more remains to be done, both at a mathematical and an algorithmic level. The potential payoff for any development could be, however, significant. To understand this, it is sufficient, for example, to consider the scaling argument presented at the beginning of this chapter, which implied that a brute force approach to simulating the boundary layers has prohibitive computational cost for many flows of practical interest. Considering alternative approaches appears the only reasonable path.

We end this final note by mentioning two interesting attempts to finesse the boundary commutation error question: defining averages by projection in Hughes' Variational Multiscale Method [160, 161, 162] and defining averages by differential filters [127, 126] (both treated in other chapters).

Near Wall Models in LES

10.1 Introduction

As we saw in the previous chapter, one basic problem in LES is turbulence driven by interaction of a flow with a wall. Mathematically, this is the problem of *specifying boundary conditions for flow averages*. Flow averages (with constant averaging radius δ) are *inherently nonlocal*: they depend on the behavior of the unknown, underlying turbulent flow near the boundary. On the other hand, to be guided by the mathematical theory of the equations of fluid motion and seek boundary conditions that have hope of leading to a well-posed problem, those boundary conditions should be *local*.

One key seems to be the work on the commutation error (presented in the previous chapter), which accounts for a significant part of the nonlocal effects near the walls. Thus, a reliable model for the commutation error appears to be an essential ingredient in the development of appropriate local boundary conditions for the flow averages.

At this point, some comments are necessary. In LES, the question of finding boundary conditions when using a constant averaging radius δ is known as *Near Wall Modeling* and a boundary condition is known as *Near Wall Model (NWM)*. This is related to the extensive literature in Conventional Turbulence Modeling (CTM) on "wall-laws." CTM seeks to approximate long time averages of flow quantities and, conveniently for CTM, there is a lot of experimental and asymptotic information available about time averaged turbulent boundary layers. One common approach in CTM is to place an artificial boundary *inside* the flow domain and *outside* the boundary layers. A boundary condition is given for the CTM on this artificial boundary by a Dirichlet condition for the stresses: they are required to match the stress at the edge of the layer given by, *e.g.* a log-law of the wall profile.

There are some interesting differences between CTM and the problem of near wall modeling in LES. First, with constant averaging radius, there is no structure in $\overline{\mathbf{u}}$ smaller than $O(\delta)$. Thus, there is no need to try to guess the edge of any layer and construct artificial boundaries inside Ω. Understandably, early LES studies used the extensive experience in CTM and tried NWM with the same approach as wall laws in CTM. Nevertheless, it seems clear now that in NWM applied to LES *the boundary condition can (and should) be imposed at the physical boundary.* The second distinction is that LES describes inherently dynamic phenomena, so imposing a condition that $\overline{\mathbf{u}}$ should match some equilibrium profile cannot be correct. At this point, one challenge in LES is how to use the extensive information on *time averaged turbulent* boundary layers to generate NWMs that allow *time fluctuating* solution behavior near the wall. We feel that the solution outlined in this chapter is a step along the correct path for this problem.

The last issue is how to reflect the fact that $\overline{\mathbf{u}}$ is inherently nonlocal near the boundary. As we stated earlier, we believe that the right approach is to separate the issue of nonlocality, which we believe is due predominantly to the commutation error term (Chap. 9), from the question of appropriate, well-posed boundary conditions, and then to study each carefully and combine their solutions.

10.2 Wall Laws in Conventional Turbulence Modeling

In this section, we present some of the wall laws used in devising physically reasonable boundary conditions in CTM, even if they have been used also in LES. We will focus mainly on the mathematical properties of this topic, and we refer the reader to Cousteix [79] and to Chap. 9 in Sagaut [267] for a more detailed physical statement of the problem.

A classical approach, introduced for the $k-\varepsilon$ model, consists in eliminating part of the boundary layer; see Launder and Spalding [200]. The boundary that is considered is not the real boundary $\partial\Omega$, but an artificial one $\partial\Omega_1$, lying inside the volume of the flow. If the boundary is smooth, we can impose the following boundary condition:

$$\begin{cases} \overline{\mathbf{u}} \cdot \mathbf{n} = \phi(\mathbf{x}), \\ \\ \mathbf{n} \cdot \sigma(\overline{\mathbf{u}}, \overline{p}) \cdot \boldsymbol{\tau}_i + \dfrac{u_\tau^2}{|\overline{\mathbf{u}}|}\overline{\mathbf{u}} \cdot \boldsymbol{\tau}_i = 0, \quad i = 1, \ldots, d-1, \end{cases} \quad \text{for } (\mathbf{x}, t) \in \partial\Omega_1 \times [0, T].$$

In the above formula $\boldsymbol{\tau}_i$ is an orthonormal set of tangent vectors, while σ is the stress tensor[1]. In particular, in the Smagorinsky model (this is the one studied with the above artificial boundary conditions by Parés [249]) the turbulent stress tensor is given by

$$\sigma(\overline{\mathbf{u}}, \overline{p}) = -\overline{p}\,\mathbb{I} + (\nu + \nu_T)\,\nabla^s\overline{\mathbf{u}},$$

where ν is the usual kinematic viscosity, while $\nu_T = \nu_T(\delta, \nabla^s\overline{\mathbf{u}})$ is the turbulent viscosity. The quantity u_τ appearing in the formula is the so-called

---

[1] In this section, and just in this one, we use the dimensional form of the equations.

wall shear velocity (or skin friction velocity). It has the dimensions of length divided by time and acts as a characteristic velocity for the turbulent flow. The reader can find a detailed presentation of the formulas involving u_τ in Chap. 12, p. 299. For more details, the reader is referred to Sects. 42–44 of Landau and Lifshitz [199], where there is an overview of results obtained mainly by von Kármán and Prandtl; see also Sect. 7.1.3 in Pope [258].

The particular case in which $u_\tau^2/|\overline{\mathbf{u}}|$ is a nonnegative constant corresponds to a rough surface. An analysis, together with a numerical implementation of this condition can be found in John [174, 175], for some classes of LES models.

Generally, the *mean* velocity profile of the flow in a boundary layer may be approximated by

$$u^+ = f(y^+), \tag{10.1}$$

where f is the so-called *law-of-the-wall*. In (10.1),

$$u^+ = \frac{\langle \overline{u} \rangle}{u_\tau} \quad \text{and} \quad y^+ = \frac{u_\tau y}{\nu},$$

where y is the distance from the wall and a $^+$ superscript denotes the quantities measured in wall-units. For more details on the significance and importance of measuring flow variables in wall-units, the reader is again referred to Chap. 12 and Sect. 7.1 in Pope [258]. Many different expressions for f may be found in the literature, however all of them are monotonic, and some are linear near 0 (in the so-called *viscous sublayer*), and with logarithmic growth at infinity. We report, see [271], two of them:

(a) Prandtl–Taylor law

$$f(y^+) = \begin{cases} y^+ & \text{if } 0 \le y^+ \le y_0^+ \\ 2.5 \log(y^+) + 5.5 & \text{if } y_0^+ < y^+, \end{cases}$$

where y_0^+ is chosen such that f be continuous.

(b) Reichardt law

$$f(y^+) = 2.5 \log(1 + 0.4 y^+) + 7.8 \left(1 - e^{-y^+/11} - \frac{y^+}{11} e^{-y^+/3}\right),$$

which is smoother than the Prandtl–Taylor law and is used if higher regularity of the solution is desired.

Remark 10.1. The above laws have been used successfully in the analytical treatment of the LES equations, see for instance Parés [249], even if the statistical description of the canonical boundary layer is slightly different. In the case of the canonical boundary layer, there are three layers in the inner region (the region whose distance from the boundary is less than or equal to $0.2\,\delta$), where dynamics is controlled by viscous effects: in the *viscous sublayer* (the region such that $y^+ \leq 5$) the mean velocity is linear. This means that the mean velocity is distributed according to the same law as the true velocity would be for a laminar flow, under the same conditions. In the *buffer layer* ($5 < y^+ \leq 30$) and in the *logarithmic inertial layer* ($30 < y^+$) the mean average velocity is controlled by log-like laws. On the contrary, the outer region (*i.e.* with distance greater than $0.2\,\delta$) is controlled by turbulence.

In the inner region, the correct length scale needed to describe the dynamics is the viscous length $l_\tau = \nu/u_\tau$. In the outer region the characteristic length is δ and the mean velocity is logarithmic in the *logarithmic inertial region*, while it is controlled by a logarithm added to a linear function in the *wake region*.

To implement the boundary condition related to the Prandlt law (or to the Reichardt one) we have to consider then the *no-penetration* condition $\bar{\mathbf{u}} \cdot \mathbf{n} = 0$ together with

$$\mathbf{n} \cdot \sigma(\bar{\mathbf{u}}, \bar{p}) \cdot \tau_i + G(\bar{\mathbf{u}}) \cdot \tau_i = 0, \quad \text{with} \quad G(\bar{\mathbf{u}}) = \begin{cases} \dfrac{h(|\bar{\mathbf{u}}|)}{|\bar{\mathbf{u}}|} \, \bar{\mathbf{u}} & \text{if } |\bar{\mathbf{u}}| > 0 \\ 0 & \text{if } |\bar{\mathbf{u}}| = 0, \end{cases}$$

where $h : \mathbb{R}^+ \to \mathbb{R}$ is the function defined by $h(|\bar{\mathbf{u}}|) = u_\tau^2$, and u_τ is calculated by inverting the law-of-the-wall

$$|\bar{\mathbf{u}}| = u_\tau \, f\left(\frac{u_\tau \, y}{\nu}\right).$$

Since the real function $s \mapsto s\,f(s\,\delta/\nu)$ is strictly increasing and continuous, h is strictly increasing and continuous too. Roughly speaking, the function $G(s)$ is nonnegative and behaves as $o(s^2)$, for $|s| \to \infty$. This is the basic property that such a function should satisfy to produce a boundary value problem that can be treated with the usual monotone operators technique, see again Parés [249].

10.3 Current Ideas in Near Wall Modeling for LES

Near Wall Resolution, in which the averaging radius δ is reduced to 0 near the boundary, besides the well-documented mathematical challenges, involves high computational cost which makes it impractical for most applications of interest. Thus, reflecting the fundamental importance of the topic, there have been correspondingly many NWMs tested in LES. The reader is referred to Sagaut [267], Piomelli and Balaras [253], and Werner and Wengle [312] for detailed surveys of the NWM. Next, we will only sketch the main directions in the development of NWM.

The first paper in LES by Deardorff [87] also used the first NWM model, while Schumann [272] was the first to impose a nonlocal condition on the wall shear stress. He assumed that the stream-wise (span-wise) stress is in phase with the stream-wise (span-wise) velocity at the first grid point away from the wall. The constant of proportionality was obtained from the logarithmic law of the wall.

Grötzbach [141] and Piomelli *et al.* [255] proposed improvements to the basic idea of Schumann, in which a simple algebraic relationship is assumed between the wall stress and the velocity at the first grid point away from the wall. Such NWMs are nonlocal in nature and thus difficult to study as boundary conditions for an LES model. Alternately, they can be viewed as involving a normal derivative of the wall stress – again a "difficult" condition since this imposes boundary conditions of higher order than the equations. (Thus, there are many interesting opportunities for mathematical understanding of existing NWMs.)

A different approach, similar in spirit to the domain decomposition techniques, has led to the two-layer model. In this approach, the three-dimensional boundary layer equations are integrated on an embedded near-wall grid to estimate the wall stresses, see Cabot [48, 49]. While incorporating more physics than the previous approach, the two-layer model is still computationally expensive. Furthermore, it does not produce better results than simpler algebraic wall models for coarse LES at high Reynolds numbers, see Nicoud *et al.* [246].

Bagwell *et al.* [11, 10], developed a different approach in which linear stochastic estimation is used to find the least squares estimate of the wall stresses, given the LES velocities on some plane or planes parallel to the wall. Bagwell used the resulting model in channel flow simulations at $Re_\tau = 180$, the Reynolds number based on u_τ (see Chap. 12, p. 298, for the definition of Re_τ.) He also attempted to rescale the model for the $Re_\tau = 640$ case, but the results were not encouraging. While this approach does not rely on the underlying physics, the two-point correlation tensor of the flow must be known to form the linear stochastic coefficients.

In experimental tests of Marusic *et al.* [233], it was noted that these (and other) commonly used NWMs degrade seriously in presence of complex geometries and at realistic, high Reynolds numbers. In [233] the authors considered a turbulent boundary layer at $Re_\tau = 1350$ and found *overall significant discrepancies* in all three models investigated: the Schumann model with the Grötzbach modification (SG) [141], the shifted SG model of Piomelli *et al.* [255], and the ejection model [255].

One recurring theme in these attempts is the use of *nonlocal* boundary conditions to incorporate solution behavior in a strip near $\partial\Omega$. From the results in Chap. 9, it appears that one essential way to incorporate it is via a discrete model for the boundary commutation error term $A_\delta(\sigma)$ (see Sect. 9.2.1) as an extra forcing function in the strip along $\partial\Omega$.

If a discrete model of $A_\delta(\sigma)$ is used, the problem of NWM modeling simplifies considerably. We can seek *local* boundary conditions for the fluid averages and thus be guided by a large body of mathematical and physical studies of well-posed boundary conditions for flow problems. With that said, how-

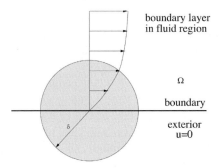

Fig. 10.1. Averaging the velocity at the boundary does not yield homogeneous Dirichlet conditions

ever, the problem remains difficult because the behavior of $\overline{\mathbf{u}}$ on $\partial\Omega$ depends on the behavior $\mathbf{u}$ in a δ-neighborhood of $\partial\Omega$, as illustrated in Fig. 10.1. Recently, there have been some interesting developments along these lines. Recognizing the importance of the commutation error $A_\delta(\sigma)$, Borggaard and Iliescu [39] proposed a numerical implementation of the commutation error by using an *Approximate Deconvolution* (AD) approach. The AD was presented at length in Chap. 7, and can be summarized as follows: by using the mathematical properties of the particular spatial filter g_δ and the numerical approximation of $\overline{\mathbf{u}}$, one can obtain an *approximation* of (some of) the subfilter-scale information contained in $\mathbf{u} - \overline{\mathbf{u}}$. AD was combined with physical insight and was successfully used in developing improved models for the stress tensor τ, yielding the so-called mixed models where the subfilter-scale tensor – due to the loss of information in the filtering process – was modeled through AD, while the subgrid-scale tensor – due to the loss of information in the discretization process – was modeled by using physical insight (eddy viscosity).

It is only natural to pursue the same approach in developing NWMs. The applications that would probably benefit most from this approach would be those in which *(i)* the boundary layer theory is not valid and physical insight is scarce (such as in complex geometries), and *(ii)* the boundary conditions are time dependent (such as in a flow control setting).

In [39], the authors modeled both the commutation error $A_\delta(\sigma)$ and the boundary conditions for $\overline{\mathbf{u}}$ using an AD approach. As a first step, they illustrated their *Approximate Deconvolution Boundary Conditions* (ADBC) algorithm for the heat equation. This linear problem was chosen to decouple the boundary treatment from the closure problem. The first conclusion of these tests was that the commutation error should be included in the numerical model: without it, the error increased by *three orders of magnitude*. The numerical tests also indicated that the ADBC algorithm yielded appropriate numerical approximations for the commutation error and the boundary conditions for $\overline{\mathbf{u}}$.

These first tests were encouraging. Obviously, the algorithm should be tested on realistic turbulent flows (at the time of writing this book, the ADBC algorithm is being tested on channel flows with time-dependent boundary conditions.)

Das and Moser recently proposed in [83] a different approach to account for the commutation error $A_\delta(\sigma)$: to estimate the shear stresses, the authors included in the computational domain a buffer region outside the wall. In this region, the velocities are set to zero, and the wall stresses are determined to minimize the kinetic energy in the buffer region. The resulting system can be thought of as an LES version of embedded boundary techniques. The approach has been tested on several model problems, including the heat equation, Burgers equation, and turbulent channel flow.

10.4 New Perspectives in Near Wall Models

Our intuition of large structures touching a wall is that they do *not penetrate* the wall and *slide* along the wall losing energy as they slide, *e.g.* Navier [244] and Galdi and Layton [122]. This is in accord with Fig. 10.1 (in which $\overline{\mathbf{u}} \cdot \mathbf{n} \cong 0$ while $\overline{\mathbf{u}} \cdot \boldsymbol{\tau}_i \neq 0$) and also with Maxwell's derivation [234] of slip with resistance boundary conditions for gases from the kinetic theory of gases [178]. Thus, as a first approximation of a good NWM, consider the local, well-posed boundary condition for $\overline{\mathbf{u}}$:

$$\overline{\mathbf{u}} \cdot \mathbf{n} = 0 \qquad \text{and} \qquad \beta \overline{\mathbf{u}} \cdot \boldsymbol{\tau}_i + \mathbf{n} \cdot \sigma(\overline{\mathbf{u}}, \overline{p}) \cdot \boldsymbol{\tau}_i = 0, \text{ on } \partial\Omega. \qquad (10.2)$$

The above boundary conditions give rise to a well-posed boundary value problem and this can be seen at least from the point of view of basic energy estimates. We can see this fact at least in the case of the stress tensor corresponding to the NSE, *i.e.*, $\sigma(\mathbf{w}, q) = 2Re^{-1}\nabla^s\mathbf{w} - q\,\mathbb{I}$. Multiplying by $\mathbf{w}$ the term $-\nabla \cdot \sigma(\mathbf{w}, q)$ and integrating by parts, we get (with the convention of summation over repeated indices)

$$\int_\Omega -\nabla \cdot \left(\frac{2}{Re}\nabla^s\mathbf{w} - q\,\mathbb{I}\right)\mathbf{w}\,d\mathbf{x}$$

$$= -\int_{\partial\Omega} \mathbf{n} \cdot \left(\frac{2}{Re}\nabla^s\mathbf{w} - q\,\mathbb{I}\right)\mathbf{w}\,d\sigma + \frac{2}{Re}\int_\Omega |\nabla^s\mathbf{w}|^2\,d\mathbf{x}$$

$$= -\int_{\partial\Gamma} \mathbf{n} \cdot \left(\frac{2}{Re}\nabla^s\mathbf{w} - q\,\mathbb{I}\right)[(\mathbf{w} \cdot \mathbf{n})\mathbf{n} + (\mathbf{w} \cdot \boldsymbol{\tau}_i)\boldsymbol{\tau}_i]\,d\sigma + \frac{2}{Re}\int_\Omega |\nabla^s\mathbf{w}|^2\,d\mathbf{x}$$

and, supposing the velocity $\mathbf{w}$ to satisfy both boundary conditions in (10.2),

$$= \int_\Gamma \beta|\mathbf{w}_\tau|^2\,d\sigma + \frac{2}{Re}\int_\Omega |\nabla^s\mathbf{w}|^2\,d\mathbf{x},$$

where $\mathbf{w}_\tau$ denotes[2] the "tangential part" of the velocity. The boundary integral is then nonnegative, provided that $\beta \geq 0$. This can be used to employ

---

[2] This $\mathbf{w}_\tau = \mathbf{w} - (\mathbf{w} \cdot \mathbf{n})\,\mathbf{n}$ should not be confused with the wall shear velocity.

the usual variational techniques needed to prove existence and H^2 regularity of weak solutions (see Beirão da Veiga [20]), at least in the linear case. A generalized Stokes problem has been also studied in [20]. For more details on the physics of this type of boundary conditions for the NSE see the wonderful introduction to this topic in Sect. 64 of Serrin [274] and also the recent analytical results in Fujita [118], Consiglieri [71], and references therein.

Herein the friction parameter β should satisfy the two consistency conditions

$$\beta \to \infty \ \text{ as } \delta \to 0 \text{ for } Re \text{ fixed,}$$

i.e. (10.2) becomes the no-slip condition;

$$\beta \to 0 \text{ as } Re \to \infty \text{ for } \delta \text{ fixed,}$$

i.e. (10.2) becomes the free slip condition.

Remark 10.2. It is clear that the energy estimate that can be derived from the above calculations remains essentially the same for all EV methods. In addition, we recall that similar boundary conditions have been studied for the Stokes problem by Solonnikov and Ščadilov [278] and Beirão da Veiga [20]. In fact, they studied the well-posedness of a more general version of (10.2) in which the tangential part of the velocity $\mathbf{u} - (\mathbf{u} \cdot \mathbf{n}) \mathbf{n}$ is supposed proportional to $\mathbf{n} \cdot \sigma - (\mathbf{n} \cdot \sigma \cdot \mathbf{n}) \mathbf{n}$, that is the tangential part of the normal stress tensor, or the tangential part of the Cauchy stress vector.

As we have seen, the 1879 work of Maxwell gives also insight into the correct scaling of β. In LES the microlength scale is δ. Thus, the natural interpretation of Maxwell's calculation is the scaling

$$\beta \sim \frac{L}{Re\,\delta}.$$

Since $\overline{\mathbf{u}}$ depends on $\mathbf{u}$ *near* $\partial\Omega$, so must β and supposing β to be a constant may be restrictive. Thus, the best available tools to determine β analytically come from boundary layer theory. Maxwell also accompanied his analysis with the disclaimer

> It is almost certain that the stratum of gas next to a solid body is in a very different state from the rest of the gas (J.C. Maxwell, 1879).

An analytic formula for β can be calculated (within the limits of accuracy and validity of boundary layer theory) by the following procedure, see [122, 269, 178]. Let $\widetilde{\mathbf{u}}$ denote a boundary layer approximation of $\mathbf{u}$ [271, 16]. Then, starting from the two-dimensional case, $g_\delta * \widetilde{\mathbf{u}}$ and $g_\delta * (\nabla^s \widetilde{\mathbf{u}})$ can be explicitly calculated (by using a symbolic mathematics program, for example, if the modeler is not a maestro in special functions) and β calculated via

$$\beta \doteq \frac{-\mathbf{n} \cdot (g_\delta * \nabla^s \widetilde{\mathbf{u}}) \cdot \boldsymbol{\tau}}{(g_\delta * \widetilde{\mathbf{u}}) \cdot \boldsymbol{\tau}}\bigg|_{\partial\Omega}.$$

NWMs of the form (10.2) have the advantage of being in accord with the physics of fluids near walls. They also have the advantage of allowing time fluctuating behavior in $\bar{\mathbf{u}}$ on the wall. Indeed, any time fluctuation in the wall stress in (10.2) results in a fluctuation in the wall slip velocity (via (10.2)) and vice versa. Furthermore, the problem of near wall modeling reduces now to determining the effective friction coefficient $\beta = \beta(\delta, Re, \dots)$. Since the essence of the formulation (10.2) allows time fluctuating behavior on the wall, the extensive information on time-averaged turbulent boundary layers can be used to get insight into β, *without constraining the near wall motion to be quasi-static.*

10.4.1 The 1/7th Power Law in 3D

Consider the case of a turbulent boundary layer. We recall that there are various theories for turbulent boundary layers, *e.g.* Barenblatt and Chorin [16], Schlichting [271], and Pope [258]. Although the following calculation can be done for other descriptions, we perform it herein for power law layers (which is in accord with current views on the subject [16]).

Consider the flat plane $\{(x, y, z) : y = 0\} \subset \mathbb{R}^3$. We say that the velocity $\mathbf{u} = (u, v, w)$ obeys the 1/7th power law, see Schlichting [271], Sect. 21, provided the time (or ensemble) average of the velocity is given by

$$
u = \begin{cases} U_\infty \left(\dfrac{y}{\eta}\right)^{1/7} & \text{for } 0 \le y \le \eta, \\ U_\infty & \text{for } \eta < y, \end{cases}
$$

$$
v = w = 0 \qquad \text{for} \quad 0 \le y,
$$

where the boundary layer thickness $\eta = \eta(x)$ is given by (21.8) in [271],

$$
\eta(x) = 0.37x \, (U_\infty \, x \, Re)^{-1/5},
$$

and U_∞ is the free stream velocity.

Remark 10.3. This power law formula is only valid away from the very thin region near the wall called the viscous sublayer, in which a different asymptotic profile holds. Using it at the wall in a pointwise sense is incorrect; it is easy to see that *without* the viscous sublayer correction, the power law formula predicts infinite stresses at the wall. This section presents a first step in the derivation of near wall models. In this first step, we shall calculate the time average of the average stress in an $O(\delta)$ radius near the wall and ignore the viscous sublayer in the calculation to simplify it significantly. (Thus, incorporating the viscous sublayer's effects into $\beta(\cdot)$ and testing the difference with and without them accounted for is an important open problem!) At this point, we conjecture that the influence of these on the computed slip velocity $\mathbf{w} \cdot \boldsymbol{\tau}$ is small but if it is used to predict wall stresses, the effect of the (herein ignored) viscous sublayer effects can be very large.

We consider the model situation of a reference plate of nondimensional length one. Let $\Omega \subset \mathbb{R}^3$ be the half space

$$\Omega = \{(x, y, z) \in \mathbb{R}^3 : y > 0\}$$

and $\partial\Omega$ the flat plane $\{y = 0\}$. In order to handle this situation, we have to eliminate the x-dependence in η by averaging in the x-direction. Since the problem is nondimensional, the x-length is thus one. Define an averaged boundary layer thickness by

$$\bar{\eta} = \int_0^1 \eta(x)\, dx = \frac{185}{900} U_\infty Re^{-1/5} = c_\eta Re^{-1/5}, \qquad (10.3)$$

and, by direct calculation, the x-averaged velocity obeys the following law:

$$u = \begin{cases} U_\infty \left(\dfrac{y}{\bar{\eta}}\right)^{1/7} & \text{for } 0 \leq y \leq \bar{\eta}, \\ U_\infty & \text{for } \bar{\eta} < y, \end{cases}$$

$$v = w = 0 \qquad \text{for } 0 \leq y.$$

Let $\mathbf{n} = (0, -1, 0)$ be the outward pointing normal vector with respect to Ω on $\{y = 0\}$ and $\boldsymbol{\tau}_1 = (1, 0, 0)$, $\boldsymbol{\tau}_2 = (0, 0, 1)$ be an orthonormal system of tangential vectors. All velocity components are extended by zero outside Ω. We

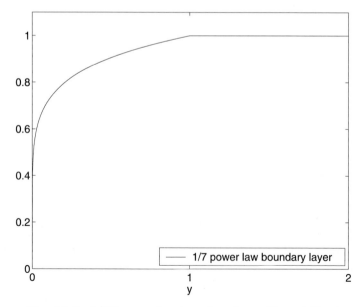

Fig. 10.2. 1/7th power law boundary layer, $U_\infty = 1, \bar{\eta} = 1$

have obviously $\bar{v} = \bar{w} = 0$ and the slip-with-friction boundary condition (10.2) thus simplifies to

$$\beta(\delta, Re)\,\bar{u} + \frac{1}{Re}\frac{\partial \bar{u}}{\partial y} = 0 \qquad \text{on } \{y = 0\}.$$

Thus,

$$\beta(\delta, Re) = \frac{1}{Re}\frac{\dfrac{\partial \bar{u}}{\partial y}(x,0)}{\bar{u}(x,0)}. \tag{10.4}$$

In the case of a filter g_δ given by the usual Gauss kernel, we obtain, by using explicit formulas involving Gaussian integrals,

$$\bar{u}(x, 0, z) = (g_\delta * u)(x, 0, z)$$

$$= U_\infty \left(\frac{\gamma}{\delta^2 \pi}\right)^{3/2}$$

$$\times \left[\int_0^{\bar{\eta}} \left(\frac{y'}{\eta}\right)^{1/7} \exp\left(-\frac{\gamma}{\delta^2}(y')^2\right) dy' \int_{-\infty}^{\infty} \exp\left(-\frac{\gamma}{\delta^2}(x')^2\right) dx'\right.$$

$$\times \int_{-\infty}^{\infty} \exp\left(-\frac{\gamma}{\delta^2}(z')^2\right) dz' + \int_{\bar{\eta}}^{\infty} \exp\left(-\frac{\gamma}{\delta^2}(y')^2\right) dy'$$

$$\left.\times \int_{-\infty}^{\infty} \exp\left(-\frac{\gamma}{\delta^2}(x')^2\right) dx' \int_{-\infty}^{\infty} \exp\left(-\frac{\gamma}{\delta^2}(z')^2\right) dz'\right]$$

$$= \frac{U_\infty}{2}\left\{\left(\frac{1}{\pi}\right)^{1/2}\left(\frac{\delta}{\sqrt{\gamma}\,\bar{\eta}}\right)^{1/7}\left[\Gamma\left(\frac{4}{7}\right) - \Gamma\left(\frac{4}{7}, \left(\frac{\sqrt{\gamma}\,\bar{\eta}}{\delta}\right)^2\right)\right]\right.$$

$$\left. + \left[1 - \operatorname{erf}\left(\frac{\sqrt{\gamma}\,\bar{\eta}}{\delta}\right)\right]\right\},$$

where $\Gamma(z)$ is the usual Gamma function

$$\Gamma(z) = \int_0^{\infty} t^{z-1}\exp(-t)\, dt,$$

while $\Gamma(z, y)$ denotes the incomplete Gamma function (see Abramowitz and Stegun [1]) defined by

$$\Gamma(z, y) = \Gamma(z) - \int_0^y t^{z-1}\exp(-t)\, dt.$$

To compute the numerator in (10.4), first we note that differentiation and convolution commute because functions have been extended off the flow domain so as to retain one weak L^2-derivative, $i.e.$

$$\frac{\partial \bar{u}}{\partial y} = g_\delta * \frac{\partial u}{\partial y}.$$

A straightforward computation (using a symbolic mathematics package) gives

$$\frac{\partial \overline{u}}{\partial y}(x,0,z) = g_\delta * \frac{\partial u}{\partial y}(x,0,z)$$

$$= U_\infty \left(\frac{\gamma}{\delta^2 \pi}\right)^{3/2}$$

$$\times \left[\int_0^{\overline{\eta}} \frac{1}{7} \left(\frac{1}{\overline{\eta}}\right)^{1/7} (y')^{-6/7} \exp\left(-\frac{\gamma}{\delta^2}(y')^2\right) dy' \int_{-\infty}^{\infty} \exp\left(-\frac{\gamma}{\delta^2}(x')^2\right) dx' \right.$$

$$\left. \times \int_{-\infty}^{\infty} \exp\left(-\frac{\gamma}{\delta^2}(z')^2\right) dz' \right]$$

$$= \frac{U_\infty}{14} \left(\frac{\gamma}{\delta^2 \pi}\right)^{1/2} \left(\frac{\delta}{\sqrt{\gamma}\,\overline{\eta}}\right)^{1/7} \left[\Gamma\left(\frac{1}{14}\right) - \Gamma\left(\frac{1}{14}, \left(\frac{\sqrt{\gamma}\,\overline{\eta}}{\delta}\right)^2\right) \right].$$

The friction coefficient $\beta(\delta, Re)$ given in (10.4) can now be computed by using the above expressions for $\overline{u}(x,0,z)$ and $\partial_y \overline{u}(x,0,z)$:

$$\beta(\delta, Re) \tag{10.5}$$

$$= \frac{\gamma^{1/2}(7\delta\, Re)^{-1} \left[\Gamma\left(\frac{1}{14}\right) - \Gamma\left(\frac{1}{14}, \left(\sqrt{\gamma}\frac{\overline{\eta}}{\delta}\right)^2\right) \right]}{\left[\Gamma\left(\frac{4}{7}\right) - \Gamma\left(\frac{4}{7}, \left(\sqrt{\gamma}\frac{\overline{\eta}}{\delta}\right)^2\right) \right] + \pi^{1/2}\left(\frac{\sqrt{\gamma}\,\overline{\eta}}{\delta}\right)^{1/7}\left[1 - \mathrm{erf}\left(\frac{\sqrt{\gamma}\,\overline{\eta}}{\delta}\right) \right]}.$$

Remark 10.4. Considering the 1/7th power law in 2D under the same geometric situation as in Sect. 10.4.2 gives the same results as in 3D, *i.e.* $\overline{u}(x,0)$ turns out to be equal to $\overline{u}(x,0,z)$, while $\partial_y \overline{u}(x,0)$ is equal to $\partial_y \overline{u}(x,0,z)$.

From John, Layton, and Sahin [178] we have the following proposition:

Proposition 10.5. *Let $\beta(\delta, Re)$ be given as in (10.5). We have the following asymptotic results: if Re is constant, then*

$$\lim_{\delta \to 0} \beta\,(\delta, Re) = \infty, \qquad \lim_{\delta \to 0} \delta\,\beta(\delta, Re) = \frac{\sqrt{\gamma}}{7Re} \frac{\Gamma\left(\frac{1}{14}\right)}{\Gamma\left(\frac{4}{7}\right)}. \tag{10.6}$$

If δ is constant, then

$$\lim_{Re \to \infty} \beta(\delta, Re) = 0, \qquad \lim_{Re \to \infty} Re\,\beta(\delta, Re) = \frac{2\sqrt{\gamma}}{\delta\sqrt{\pi}}. \tag{10.7}$$

Proof. From the definition of the Gamma functions it follows that

$$\lim_{x \to 0} \left(\Gamma(z) - \Gamma(z,x)\right) = \lim_{x \to 0} \int_0^x \exp(-t)t^{z-1}\,dt = 0, \tag{10.8}$$

$$\lim_{x \to \infty} \left(\Gamma(z) - \Gamma(z,x)\right) = \lim_{x \to \infty} \int_0^x \exp(-t)t^{z-1}\,dt = \Gamma(z). \tag{10.9}$$

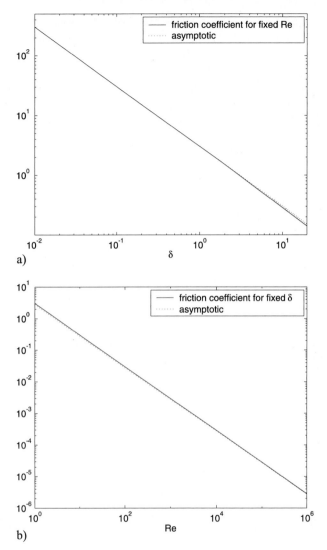

Fig. 10.3. 1/7th power law boundary layer : *top* – behavior of $\beta(\delta, Re)$ with respect to δ for constant $Re(=1), \overline{\eta}(=1)$; *bottom* – behavior of $\beta(\delta, Re)$ with respect of Re for constant $\delta(=1)$; $(\gamma = 6)$

Let Re be fixed and consider the numerator in the last factor of (10.5). The application of (10.9) proves that the numerator tends to $\Gamma(1/14)$ as $\delta \to 0$ and the first term of the denominator tends to $\Gamma(4/7)$. By applying three times the rule of (Johann) Bernoulli–de L'Hôpital, we obtain

$$\lim_{x \to 0} \left(\frac{1}{x}\right)^{1/7} \left[1 - \operatorname{erf}\left(\frac{z}{x}\right)\right] = 0, \quad z > 0.$$

Thus, the second term in the denominator tends to zero and hence the last factor in (10.5) tends to $\Gamma(1/14)/\Gamma(4/7)$. From these considerations follows the second limit in (10.6). In addition, it also follows that $\beta(\delta, Re)$ tends to infinity for $\delta \to 0$, due to the second factor in (10.5).

To prove the first limit in (10.7), note first that the numerator in the last factor in (10.5) tends to zero for $Re \to \infty$ by (10.8). The denominator will be multiplied by the leading factor. Inserting the definition (10.3) of $\overline{\eta}$, we find for the second term in the denominator,

$$\lim_{Re \to \infty} Re^{34/35} \left[1 - \mathrm{erf}\left(a\, Re^{-1/5} \right) \right] = \infty \qquad a > 0.$$

The rule of Bernoulli–de L'Hôpital gives for the first term in the denominator,

$$\lim_{Re \to \infty} Re \left(\Gamma(a) - \Gamma(a, b\, Re^{-2/5}) \right) = \infty \qquad a, b > 0.$$

Thus, the denominator multiplied by the leading term tends to infinity, which proves the first limit in (10.7).

The second limit in (10.7) can be obtained also by the rule of Bernoulli–de L'Hôpital. For details, see [178].

Remark 10.6. It is interesting that the limiting forms of the optimal linear friction coefficient are similar in the 3D turbulent case to those in the 2D laminar case. In some sense, this dimension independence indicates that δ and Re are the correct variables for the analysis.

10.4.2 The $1/n$th Power Law in 3D

When considering the α-power law

$$u = \begin{cases} U_\infty \left(\dfrac{y}{\eta} \right)^\alpha & \text{for } 0 \leq y \leq \eta, \\ U_\infty & \text{for } \eta < y, \end{cases}$$

$$v = w = 0 \qquad \text{for} \quad 0 \leq y,$$

the case $\alpha = 1/7$ is the most commonly used. However, the best available data on turbulent boundary layers suggest that the value $\alpha = 1/7$ is not universal, but should vary slowly with Reynolds number via [258]

$$\alpha = 1/n = \frac{1.085}{\ln(Re)} + \frac{6.535}{\ln(Re)^2},$$

see also Fig. 7.32 in [258]. Thus, it is important to employ here the analysis in [178], Sect. 3, by treating the general case $\alpha = 1/n$. As in the previous section, this formula does not actually hold up to the wall (*i.e.* down to $y = 0$).

Using it up to the wall ignores viscous sublayer effects and leaves an important open problem.

Let the geometric situation be the same as in Sect. 10.4.1 and let, for simplicity, $n \in \mathbb{N}$. The $1/n$th power law in 3D has the form

$$u = \begin{cases} U_\infty \left(\dfrac{y}{\overline{\eta}} \right)^{1/n} & \text{for } 0 \le y \le \overline{\eta}, \\ U_\infty & \text{for } \overline{\eta} < y, \end{cases}$$

$$v = w = 0 \qquad \text{for } 0 \le y,$$

where $\overline{\eta}$ is given as in (10.3).

The computation of the friction coefficient $\beta(\delta, Re)$ proceeds along the same lines as Sect. 10.4.1. One obtains

$$\overline{u}(x, 0, z) = (g_\delta * u)(x, 0, z)$$
$$= \frac{U_\infty}{2} \left\{ \left(\frac{1}{\pi} \right)^{1/2} \left(\frac{\delta}{\sqrt{\gamma}\,\overline{\eta}} \right)^{1/n} \left[\Gamma \left(\frac{n+1}{2n} \right) - \Gamma \left(\frac{n+1}{2n}, \left(\frac{\sqrt{\gamma}\,\overline{\eta}}{\delta} \right)^2 \right) \right] \right.$$
$$\left. + \left[1 - \mathrm{erf} \left(\frac{\sqrt{\gamma}\,\overline{\eta}}{\delta} \right) \right] \right\},$$

while

$$\frac{\partial \overline{u}}{\partial y}(x, 0, z) = g_\delta * \frac{\partial u}{\partial y}(x, 0, z)$$
$$= \frac{U_\infty}{2n} \left(\frac{\gamma}{\delta^2 \pi} \right)^{1/2} \left(\frac{\delta}{\sqrt{\gamma}\,\overline{\eta}} \right)^{1/n} \left[\Gamma \left(\frac{1}{2n} \right) - \Gamma \left(\frac{1}{2n}, \gamma \left(\frac{\overline{\eta}}{\delta} \right)^2 \right) \right],$$

and finally

$$\beta(\delta, Re) \qquad\qquad\qquad\qquad\qquad\qquad\qquad\qquad\qquad\qquad (10.10)$$

$$= \frac{\gamma^{1/2}(Re\, n\, \delta)^{-1} \left[\Gamma \left(\frac{1}{2n} \right) - \Gamma \left(\frac{1}{2n}, \left(\frac{\sqrt{\gamma}\,\overline{\upsilon}\,\overline{\eta}}{\delta} \right)^2 \right) \right]}{\left[\Gamma \left(\frac{n+1}{2n} \right) - \Gamma \left(\frac{n+1}{2n}, \left(\sqrt{\gamma}\,\frac{\overline{\upsilon}}{\delta} \right)^2 \right) \right] + \pi^{1/2} \left(\frac{\sqrt{\gamma}\,\overline{\upsilon}\,\overline{\eta}}{\delta} \right)^{1/n} \left[1 - \mathrm{erf} \left(\frac{\sqrt{\gamma}\,\overline{\upsilon}\,\overline{\eta}}{\delta} \right) \right]}.$$

Along the same lines as the proof of Proposition 10.5, one can prove the following double asymptotics of the friction coefficient, see [178]:

Proposition 10.7. *Let* $\beta(\delta, Re)$ *be given as in* (10.10). *If* Re *is constant, then*

$$\lim_{\delta \to 0} \beta(\delta, Re) = \infty, \quad \lim_{\delta \to 0} \delta\, \beta(\delta, Re) = \frac{\sqrt{\gamma}}{n\, Re} \frac{\Gamma \left(\frac{1}{2n} \right)}{\Gamma \left(\frac{n+1}{2n} \right)}.$$

If δ is constant, then

$$\lim_{Re \to \infty} \beta(\delta, Re) = 0, \quad \lim_{Re \to \infty} Re \, \beta(\delta, Re) = \frac{2\sqrt{\gamma}}{\delta\sqrt{\pi}}.$$

The basic idea of the simple (linear) slip-with-friction model introduced in (10.2) is sound but the derivation of the model places severe limitations on the flow (such as no recirculation regions, no reattachment points, ...). Motivated by some of these limitations, we will survey some elaborations of (10.2) proposed in order to extend its applicability.

10.4.3 A Near Wall Model for Recirculating Flows

In the previous sections we studied linear near wall models, *i.e.* with a friction coefficients β based upon a global Reynolds number. In recirculating flows, there are usually large differences between reference velocities in the freestream and in the recirculation regions. Thus, a linear NWM will tend to overpredict the friction in attached eddies and underpredict it away from attached eddies. A solution of this difficulty is to base the NWM upon the *local* Reynolds number as follows.

The analysis performed in the previous sections reveals that the predicted *local* slip velocity, $\overline{\mathbf{u}} \cdot \boldsymbol{\tau}$, is a monotone function of Re. Thus, the relationship can be inverted and inserted into the appropriate place in the derivation of the NWM to give a β dependent on the local slip speed,

$$\beta = \beta(\delta, |\overline{\mathbf{u}} \cdot \boldsymbol{\tau}|).$$

To carry out this program, we assume the 1/7th power law holds. The 2D calculations in Sect. 10.4.1 reveal that the tangential velocity (10.5) can be written in the following form:

$$\overline{\mathbf{u}} \cdot \boldsymbol{\tau}_1 = \frac{U_\infty}{2} \left\{ \left(\frac{1}{\pi}\right)^{1/2} \left(\frac{1}{\xi}\right)^{1/7} \left[\Gamma\left(\frac{4}{7}\right) - \Gamma\left(\frac{4}{7}, \xi^2\right) \right] + [1 - \text{erf}(\xi)] \right\} = g(\xi), \tag{10.11}$$

with

$$\xi = \frac{\sqrt{\gamma}\,\overline{\eta}}{\delta} = \frac{\sqrt{\gamma}\,c_\eta}{\delta \, Re^{1/5}} > 0.$$

Consequently one finds, by direct evaluation,

$$\frac{d\,\overline{\mathbf{u}} \cdot \boldsymbol{\tau}_1}{d\xi} = g'(\xi) = -\frac{U_\infty}{14\sqrt{\pi}} \left(\frac{1}{\xi}\right)^{8/7} \left[\Gamma\left(\frac{4}{7}\right) - \Gamma\left(\frac{4}{7}, \xi^2\right) \right] < 0,$$

and this calculation proves the following lemma.

Lemma 10.8. *Let $\overline{\mathbf{u}} \cdot \boldsymbol{\tau}_1$ be given by (10.11). Then, $\overline{\mathbf{u}} \cdot \boldsymbol{\tau}_1$ is a strictly monotone, decreasing function of ξ, hence a strictly monotone increasing function of Re. Thus, an inverse function $\xi = g^{-1}(\overline{\mathbf{u}} \cdot \tau_1)$ exists.*

An ideal NWM can thus be obtained by using this inverse function for Re in (10.5):

$$\beta = \beta(\delta, g^{-1}(\overline{\mathbf{u}} \cdot \boldsymbol{\tau})), \qquad (10.12)$$

and β given by (10.5).

However, this cannot easily be used in practical calculations. Thus, we shall develop an accurate and simple approximate inverse to $g(\xi)$ which still captures the correct double asymptotics. The idea to obtain a usable non-linear friction coefficient consists in: (i) finding an approximation $\widetilde{g}(\xi)$ of $g(\xi)$ which can be easily inverted, and (ii) replacing ξ and Re in (10.5) by $\widetilde{g}^{-1}(\overline{\mathbf{u}} \cdot \boldsymbol{\tau}_1)$.

A careful examination of $g(\xi)$ reveals that an appropriate approximation over $0 \le \xi < \infty$ is that of the form

$$\overline{\mathbf{u}} \cdot \boldsymbol{\tau}_1 \approx \frac{U_\infty}{2} \exp\left(-a\,\xi^b\right) \qquad \text{with } a,\, b \in \mathbb{R}^+.$$

This gives

$$\xi = \left(-\frac{1}{a} \ln\left(\frac{2\,\overline{\mathbf{u}} \cdot \boldsymbol{\tau}_1}{U_\infty}\right)\right)^{1/b} \quad \text{and} \quad Re = \left(\frac{\sqrt{\gamma}\, c_\eta}{\delta\xi}\right)^5.$$

The constants a and b must be chosen such that the approximation is the best in a least squares sense: find $a, b > 0$ that minimize the expression

$$\int_{\xi_l}^{\xi_r} \left[\left(\frac{1}{\pi}\right)^{1/2} \left(\frac{1}{\xi}\right)^{1/7} \left(\Gamma\left(\frac{4}{7}\right) - \Gamma\left(\frac{4}{7}, \xi^2\right)\right) + [1 - \text{erf}\,(\xi)] - \exp\left(-a\xi^b\right)\right]^2 d\xi.$$

$$(10.13)$$

In the above formula, the left boundary ξ_l and the right boundary ξ_r of the integral must be specified by using the data of the given problem. If they are given, the optimal parameters can be approximated numerically. Such an approximation can be obtained in the following way: the interval $[\xi_l, \xi_r]$ is divided into N equal subintervals $[\xi_i, \xi_{i+1}]$ with $\xi_0 = \xi_l$ and $\xi_N = \xi_r$. Then, the continuous minimization (10.13) is replaced by its discrete counterpart: find $a, b > 0$ that minimize the expression

$$\sum_{i=0}^{N} \left[\left(\frac{1}{\pi}\right)^{1/2} \left(\frac{1}{\xi_i}\right)^{1/7} \left(\Gamma\left(\frac{4}{7}\right) - \Gamma\left(\frac{4}{7}, \xi_i^2\right)\right) + [1 - \text{erf}\,(\xi_i)] - \exp\left(-a\xi_i^b\right)\right]^2.$$

The necessary condition for a minimum, *i.e.* that the partial derivatives with respect to a and b vanish, leads to a nonlinear system of two equations. This can be solved iteratively, *e.g.* by Newton's method. We give some examples of optimal parameters for some intervals in Table 10.1. These parameters were computed with $N = 50\,000$ using Newton's method. An illustration of the approximation is presented in Fig. 10.4.

Preliminary testing of NWM of the above type, performed by John, Layton, and Sahin [178] and Sahin [269] on flow over a step, seems to suggest that they improve the estimation of the reattachment point *before* separation occurs.

Table 10.1. Optimal parameters in (10.13) for different intervals $[\xi_l, \xi_r]$

ξ_l	ξ_r	a	b
0	0.1	0.142864	1.00312
0	1	0.137149	0.961851
0	10	0.154585	0.497275
0	100	0.238036	0.268180
0	1 000	0.342360	0.174579
0	10^6	0.689473	0.0812879
1	10	0.170289	0.444825

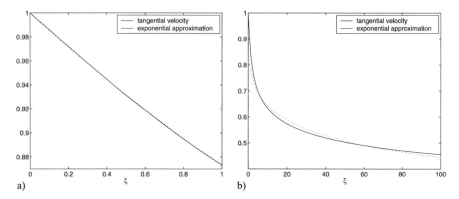

a)   b)

Fig. 10.4. The function (10.11) and its exponential approximation according to Table 10.1, $[\xi_l, \xi_r] = [0, 1]$ (*left*), $[\xi_l, \xi_r] = [0, 100]$ (*right*), $U_\infty = 2$

Remark 10.9. We stress that the above model, though being nonlinear, is only one step along the required path of developing NWMs for practical flows. In the following subsections we will survey the steps that (to us at this point in time at least) seem necessary.

10.4.4 A Near Wall Model for Time-averaged Modeling of Time-fluctuating Quantities

Boundary layer theory (*e.g.* Schlichting [271]) describes *time-averaged* flow profiles near walls. Thus, time-fluctuating information is not incorporated into NWM like (10.2). This (necessarily omitted) fluctuating information in the wall-normal direction can play an important role in triggering separation and detachment, as pointed out in Layton [205].

One attempt to mimic these effects is to introduce noise into the wall-normal condition, aiming to trigger separation and detachment when attached eddies become sufficiently unstable:

$$\overline{\mathbf{u}} \cdot \mathbf{n} = \delta^2\, \omega(\mathbf{x}, t) \quad \text{and} \quad \beta\, \overline{\mathbf{u}} \cdot \boldsymbol{\tau}_j + \frac{2}{Re}\mathbf{n} \cdot \nabla^s \overline{\mathbf{u}} \cdot \boldsymbol{\tau}_j = 0 \quad \text{on } \partial\Omega \times [0, T], \quad (10.14)$$

where $\omega(\mathbf{x}, t)$ is highly fluctuating and satisfies

$$0 = \int_{\partial\Omega} \omega(\mathbf{x}, t)\, d\sigma(\mathbf{x}) \qquad \text{for each } t \in [0, T]. \tag{10.15}$$

The compatibility condition (10.15) is required by the incompressibility condition $\nabla \cdot \overline{\mathbf{u}} = 0$, which implies

$$0 = \int_{\Omega} \nabla \cdot \overline{\mathbf{u}}(\mathbf{x})\, d\mathbf{x} = \int_{\partial\Omega} \overline{\mathbf{u}} \cdot \mathbf{n}\, d\sigma(\mathbf{x}) = \delta^2 \int_{\partial\Omega} \omega(\mathbf{x}, t)\, d\sigma(\mathbf{x}).$$

The *ad hoc* modification (10.14) is similar in spirit to so-called "vorticity seeding" methods. A preliminary analytical result, at least in the case of the 2D NSE, has been recently obtained by Berselli and Romito in [35], where nearly optimal conditions on ω ensuring the existence of weak solutions in the sense of Leray–Hopf are found. In addition, it is proved that as $\delta \to 0$ the solutions to the vorticity seeding model converge (in appropriate norms) to those of the NSE with the usual no-slip boundary condition.

10.4.5 A Near Wall Model for Reattachment and Separation Points

In the previous Sects. 10.4.1 and 10.4.2 the friction coefficient β has been derived using asymptotics of time averages of attached turbulent boundary layers along flat plates. Thus, it is not applicable when the curvature of the boundary becomes large relative to other physical parameters (such as at a corner) and it fails completely at a reattachment or separation point. The geometry of flows at such points suggests that at such points $\overline{\mathbf{u}} \cdot \mathbf{n} \neq 0$ but $\overline{\mathbf{u}} \cdot \boldsymbol{\tau} \doteq 0$. Thus, at reattachment/separation points, a wall-normal condition of the form

$$\gamma\, \overline{\mathbf{u}} \cdot \mathbf{n} + \mathbf{n} \cdot \left(\frac{2}{Re} \nabla^s \overline{\mathbf{u}} - \overline{p}\,\mathbb{I} \right) \cdot \mathbf{n} = 0$$

should be investigated for the NSE[3]. Much less is known about flow near such points. There is one known exact solution $(\widetilde{\mathbf{u}}, \widetilde{p})$ for a jet impinging upon a wall, Schlichting [271], which has an analogous flow pattern. From this, the resistance coefficient γ could be calculated via

$$\gamma = \left. \frac{-\mathbf{n} \cdot \left[2Re^{-1}(g_\delta * \nabla^s \widetilde{\mathbf{u}}) - g_\delta * \widetilde{p}\,\mathbb{I} \right] \cdot \mathbf{n}}{(g_\delta * \widetilde{\mathbf{u}}) \cdot \mathbf{n}} \right|_{\partial\Omega}. \tag{10.16}$$

So far, this calculation has not been performed in significant cases and the correct double asymptotics of $\gamma = \gamma(\delta, Re)$ are still unclear.

---

[3] Clearly in the case of the Smagorinsky model this will become

$$\gamma\, \overline{\mathbf{u}} \cdot \mathbf{n} + \mathbf{n} \cdot \left(\frac{2}{Re} \nabla^s \overline{\mathbf{u}} + (C_s \delta)^2 |\nabla^s \overline{\mathbf{u}}| \nabla^s \overline{\mathbf{u}} - \overline{p}\,\mathbb{I} \right) \cdot \mathbf{n} = 0.$$

10.5 Conclusions

To summarize an admittedly speculative program for NWMs, we propose local boundary conditions for an LES average velocity $\overline{\mathbf{u}}$ on walls of the general form:

$$
\begin{cases}
\beta\left(\delta, |\overline{\mathbf{u}} \cdot \boldsymbol{\tau}|\right)\overline{\mathbf{u}} \cdot \boldsymbol{\tau}_i + \mathbf{n} \cdot \sigma(\overline{\mathbf{u}}, \overline{p}) \cdot \boldsymbol{\tau}_i = 0 & \text{on } \partial\Omega \times [0, T] \\[2mm]
\gamma\,\overline{\mathbf{u}} \cdot \mathbf{n} + \mathbf{n} \cdot \sigma(\overline{\mathbf{u}}, \overline{p}) \cdot \mathbf{n} = \delta^2\,\omega(\mathbf{x}, t) & \text{on } \partial\Omega \times [0, T].
\end{cases}
\tag{10.17}
$$

The nonlinear friction coefficient $\beta = \beta(\delta, |\overline{\mathbf{u}} \cdot \boldsymbol{\tau}|)$ can be calculated following (10.12) and the linear filtration-resistance coefficient γ is calculated by (10.16). The wall-normal forcing $\omega(\mathbf{x}, t)$ is a perturbation satisfying the consistency condition (10.15).

So far, only preliminary tests have been performed with the simplest, first step (10.2) in the direction of (10.17), with moderate success, Sahin [269]. The form we are seeking for the NWMs will ensure the combined LES model plus NWM has a chance at robustness: the conditions (10.17) make mechanical sense and mathematical tools exist for studying the well-posedness of (10.17) with an LES model. Even in this simple approach, the important effect of the viscous sublayer has been omitted to make the calculations tractable. Thus, finding and testing this correction is an interesting open problem.

The quest for the "right" boundary conditions (NWMs) for LES models represents one of the most important challenges in LES. This is a very active area of research, and giving a detailed presentation of existing NWMs is a challenge in itself and, clearly, beyond the scope of this book. The reader is referred to Sagaut [267], and Piomelli and Balaras [253] for detailed surveys.

In this chapter, we just tried to sketch the general framework and list some of the main directions in the development of NWM. Thus, unfortunately, we had to leave out some of these NWMs. We preferred to focus instead on one promising direction that we have been exploring lately.

Part V

Numerical Tests

11

Variational Approximation of LES Models

11.1 Introduction

In the approximation of underresolved flow problems, one question that reappears is: What are the correct variables to seek to compute? In LES the "answer" is the large, spatial, coherent structures. The traditional definition of the structures $(\overline{\mathbf{u}}, \overline{p})$ is by convolution or space filtering:

$$\overline{\mathbf{u}}(\mathbf{x}, t) := \int_{\mathbb{R}^d} \mathbf{u}(\mathbf{x} - \mathbf{x}', t)\, g_\delta(\mathbf{x}')\, d\mathbf{x}' \qquad \overline{p}(\mathbf{x}, t) := \int_{\mathbb{R}^d} p(\mathbf{x} - \mathbf{x}', t)\, g_\delta(\mathbf{x}')\, d\mathbf{x}'.$$

Averaging/filtering the Navier–Stokes equations is the traditional approach in LES. As we have seen, it leads to problems of closure (a closure model must be selected), near wall modeling (boundary conditions for flow averages must be provided), and noncommutativity of filtering and differentiation on bounded domains. The resulting continuum model must still be discretized and an approximate solution calculated. Assessing the reliability of the computed solutions inevitably leads to classic numerical analysis issues of stability, consistency, and convergence of an algorithm as well as the questions of well posedness of the continuum model.

In the following sections, we give a description of the variational methods that are used in the experiments presented in the next chapter. At this point, there are more open questions than clear theoretical answers in the numerical analysis of LES models. In particular, the classic approaches to stability, consistency, and convergence do not give predictions for the most important outputs of turbulent flow calculations, namely, time-averaged flow statistics. A new numerical analysis needs to be developed studying the accuracy of flow statistics for problems in which the actual flow predictions may not be accurate!

One of the most interesting recent approaches to LES is the *Variational Multiscale Method* (VMM), developed by Hughes and his co-workers [160]. Related ideas have been pursued by Temam and co-workers as the "dynamic multilevel method" [96] and Brown *et al.* [47] and Hylin *et al.* [163] as the "additive turbulent decomposition." Each approach has its own interesting and unique features. In the VMM, the above "answer" is that the large solution scales (large eddies) are defined by orthogonal projection onto functions which can be represented on a given mesh. A simple approximation of the first unresolved scale is made (and used as a closure model) and only the effects of further unresolved scales on the first unresolved scales are modeled. The new and interesting point is that all this occurs in a variational framework. The exact coupling between large and small scales in the NSE then acts as a type of "expert system" to determine the effective LES model. Although the VMM is outside the classical approach to LES, which we focus on in this book, because of its great promise, we give a synopsis of one approach to VMM in Sect. 11.5.

One of the most promising approaches to EV models consists of models whose action is restricted to either the fluctuations or the smallest resolved scales. In Sect. 11.6, we consider one approach to such methods. Interestingly, we will also show in Sect. 11.6 that, by simple choices of the model fluctuations and large scales, this method is equivalent to a Variational Multiscale Method! This leads to the idea that the VMM framework might be universal. Is every consistently stabilized variational approximation a Variational Multiscale Method? Is every LES model that (given the approximation of small scales) uses exact equations for large-scale motion also a Variational Multiscale Method? The answers to these questions are unknown.

11.2 LES Models and their Variational Approximation

The traditional path (which we study in this section) is to average the NSE, giving

$$\overline{\mathbf{u}}_t + \nabla \cdot (\overline{\mathbf{u}}\,\overline{\mathbf{u}}^T) - \frac{1}{Re}\Delta\overline{\mathbf{u}} + \nabla\overline{p} + \nabla\cdot(\overline{\mathbf{u}\,\mathbf{u}^T} - \overline{\mathbf{u}}\,\overline{\mathbf{u}}^T) = \overline{\mathbf{f}} + A_\delta(\mathbf{u}, p) \quad (11.1)$$

$$\nabla \cdot \overline{\mathbf{u}} = 0. \quad (11.2)$$

Next a closure model is chosen: the subfilter-scale stress tensor τ is replaced by one depending only on $\overline{\mathbf{u}}$;

$$\tau = \overline{\mathbf{u}\mathbf{u}} - \overline{\mathbf{u}}\,\overline{\mathbf{u}} \approx \mathcal{S}(\overline{\mathbf{u}}, \overline{\mathbf{u}}),$$

and approximate boundary conditions are selected. (Recall that actually what is needed is a trace-free approximation of the trace-free part of $\overline{\mathbf{u}\,\mathbf{u}^T} - \overline{\mathbf{u}}\,\overline{\mathbf{u}}^T$.) Picking a simple example, we choose no-penetration and slip-with-friction (see Chap. 9)

$$\overline{\mathbf{u}} \cdot \mathbf{n} = 0 \quad \text{and} \quad \beta\overline{\mathbf{u}} \cdot \boldsymbol{\tau}_j - \mathbf{n} \cdot \boldsymbol{\sigma}(\overline{\mathbf{u}}, \overline{p}) \cdot \boldsymbol{\tau}_j = 0 \quad \text{on } \partial\Omega,$$

where $\mathbf{n}, \boldsymbol{\tau}_j$ are respectively the normal and tangent vectors to $\partial\Omega$, while $\boldsymbol{\sigma}$ is the total stress for the model

$$\boldsymbol{\sigma}(\overline{\mathbf{u}}, \overline{p}) := \overline{p}\,\mathbb{I} - \frac{2}{Re}\,\nabla^s\overline{\mathbf{u}} + \mathcal{S}(\overline{\mathbf{u}}, \overline{\mathbf{u}}).$$

Finally, approximation to the commutation error $A_\delta(\mathbf{u}, p)$ (see Chap. 9 for a detailed presentation) is needed and, although there are ideas under development, an acceptable one is not yet known. Thus, we shall drop it for the moment.

   With these choices, we then have a boundary value problem for a velocity $\mathbf{w}(\mathbf{x}, t)$ and a pressure $q(\mathbf{x}, t)$ which model $\overline{\mathbf{u}}(\mathbf{x}, t)$ and $\overline{p}(\mathbf{x}, t)$ and which are given

$$\mathbf{w}_t + \nabla \cdot (\mathbf{w}\,\mathbf{w}^T) - \nabla \cdot \left(\frac{2}{Re}\nabla^s\mathbf{w} - \mathcal{S}(\mathbf{w}, \mathbf{w})\right) + \nabla q = \overline{\mathbf{f}} \text{ in } \Omega \times (0, T]$$

$$\tag{11.3}$$

$$\nabla \cdot \mathbf{w} = 0 \quad \text{in } \Omega \times (0, T] \tag{11.4}$$

$$\mathbf{w}(\mathbf{x}, 0) = \overline{\mathbf{u}}_0(\mathbf{x}) \quad \text{in } \Omega \tag{11.5}$$

$$\overline{\mathbf{w}} \cdot \mathbf{n} = 0 \quad \text{and} \quad \beta\,\overline{\mathbf{w}} \cdot \boldsymbol{\tau}_j - \mathbf{n} \cdot \boldsymbol{\sigma}(\overline{\mathbf{w}}, \overline{p}) \cdot \boldsymbol{\tau}_j = 0 \text{ on } \partial\Omega \times (0, T],$$

$$\tag{11.6}$$

where $\boldsymbol{\sigma}(\mathbf{w}, q) := q\,\mathbb{I} - \frac{2}{Re}\,\nabla^s\mathbf{w} + \mathcal{S}(\mathbf{w}, \mathbf{w})$, as above.

   The system (11.3)–(11.6) must still be discretized and an approximate solution calculated by good algorithms on computationally feasible meshes. Further, since the only real data and solutions to (11.3) that are available are based on these simulations, it is very difficult to distinguish in practice between modeling errors (the error going from (11.1) to (11.3)) and numerical errors (the error between (11.3) and its computational realization). This introduces in an essential way the classical numerical analysis questions of accuracy, stability, convergence, and robustness (meaning behavior of algorithms as $h \to 0$ and $\delta \to \infty$) for discretizations of (11.3). As usual though, the answers to such universal questions depend on specific features of each choice made and, in particular, on a clear understanding of the mathematical foundation of the specific model (11.3) chosen. This is a topic with enough scope for a series of books (in particular, we refer to the ground breaking work in John [175]) and beyond the goals of the present treatment. However, some results of variational approximation are reported and the ideas of the algorithms behind those results will be described herein.

11.2.1 Variational Formulation

The variational formulation of (11.3)–(11.6) is obtained in the usual way: multiply by a test function $\mathbf{v}$ (not necessarily divergence-free), integrate over

Ω, and integrate by parts. This gives

$$(\mathbf{w}_t, \mathbf{v}) + (\mathbf{w} \cdot \nabla \mathbf{w}, \mathbf{v}) + \left(\frac{2}{Re}\nabla^s \mathbf{w} - \mathcal{S}(\mathbf{w}, \mathbf{w}), \nabla \mathbf{v}\right) + \Gamma - (q, \nabla \cdot \mathbf{v}) = (\bar{\mathbf{f}}, \mathbf{v}),$$

where Γ denotes the boundary terms arising from all the integration by parts. Many of these terms vanish if $\mathbf{v} \cdot \mathbf{n} = 0$ on $\partial\Omega$ (which we shall assume) and the term $\nabla \mathbf{v}$ may be replaced by $\nabla^s \mathbf{v}$ if $\mathcal{S}(\mathbf{w}, \mathbf{w})$ is a symmetric tensor.

Assuming $\mathbf{v} \cdot \mathbf{n} = 0$, the boundary term Γ reduces to

$$\Gamma = -\int_{\partial\Omega} \mathbf{n} \cdot \left(\frac{2}{Re}\nabla^s \mathbf{w} - \mathcal{S}(\mathbf{w}, \mathbf{w})\right) \cdot \mathbf{v}\, dS.$$

The vector function $\mathbf{v}$ can be decomposed as

$$\mathbf{v} = (\mathbf{v} \cdot \mathbf{n})\,\mathbf{n} + (\mathbf{v} \cdot \boldsymbol{\tau}_j)\,\boldsymbol{\tau}_j = (\mathbf{v} \cdot \boldsymbol{\tau}_j)\,\boldsymbol{\tau}_j$$

(since $\mathbf{v} \cdot \mathbf{n} = 0$ on the boundary). Thus, Γ becomes

$$\Gamma = \int_{\partial\Omega} (\mathbf{n} \cdot \boldsymbol{\sigma}(\mathbf{w}, q) \cdot \boldsymbol{\tau}_j)\,(\mathbf{v} \cdot \boldsymbol{\tau}_j)\, dS,$$

which, due to the boundary conditions imposed, becomes

$$\Gamma = \int_{\partial\Omega} \beta(\mathbf{w})\,\mathbf{w} \cdot \boldsymbol{\tau}_j\, \mathbf{v} \cdot \boldsymbol{\tau}_j\, dS.$$

Thus, if we define

$$\mathbf{X} := \left\{\mathbf{v} \in [H^1(\Omega)]^d : \mathbf{v} \cdot \mathbf{n} = 0 \text{ on } \partial\Omega\right\}$$

$$Q := \left\{q \in L^2(\Omega) : \int_\Omega q\, d\mathbf{x} = 0\right\},$$

one natural *mixed* variational formulation is to seek a velocity $\mathbf{w} : [0, T] \to \mathbf{X}$ and a pressure $q : [0, T] \to Q$, satisfying

$$\left\{ \begin{aligned} &(\mathbf{w}_t, \mathbf{v}) + (\mathbf{w} \cdot \nabla \mathbf{w}, \mathbf{v}) + \left(\frac{2}{Re}\nabla^s \mathbf{w} - \mathcal{S}(\mathbf{w}, \mathbf{w}), \nabla^s \mathbf{v}\right) \\ &\quad + \int_{\partial\Omega} \beta(\mathbf{w})\,\mathbf{w} \cdot \boldsymbol{\tau}_j\, \mathbf{v} \cdot \boldsymbol{\tau}_j\, dS - (q, \nabla \cdot \mathbf{v}) = (\bar{\mathbf{f}}, \mathbf{v}), \quad \forall \mathbf{v} \in \mathbf{X}, \\ &\qquad\qquad\qquad\qquad\qquad\qquad\quad (\nabla \cdot \mathbf{w}, \lambda) = 0, \quad \forall \lambda \in Q. \end{aligned} \right.$$

$$(11.7)$$

Even if $\mathcal{S}(\mathbf{w}, \mathbf{w}) = \mathbf{0}$, the difference in the boundary conditions requires an extension of the usual mathematical architecture surrounding the analysis and numerical analysis of the Navier–Stokes equations (Girault and Raviart [137] and Gunzburger [146]). This extension has been successfully carried out in, for example, [202, 50, 51, 52] and tested in [178, 175]. Thus, it is safe to suppress some technical points related to the boundary conditions and $\mathbf{X}$ vs. $[H_0^1(\Omega)]^d$ (the usual velocity space).

A variational approximation to the model is simply a finite-dimensional approximation to the variational form of the model (11.7) rather than to its strong form (11.3). Many choices are possible here as well; the most fundamental is the Galerkin method (which we have already used as a theoretical tool to prove existence of solutions to several LES models). In the Galerkin method, finite-dimensional conforming velocity–pressure subspaces $\mathbf{X}_h \subset \mathbf{X}$ and $Q_h \subset Q$ are chosen, and approximate solutions

$$\mathbf{w}^h : [0, T] \to \mathbf{X}_h, \qquad q^h : (0, T] \to Q_h$$

are found satisfying (11.7) restricted to $\mathbf{X}_h \times Q_h$:

$$\begin{cases} (\mathbf{w}_t^h, \mathbf{v}^h) + b^*(\mathbf{w}^h, \mathbf{w}^h, \mathbf{v}^h) + \left(\dfrac{2}{Re} \nabla^s \mathbf{w}^h - \mathcal{S}(\mathbf{w}^h, \mathbf{w}^h), \nabla^s \mathbf{v}^h \right) \\[2mm] \qquad + \displaystyle\int_{\partial\Omega} \beta(\mathbf{w}^h)\, \mathbf{w}^h \cdot \boldsymbol{\tau}_j\, \mathbf{v}^h \cdot \boldsymbol{\tau}_j\, dS - (q^h, \nabla \cdot \mathbf{v}^h) = (\overline{\mathbf{f}}, \mathbf{v}^h), \quad \forall \mathbf{v}^h \in \mathbf{X}_h, \\[3mm] \qquad\qquad\qquad\qquad\qquad\qquad (\nabla \cdot \mathbf{w}^h, \lambda^h) = 0, \quad \forall \lambda^h \in Q_h. \end{cases}$$
$$(11.8)$$

For stability reasons, the second term in (11.7) is usually replaced by its explicit skew-symmetrization $b^*(\cdot, \cdot, \cdot)$ in (11.8) given by

$$b^*(\mathbf{u}, \mathbf{v}, \mathbf{w}) := \frac{1}{2}(\mathbf{u} \cdot \nabla \mathbf{v}, \mathbf{w}) - \frac{1}{2}(\mathbf{u} \cdot \nabla \mathbf{w}, \mathbf{v}).$$

Also, for stability of the pressure q^h, the spaces $\mathbf{X}_h$ and Q_h must either satisfy the inf-sup (Ladyžhenskaya–Babuška–Brezzi) compatibility condition

$$\inf_{q^h \in Q_h} \sup_{\mathbf{v}^h \in \mathbf{X}_h} \frac{(q^h, \nabla \cdot \mathbf{v}^h)}{\|q^h\| \, \|\nabla \mathbf{v}^h\|} \geq C > 0 \qquad \text{uniformly in } h,$$

which we will assume, or include extra stabilization of the pressure employed in (11.8).

Further choices must be made of $\mathbf{X}_h \times Q_h$, determining different methods such as spectral, finite element or spectral element methods. Also, a further discretization of the time variable must be selected for (11.9) giving yet more algorithmic options.

Proposition 11.1 (Stability of (11.8)). *Let $(\mathbf{w}^h, q^h)$ satisfy (11.8). Then,*

$$\frac{1}{2}\|\mathbf{w}^h(t)\|^2$$
$$+ \int_0^t \left[\frac{1}{Re}\|\nabla^s \mathbf{w}^h\|^2 - (\mathcal{S}(\mathbf{w}^h, \mathbf{w}^h), \nabla^s \mathbf{v}^h) + \int_{\partial\Omega} \beta(\mathbf{w}^h)\, |\mathbf{w}^h \cdot \boldsymbol{\tau}_j|^2\, dS \right] dt'$$
$$\leq \frac{1}{2}\|\overline{\mathbf{u}}_0\|^2 + C\, Re \int_0^t \|\overline{\mathbf{f}}\|_{-1}^2\, dt'. \qquad (11.9)$$

If, additionally, $\beta(\cdot) \geq \beta_0 > 0$ and the model in dissipative in the sense that

$$(S(\mathbf{v}, \mathbf{v}), \nabla^s \mathbf{v}) \leq 0 \qquad \forall \, \mathbf{v} \in \mathbf{X},$$

then the method (11.8) *is stable.*

Proof. Set $\mathbf{v}^h = \mathbf{w}^h, \lambda^h = q^h$, and add the two equations in (11.8). This gives

$$\frac{1}{2}\|\mathbf{w}^h(t)\|^2 + \frac{2}{Re}\|\nabla^s \mathbf{w}^h\|^2 - (S(\mathbf{w}^h, \mathbf{w}^h), \nabla^s \mathbf{w}^h) + \int_{\partial\Omega} \beta(\mathbf{w}^h)\,|\mathbf{w}^h \cdot \boldsymbol{\tau}_j|^2\,dS$$

$$= (\bar{\mathbf{f}}, \mathbf{w}^h)$$

$$\leq C\,Re\,\|\bar{\mathbf{f}}\|_{-1}^2 + \frac{1}{Re}\,\|\nabla^s \mathbf{w}^h\|^2,$$

where we applied Körn's inequality to get the right-hand side. Integrating the result from 0 to t, yields (11.9). $\square$

On the Stability of the Method

The two conditions for stability in Proposition 11.1 are worth examining. The first is that

$$\beta(\mathbf{w}) = \beta(\mathbf{w}, \delta, Re) \geq \beta_0 = \beta_0(\delta, Re) > 0.$$

This should be true of any reasonable boundary condition (keeping in mind the typical limiting behavior of β: $\beta \to +\infty$ as $\delta \to 0$ and $\beta \to 0^+$ as $Re \to +\infty$).

The second condition is dissipativity:

$$\int_\Omega S(\mathbf{v}, \mathbf{v}) : \nabla^s \mathbf{v}\,d\mathbf{x} \leq 0 \qquad \forall \, \mathbf{v} \in \mathbf{X}. \tag{11.10}$$

This condition is not universally true for models in use. For EV models

$$S^*(\mathbf{v}, \mathbf{v}) = -\nu_T(\delta, \mathbf{v})\,\nabla^s \mathbf{v}, \qquad \text{where } \nu_T \geq 0,$$

(11.10) does hold. However, EV models have large modeling errors. For other models with asymptotically smaller modeling errors it is more problematic. For example, both the Gradient LES model (7.3) and the Rational LES (7.18) model in Chap. 7 fail (11.10), as does the Bardina and, in fact, most scale similarity models (Chap. 8). Thus, the stability of discretizations of non-eddy viscosity LES models must also be treated on a case-by-case basis exploiting the particular features of each model: a universal analysis of discretization errors in LES models is not yet achievable.

Furthermore, the dissipativity assumption (11.10) is too restrictive: many important models and interesting physical behaviors are eliminated by (11.10). The question remains open: what is the correct one? One (speculative) possibility is to ask that $S(\mathbf{v}, \mathbf{v})$ act in a dissipative manner on fluctuations:

$$\int_{\Omega} \mathcal{S}(\mathbf{v}', \mathbf{v}') : \nabla^s \mathbf{v}' \, d\mathbf{x} \leq 0 \qquad \text{for } \mathbf{v}' = \mathbf{v} - \overline{\mathbf{v}}, \text{ and } \forall \, \mathbf{v} \in \mathbf{X},$$

while it acts as a sort of reaction term on the resolved scales. The correct formulation of this second condition is not yet clear.

Another (speculative) possibility is to connect the kinetic energy balance in the discrete equations to the kinetic energy balance in the model by deconvolution to that of the continuous NSE. Briefly, let A denote an Approximate Deconvolution operation. This means that, for smooth enough $\mathbf{v}$,

$$A \, \overline{\mathbf{v}} = \mathbf{v} + o(1) \quad \text{as } \delta \to 0.$$

If $\mathbf{w} \approx \overline{\mathbf{u}}$, then

$$(\mathbf{w}, A^* A \mathbf{w}) \approx (A\mathbf{w}, A\mathbf{w}) \approx \|\mathbf{u}\|^2.$$

Further, it should be hoped that since $(\nabla \cdot (\mathbf{u}\,\mathbf{u}), \mathbf{u}) = 0$, then $(\nabla \cdot (\overline{\mathbf{u}}\,\overline{\mathbf{u}}), A^*\mathbf{u}) \approx 0$. Now, when $\overline{\mathbf{u}\,\mathbf{u}}$ is modeled by $\overline{\mathbf{u}}\,\overline{\mathbf{u}} + \mathcal{S}(\overline{\mathbf{u}}, \overline{\mathbf{u}})$, one approach to try to verify stability is to construct an operator A such that

$$(\nabla \cdot (\mathbf{w}\,\mathbf{w} + \mathcal{S}(\mathbf{w}, \mathbf{w})), A^* A \mathbf{w}) = 0.$$

If such a construction is achievable, then the model is stable. Further, when achievable, it suggests that the above variational formulation is not the correct one for numerics: the equation should be tested against $A^* A \mathbf{v}^h$ instead of $\mathbf{v}^h$.

11.3 Examples of Variational Methods

The three most prominent examples of variational methods are spectral methods, finite element methods and spectral element methods. They differ only in the choices of the spaces $\mathbf{X}_h$ and Q_h.

11.3.1 Spectral Methods

The books of Peyret [252] and Canuto et al. [54] give excellent, comprehensive treatments of spectral methods in computational fluid dynamics. Briefly, spectral methods choose a basis for $\mathbf{X}_h$ and Q_h that is very close to eigenfunctions of the Stokes operator under the indicated boundary conditions. These choices simplify the equations considerably. They also ensure very high accuracy. Their computational realization for simple geometries and boundary conditions is usually very direct and easy. On the other hand, their intricacy increases rapidly with geometric complexity. In simple geometries (which often correspond to geometries for which good experimental data is available) it is often thus possible to pick basis functions and velocity spaces $\mathbf{X}_h$ that are exactly divergence-free, thereby eliminating the pressure from the discrete system.

11.3.2 Finite Element Methods

There are a number of excellent books treating Finite Element Methods
(FEM) for flow problems. The books of Pironneau [256], Gunzburger [146],
Cuvelier, Segal, and van Steenhoven [80], and the series by Gresho and
Sani [139, 140] are good beginning points, and Girault and Raviart [137]
is the definitive reference to the mathematical analysis of the method. FEM
are based on a flexible description of an unstructured finite element mesh on
the flow domain. Once such a mesh is constructed and stored in the appropriate way, the velocity and pressure finite element spaces are constructed
based upon the mesh. Typically, FEMs compute an approximate velocity and
pressure that is globally continuous across mesh edges and polynomial inside
each mesh cell. Finite element methods are very highly developed for laminar
flow problems, see again [256, 146, 140, 137]. The behavior of the methods for
turbulent flows and for approximating turbulence models is much less understood; see Mohammadi and Pironneau [239] for some first steps.

11.3.3 Spectral Element Methods

An excellent introduction to the Spectral Element Methods (SEM) is given
in the book of Deville, Fischer, and Mund [89]. SEM, introduced by Patera
and coworkers [251, 224], combine the geometric flexibility of finite element
methods with the accuracy of spectral methods. Thus, they represent an appropriate tool for the LES of turbulent flows (where the high accuracy of the
numerical method is believed to be important) in complex geometries that
would be challenging for spectral methods. SEM employ a high-order weighted
residual technique based on compatible velocity and pressure spaces that are
free of spurious modes. Locally, the spectral element mesh is structured, with
the solution, data, and geometry expressed as sums of Nth-order Lagrange
polynomials on tensor-products of Gauss or Gauss–Lobatto quadrature points.
Globally, the mesh is an unstructured array of K deformed hexahedral elements and can include geometrically nonconforming elements. For problems
having smooth solutions, the SEM achieve exponential convergence with N,
despite having only C^0 continuity (which is advantageous for parallelism). The
mathematical analysis associated with SEM was presented by Maday and Patera [224, 225]. For recent developments in SEM, including time-discretizations,
preconditioners for the linear solvers, parallel performance, and stabilization
high-order filters, the reader is directed to the papers of Fischer and collaborators [105, 107, 108].

11.4 Numerical Analysis of Variational Approximations

In this section we address some very basic facts concerning the numerical
analysis of variational equations. This topic is worthy of an entire book and we

want to focus on some specific problems and questions arising in the numerical analysis of LES equations. We essentially restrict discussion to the role of stability.

Since the stability of variational approximation depends upon the exact choice of the LES model, it is not surprising that a universal and model-independent numerical analysis of variational approximation is not possible. For specific models, the overarching goal of such a numerical analysis is to prove convergence to the solution of the model as $h \to 0$, in a natural norm (such as $L^2(\Omega \times (0,T))$) which is uniform in the Reynolds number, for δ fixed. This question is open for most interesting models (with only a few first steps, *e.g.* John and Layton [177], and M. Kaya [185].) In fact, the numerical analysis of [177] for the Smagorinsky model hardly seems extensible to most good models. Thus, it seems that a "new" numerical analysis is needed to address issues in LES. One possible avenue is to study convergence of statistics. This goal is to give analytic insight describing how well statistics computed using a given model and algorithm match the true statistics. We give one example next. Let $\langle \cdot \rangle$ denote the *time average* of the indicated quantity. For example,

$$\langle \phi(\mathbf{x},t) \rangle = \lim_{T \to \infty} \frac{1}{T} \int_0^T \phi(\mathbf{x},t)\,dt.$$

When the above limit does not exist, it is usual to replace it by a limit superior (or a Banach limit agreeing in value with the lim sup). One important statistic from turbulent flows is the time-averaged energy dissipation rate, defined by

$$\langle \epsilon(\mathbf{u}) \rangle := \limsup_{T \to \infty} \frac{1}{T} \int_0^T \frac{1}{|\Omega|} \int_\Omega \frac{1}{Re} |\nabla^s \mathbf{u}|^2\,d\mathbf{x},dt.$$

If $\mathbf{f}(\mathbf{x},t)$ is a smooth, bounded function, *e.g.*, $\mathbf{f} \in L^\infty(0,\infty; L^2(\Omega))$, then it is quite easy to show that for any weak solution of the NSE satisfying the energy inequality

$$\langle \epsilon(\mathbf{u}) \rangle \leq \left\langle \frac{1}{|\Omega|} \int_\Omega \mathbf{f} \cdot \mathbf{u}\,dx \right\rangle,$$

(*i.e.* the time-averaged energy dissipation rate is bounded by the time-averaged power input rate) and, if $\mathbf{u}$ satisfies the energy equality then equality holds in the above:

$$\langle \epsilon(\mathbf{u}) \rangle = \left\langle \frac{1}{|\Omega|} \int_\Omega \mathbf{f} \cdot \mathbf{u}\,dx \right\rangle.$$

Let us focus on the "easy" case of eddy viscosity models. Assume

$$S^*(\mathbf{v},\mathbf{v}) = -\nu_T(\delta, \mathbf{v})\,\nabla^s \mathbf{v}. \tag{11.11}$$

Here, energy is dissipated due to three effects:

1. molecular diffusion,
2. eddy diffusivity, and
3. friction large eddies encounter when contacting walls.

Including all three effects gives the computed energy dissipation rate to be given by

$$\langle \epsilon^h_{model}(\mathbf{w}^h) \rangle := \limsup_{T \to \infty} \frac{1}{T} \int_0^T \frac{1}{|\Omega|} \left[\int_\Omega \left(\frac{1}{Re} + \nu_T(\mathbf{w}^h) \right) |\nabla^s \mathbf{w}^h|^2 \, d\mathbf{x} \right.$$

$$\left. + \int_{\partial\Omega} \beta(\mathbf{w}^h) |\mathbf{w}^h \cdot \boldsymbol{\tau}_j|^2 \, dS \right] dt. \quad (11.12)$$

It is not hard to show (after some simple calculations) that for eddy viscosity models (*i.e.* models whose subgrid stress tensor satisfies (11.11)) under the same conditions on $\mathbf{f}$ as in the NSE case, that

$$\langle \epsilon^h(\mathbf{w}^h) \rangle = \left\langle \frac{1}{|\Omega|} \int_\Omega \mathbf{f} \cdot \mathbf{w}^h \, d\mathbf{x} \right\rangle.$$

Additionally, if $\mathbf{f} = \langle \mathbf{f} \rangle$ (*e.g.* if $\mathbf{f} = \mathbf{f}(\mathbf{x})$), then

$$\left\langle \int_\Omega \mathbf{f} \cdot \mathbf{w}^h \, d\mathbf{x} \right\rangle = \int_\Omega \mathbf{f} \cdot \langle \mathbf{w}^h \rangle \, d\mathbf{x}.$$

Collecting these – admittedly simple – observations into a proposition gives the following:

Proposition 11.2. *Suppose* $\mathbf{f} \in L^\infty(0, \infty; L^2(\Omega))$, *the LES model is an eddy viscosity model (i.e. (11.11) holds), and the limits in the definitions of* $\langle \epsilon(\mathbf{u}) \rangle$ *and* $\langle \epsilon^h(\mathbf{w}^h) \rangle$ *exist. Then*

$$\langle \epsilon(\mathbf{u}) \rangle - \langle \epsilon^h(\mathbf{w}^h) \rangle \leq \left\langle \frac{1}{|\Omega|}(\mathbf{f}, \mathbf{u}) \right\rangle - \left\langle \frac{1}{|\Omega|}(\mathbf{f}, \mathbf{w}^h) \right\rangle.$$

If the NSE satisfies the energy equality, then equality holds in the above.

Thus, the key to replicating this statistic (at least) is to match as accurately the time-averaged rate of power input to the flow through body force–flow interactions. One idea is to monitor an *a posteriori* estimator for the functional

$$\mathbf{w}^h \mapsto \frac{1}{t_n} \int_0^{t_n} \frac{1}{|\Omega|}(\mathbf{f}, \mathbf{w}^h) \, dt'$$

and, using that information, adaptively tune the eddy diffusivity for the following time step.

Remark 11.3. The case of a flow driven by a given body force is a very easy case. There have been some exciting developments and analytic estimates for shear flows (a much harder case). See the work of Doering and Constantin [90], Doering and Foiaş [91], Wang [311], and the references therein: [61, 109, 280, 281]. Extension of this work to LES models is an interesting and important open problem. So far only the Smagorinsky model has been considered [207].

11.5 Introduction to Variational Multiscale Methods

The VMM is naturally a *variational* method so it is most naturally presented for variational discretizations. To begin, we consider the finite element discretizations of the NSE on a polyhedral domain in $\mathbb{R}^3$, satisfying no-slip boundary conditions, driven by a body force, and with the usual (mathematically convenient form of the) pressure normalization condition $p \in Q$.

For reader convenience, we collect here all the needed definitions of functional spaces and multilinear forms.

Definition 11.4. *(a)* $\| \cdot \|$, $(\cdot, \cdot)$ *will denote the usual $L^2(\Omega)$ norm and inner product*

$$(\phi, \psi) = \int_\Omega \phi \cdot \psi \, d\mathbf{x}, \quad \|\phi\| = (\phi, \phi)^{1/2}.$$

(b) $(\mathbf{X}, Q)$ *will denote the usual velocity-pressure Sobolev spaces*

$$\mathbf{X} := \{\mathbf{v} \in [H^1(\Omega)]^d : \mathbf{v}|_{\partial\Omega} = 0\},$$

$$Q := \left\{ q \in L^2(\Omega) : \int_\Omega q \, d\mathbf{x} = (q, 1) = 0 \right\}.$$

(c) $\mathbf{V}$ *will denote the space of weakly divergence-free functions in $\mathbf{X}$:*

$$\mathbf{V} := \{\mathbf{v} \in \mathbf{X} : (\nabla \cdot \mathbf{v}, q) = 0, \ \forall q \in Q\}.$$

(d) $a(\mathbf{u}, \mathbf{v}) : \mathbf{X} \times \mathbf{X} \to \mathbb{R}$ *will denote the bilinear form*

$$a(\mathbf{u}, \mathbf{v}) := \int_\Omega \frac{1}{Re} \nabla^s \mathbf{u} : \nabla^s \mathbf{v} \, d\mathbf{x},$$

and $b(\mathbf{u}, \mathbf{v}, \mathbf{w}) : \mathbf{X} \times \mathbf{X} \times \mathbf{X} \to \mathbb{R}$ *will denote the (explicitly skew-symmetrized) nonlinear convection trilinear form*

$$b(\mathbf{u}, \mathbf{v}, \mathbf{w}) := \frac{1}{2} \int_\Omega [\mathbf{u} \cdot \nabla \mathbf{v} \cdot \mathbf{w} - \mathbf{u} \cdot \nabla \mathbf{w} \cdot \mathbf{v}] \, d\mathbf{x}.$$

It is important to recall that, due to the inequalities of classical mechanics of Poincaré and Körn, the following are all equivalent norms on $\mathbf{X}$:

$$\|\mathbf{u}\|_1 := \left[\|\nabla \mathbf{u}\|^2 + \|\mathbf{u}\|^2 \right]^{1/2}, \quad \|\nabla \mathbf{u}\|, \quad \text{and } \|\nabla^s \mathbf{u}\|.$$

Further, note that, by construction,

$$b(\mathbf{u}, \mathbf{v}, \mathbf{w}) = -b(\mathbf{u}, \mathbf{w}, \mathbf{v}) \quad \text{and} \quad b(\mathbf{u}, \mathbf{v}, \mathbf{v}) = 0, \quad \forall \mathbf{u}, \mathbf{v}, \mathbf{w} \in \mathbf{X}.$$

The usual, mixed variational formulation of the continuous NSE that is used as a first step to an approximate solution is then: find $\mathbf{u} : [0, T] \to \mathbf{X}$ and $p : (0, T] \to Q$ satisfying

$$\begin{cases} (\mathbf{u}_t, \mathbf{v}) + a(\mathbf{u}, \mathbf{v}) + b(\mathbf{u}, \mathbf{u}, \mathbf{v}) - (p, \nabla \cdot \mathbf{v}) = (\mathbf{f}, \mathbf{v}) & \forall \mathbf{v} \in \mathbf{X}, \\ (q, \nabla \cdot \mathbf{u}) = 0 & \forall q \in Q, \quad (11.13) \\ \mathbf{u}(\mathbf{x}, 0) = \mathbf{u}_0(\mathbf{x}) & \forall \mathbf{x} \in \Omega. \end{cases}$$

Following Hughes, Mazzei, and Jansen [160], in the VMM a finite element mesh is selected and a standard velocity finite element space $\mathbf{X}^h \subset \mathbf{X}$ is constructed. This finite element space is identified as the space of mean velocities ([160], p. 52). Specifically, decompose

$$\mathbf{X} = \overline{\mathbf{X}} \oplus \mathbf{X}', \quad \text{where} \quad \overline{\mathbf{X}} := \mathbf{X}^h \text{ is the chosen finite element space. } (11.14)$$

Obviously, the complement $\mathbf{X}'$ turns out to be infinite dimensional.

Corresponding to (11.14), define the decomposition of the velocity into means and fluctuations as

$$\mathbf{u} = \overline{\mathbf{u}} + \mathbf{u}', \quad \overline{\mathbf{u}} = \mathbf{u}^h := \overline{P}\mathbf{u} \in \mathbf{X}^h, \quad \mathbf{u}' = (\mathbb{I} - \overline{P})\mathbf{u} \in \mathbf{X}',$$

where $\overline{P} : \mathbf{X} \to \overline{\mathbf{X}} = \mathbf{X}^h$ is some projection operator. Insert $\mathbf{u} = \mathbf{u}^h + \mathbf{u}'$ in (11.13), then alternately set $\mathbf{v} = \mathbf{v}^h$ then $\mathbf{v} = \mathbf{v}'$ in (11.13). This yields the following two coupled, continuous systems for $\overline{\mathbf{u}}$ and $\mathbf{u}'$ which are completely equivalent[1] to the continuous problem (11.13):

$$\begin{cases} (\overline{\mathbf{u}}_t + \mathbf{u}'_t, \mathbf{v}^h) + a(\overline{\mathbf{u}} + \mathbf{u}', \mathbf{v}^h) + b(\overline{\mathbf{u}} + \mathbf{u}', \overline{\mathbf{u}} + \mathbf{u}', \mathbf{v}^h) - (p^h + p', \nabla \cdot \mathbf{v}^h) \\ \qquad\qquad\qquad\qquad\qquad\qquad = (\mathbf{f}, \mathbf{v}^h) \quad \forall \mathbf{v}^h \in \mathbf{X}^h, \\ \\ (\overline{\mathbf{u}}_t + \mathbf{u}'_t, \mathbf{v}') + a(\overline{\mathbf{u}} + \mathbf{u}', \mathbf{v}') + b(\overline{\mathbf{u}} + \mathbf{u}', \overline{\mathbf{u}} + \mathbf{u}', \mathbf{v}') - (p^h + p', \nabla \cdot \mathbf{v}') \\ \qquad\qquad\qquad\qquad\qquad\qquad = (\mathbf{f}, \mathbf{v}') \quad \forall \mathbf{v}' \in \mathbf{X}'. \end{cases}$$

These coupled systems, after algebraic rearrangement, give

$$(\overline{\mathbf{u}}_t, \mathbf{v}^h) + a(\overline{\mathbf{u}}, \mathbf{v}^h) + b(\overline{\mathbf{u}}, \overline{\mathbf{u}}, \mathbf{v}^h) - (p^h, \nabla \cdot \mathbf{v}^h) - (\mathbf{f}^h, \mathbf{v}^h) = (\mathbf{r}', \mathbf{v}^h), \quad (11.15)$$

where

$$(\mathbf{r}', \mathbf{v}^h) := (\mathbf{f}', \mathbf{v}^h) - b(\mathbf{u}', \mathbf{u}', \mathbf{v}^h)$$
$$- [(\mathbf{u}'_t, \mathbf{v}^h) + a(\mathbf{u}', \mathbf{v}^h) + b(\overline{\mathbf{u}}, \mathbf{u}', \mathbf{v}^h) + b(\mathbf{u}', \overline{\mathbf{u}}, \mathbf{v}^h - (p', \nabla \cdot \mathbf{v}^h)]$$

and

$$(\mathbf{u}'_t, \mathbf{v}^h) + a(\mathbf{u}', \mathbf{v}') + b(\mathbf{u}', \mathbf{u}', \mathbf{v}') - (p', \nabla \cdot \mathbf{v}') - (\mathbf{f}', \mathbf{v}') = (\mathbf{r}^h, \mathbf{v}'), \quad (11.16)$$

---

[1] These should also be coupled with $(\nabla \cdot (\mathbf{u}^h + \mathbf{u}'), q) = 0, \forall q \in Q$.

where

$$(\mathbf{r}^h, \mathbf{v}') := (\overline{\mathbf{f}}, \mathbf{v}') - b(\overline{\mathbf{u}}, \overline{\mathbf{u}}, \mathbf{v}')$$
$$- [(\overline{\mathbf{u}}_t, \mathbf{v}') + a(\overline{\mathbf{u}}, \mathbf{u}') + b(\mathbf{u}', \overline{\mathbf{u}}, \mathbf{v}') + b(\overline{\mathbf{u}}, \mathbf{u}', \mathbf{v}') - (\overline{p}, \nabla \cdot \mathbf{v}')].$$

Thus (as pointed out in [160]), the large scales are also driven by the projection of the small scales' residual into $\mathbf{X}^h$ and vice versa for the small scales. In the usual (continuous time) element method, the RHS of (11.15) would be identically zero so none of the effects of the unresolved scales would be incorporated.

In the *Variational Multiscale Method*, (11.15) and (11.16) are simultaneously discretized, as follows (again, following [160]):

- with $\mathbf{X}^h$ chosen, a complementary <u>finite dimensional</u> subspace, $\mathbf{X}'_b$, is chosen for the discrete fluctuations;
- since (11.16) involves reduction of an infinite dimensional problem (in $\mathbf{X}'$) into a finite-dimensional problem (in $\mathbf{X}'_b$) extra stabilization is added to the discrete fluctuation equation in the form

$$(\nu_T(\mathbf{u}) \nabla \mathbf{u}', \nabla \mathbf{v}').$$

With the above choices, the problem is to find

$$\overline{\mathbf{u}} : [0, T] \to \mathbf{X}^h, \quad \overline{p} : (0, T] \to Q^h,$$
$$\mathbf{u}'_b : [0, T] \to \mathbf{X}'_b, \quad p' : (0, T] \to Q'_b$$

satisfying

$$\mathbf{u}^h(0) = \overline{\mathbf{u}}_0 = \overline{P}\mathbf{u}_0 \text{ in } \Omega, \qquad \mathbf{u}'_b(0) = \mathbf{u}'_0 \approx (\mathbb{I} - \overline{P})\mathbf{u}_0 \text{ in } \Omega,$$

and satisfying

$$(\overline{\mathbf{u}}_t, \mathbf{v}^h) + a(\overline{\mathbf{u}}, \mathbf{v}^h) + b(\overline{\mathbf{u}}, \overline{\mathbf{u}}, \mathbf{v}^h) - (p^h, \nabla \cdot \mathbf{v}^h) + (q^h, \nabla \cdot \overline{\mathbf{u}}) - (\mathbf{f}^h, \mathbf{v}^h)$$
$$= (\mathbf{r}'_b, \mathbf{v}^h) \quad \forall \mathbf{v}^h \in \mathbf{X}^h, \ q^h \in Q^h \ (11.17)$$

where

$$(\mathbf{r}'_b, \mathbf{v}^h) := (\mathbf{f}', \mathbf{v}^h) - b(\mathbf{u}', \mathbf{u}', \mathbf{v}^h)$$

$$- [(\mathbf{u}'_{bt}, \mathbf{v}^h) + a(\mathbf{u}'_b, \mathbf{v}^h) + b(\overline{\mathbf{u}}, \mathbf{u}'_b, \mathbf{v}^h)$$

$$+ b(\mathbf{u}'_b, \overline{\mathbf{u}}, \mathbf{v}^h) - (p'_b, \nabla \cdot \mathbf{v}^h)],$$

and

$$(\mathbf{u}'_{b,t}, \mathbf{v}'_b) + a(\mathbf{u}'_b, \mathbf{v}'_b) + (\nu_T(\overline{\mathbf{u}} + \mathbf{u}'_b) \nabla^s \mathbf{u}'_b, \nabla^s \mathbf{v}'_b) + b(\mathbf{u}'_b, \mathbf{u}'_b \mathbf{v}'_b)$$

$$-(p'_b, \nabla \cdot \mathbf{v}'_b) + (q'_b, \nabla \cdot \mathbf{u}'_b) = (\mathbf{r}^h, \mathbf{v}'), \quad \forall \mathbf{v}'_b \in \mathbf{X}'_b, \ \forall q'_b \in Q'_b, (11.18)$$

where

$$(\mathbf{r}^h, \mathbf{v}_b') := (\overline{\mathbf{f}}, \mathbf{v}_b') - b(\overline{\mathbf{u}}, \overline{\mathbf{u}}, \mathbf{v}_b')$$
$$- [(\overline{\mathbf{u}}_t, \mathbf{v}_b') + a(\overline{\mathbf{u}}, \mathbf{v}_b') + b(\mathbf{u}_b', \overline{\mathbf{u}}, \mathbf{v}_b') + b(\overline{\mathbf{u}}, \mathbf{u}_b', \mathbf{v}_b') - (p^h, \nabla \cdot \mathbf{v}_b')].$$

The calculations reported in [161, 162] all employ simple variants on the Smagorinsky model, *e.g.*

$$\nu_T = (C_s \delta)^2 \, |\nabla^s(\overline{\mathbf{u}} + \mathbf{u}_b')|, \quad \nu_T = (C_s \delta)^2 \, |\nabla^s \mathbf{u}_b'|, \quad \text{etc.},$$

but *acting only on the model for the small scales!*

The above approximation actually yields $(\overline{\mathbf{u}} + \mathbf{u}_b')$ as a DNS approximation to $\mathbf{u}$ since, so far, (almost) no information is lost between the NSE and the discretization. The key to the method's computational feasibility rests in *losing the right information.*

For (11.17) for $\overline{\mathbf{u}}^h$ to be accurate all that is required is a rough approximation of $\mathbf{u}_b'$ from (11.18): only $\|\mathbf{r}_b'\|_{H^{-1}(\Omega)}$ need be small. Thus, VMM's typically *use a computational model for the fluctuations* that uncouples (11.18) into one small system per mesh cell. (This is the link to *residual free bubbles*, a finite element idea that has established a connection between Galerkin FEMs and streamline diffusion FEMs, Franca and Farhat [114], Franca, Nesliturk, and Stynes [115], and Hughes [159].)

For each mesh cell K^h a bubble function ϕ_K is chosen such that $\phi_K > 0$ in K^h but $\phi_K = 0$ on ∂K^h. This gives

$$\phi_{K^h} \in H_0^1(K^h).$$

Define then

$$\mathbf{X}_b' := \text{span}\left\{ \phi_{K^h} : \text{ all mesh cells } K^h \right\}^3.$$

Good algorithms and good computational results flow from this choice of $\mathbf{X}_b'$, see [160]. The only drawback seems to be, that, since *every function in $\mathbf{X}_b$ vanishes on all mesh lines and mesh faces*, this choice is, in essence, a computational model that the *fluctuations are quasi-stationary.*

For many LES models, the global kinetic energy balance is very murky; the kinetic energy in some can even blow up in finite time [173, 169]. However, the VMM inherits the correct energy equality from the NSE.

Proposition 11.5. *Consider* (11.17) *and* (11.18) *for* $\nu_T \geq 0$. *Let* $\widetilde{\mathbf{X}} :=$ $\mathbf{X}^h \oplus \mathbf{X}_b'$ *and let* $\widetilde{\mathbf{u}} = \overline{\mathbf{u}} + \mathbf{u}_b'$. *Then* $\widetilde{\mathbf{u}}$ *satisfies:*

$$\frac{1}{2}\|\widetilde{\mathbf{u}}(t)\|^2 + \int_0^t \left(\frac{1}{Re}\|\nabla^s\widetilde{\mathbf{u}}(t')\|^2 + \int_\Omega \nu_T(\mathbf{u})|\nabla^s\mathbf{u}'(t')|^2 \, dx \right) dt'$$
$$= \frac{1}{2}\|\mathbf{u}_0\|^2 + \int_0^t (\mathbf{f}(t'), \widetilde{\mathbf{u}}(t')) \, dt'.$$

Proof. Add (11.17) and (11.18). Then, set $\mathbf{v}^h = \overline{\mathbf{u}}$, $\mathbf{v}_b' = \mathbf{u}_b'$, $q^h = p^h$, and $q_b' = p_b'$. After that, the proof follows exactly the NSE case.   □

11.6 Eddy Viscosity Acting on Fluctuations as a VMM

The idea of subgrid-scale eddy viscosity can be thought of as being implicit in the effect of inertial forces on the resolved scales in Richardson's Energy cascade. The most natural algorithmic interpretation occurs with spectral methods in the early work of Maday and Tadmor [226] on spectral vanishing viscosity methods. Recent work of Guermond [142, 143, 144] (see also [203, 160, 204]) has shown that bubble functions can be used to give a realization of the idea in physical space as opposed to wavenumber space. Guermond [142, 143] has also shown that subgrid-scale eddy viscosity can provide good balance between accuracy and stability.

   These are exciting ideas that give mathematical structure to the physical interpretation of the action of the SFS stress tensor on the resolved scales; and, there is certainly more to be done. This section presents a third, complementary approach to subgrid-scale eddy viscosity of [204]. This third approach has the following characteristics:

(i)   It is based on a consistent variational formulation.
(ii)   It uses essentially a multiscale decomposition of the fluid *stresses* rather than the fluid velocities.
(iii) The computational model for fluctuations allows discrete fluctuations that cross mesh edges and faces.

We shall see later that, after reorganization, it is actually a VMM. One conjecture coming out of this connection is that *all consistently stabilized methods are equivalent to a VMM.*

   To present the new method, we follow the notation of Sect. 11.5 (and focus only on the simplest cases). Let $\pi^H(\Omega)$ denote a coarse finite element mesh which is refined (once, twice, ...) to produce the finer mesh $\pi^h(\Omega)$, so $h < H$. On these meshes, conforming velocity–pressure finite element spaces are constructed:

$$Q^H \subset Q^h \subset Q := L_0^2(\Omega), \text{ and } \mathbf{X}^H \subset \mathbf{X}^h \subset \mathbf{X} := [H_0^1(\Omega)]^3.$$

These are assumed to satisfy the usual inf-sup condition for stability of the pressure (explained in Gunzburger [146]):

$$\inf_{q^\mu \in Q^\mu} \sup_{\mathbf{v}^\mu \in \mathbf{X}^\mu} \frac{(q^\mu, \nabla \cdot \mathbf{v}^\mu)}{\|q^\mu\| \, \|\nabla \mathbf{v}^\mu\|} \geq \beta > 0, \quad \text{for } \mu = h, H.$$

Since the key is to construct a multiscale decomposition of the deformation tensor, $\nabla^s \mathbf{u}^h$, we need deformation spaces. Since $\mathbf{u}^h \in \mathbf{X}$, naturally

$$\nabla^s \mathbf{u}^h \in \mathbf{L} := \{\ell = \ell_{ij} : \ell_{ij} = \ell_{ji} \text{ and } \ell_{ij} \in L^2(\Omega), \ i, j = 1, 2, 3.\}$$

Accordingly, choose *discontinuous* finite element spaces on the coarse finite element mesh $\pi^H(\Omega)$:

$$\mathbf{L}^H \subset \mathbf{L}^h \subset \mathbf{L}.$$

The best example to keep in mind is: for $\mu = h$ and H

$$\mathbf{X}^\mu := \{\mathcal{C}^0 \text{ piecewise linear (vectors) on } \pi^\mu(\Omega)\},$$

$$\mathbf{L}^\mu := \{L^2 \text{ discontinuous, piecewise constant (symmetric tensors) on } \pi^\mu(\Omega)\}.$$

Note that in this example $\mathbf{L}^\mu = \nabla^s \mathbf{X}^\mu$. (Also, it is well known that various adjustments are necessary with low-order velocity spaces to satisfy the inf-sup condition.) The idea of the method is to *add* global eddy viscosity to the centered Galerkin FEM and to *subtract* its effects on the large scales as follows:

Find $\mathbf{u}^h : [0, T] \to \mathbf{X}^h$, $p^h : (0, T] \to Q^h$, and $\mathbf{g}^H : (0, T] \to \mathbf{L}^H$ satisfying

$$\begin{cases} (\mathbf{u}_t^h, \mathbf{v}^h) + a(\mathbf{u}^h, \mathbf{v}^h) + b(\mathbf{u}^h, \mathbf{u}^h, \mathbf{v}^h) - (p^h, \nabla \cdot \mathbf{u}^h) + (q^h, \nabla \cdot \mathbf{u}^h) \\ \quad + (\nu_T \nabla^s \mathbf{u}^h, \nabla^s \mathbf{v}^h) - (\nu_T \mathbf{g}^H, \nabla^s \mathbf{v}^h) = (\mathbf{f}, \mathbf{v}^h), \ \forall \mathbf{v}^h \in \mathbf{X}^h, \forall q^h \in Q^h, \\ \qquad\qquad\qquad\qquad (\mathbf{g}^H - \nabla^s \mathbf{u}^h, \ell^H) = 0, \quad \forall \ell^H \in \mathbf{L}^H \end{cases}$$

$$(11.19)$$

This form of the subgrid scale stress tensor was proposed for the first time by Layton [206]. However, the general idea of adding a global stabilization and subtracting its undesired action is quite common in viscoelastic flow simulations, *e.g.* the "EVSS-G" method, Fortin, Guénette, and Pierre [113].

The action of the extra terms in (11.19) is easy to assess. The second equation of (11.19) implies that

$$\mathbf{g}^H = P_H(\nabla^s \mathbf{u}^h), \quad \text{where } P_H' : \mathbf{L} \to \mathbf{L}^H \text{ is the } L^2 \text{ orthogonal projector.}$$

With this the extra terms in the first equation of (11.19) can be rewritten

$$\text{Extra Terms} = (\nu_T[(\nabla^s \mathbf{u}^h) - P_H(\nabla^s \mathbf{u}^h)], \nabla^s \mathbf{v}^h).$$

It is natural to think of $P_H(\nabla^s \mathbf{u}^h)$ as a mean deformation, and of $(\mathbb{I} - P_H)(\nabla^s \mathbf{u}^h)$ as a deformation fluctuation.

Definition 11.6. *With $P_H : \mathbf{L} \to \mathbf{L}^H$ the $L^2(\Omega)$ orthogonal projector, define*

$$\overline{(\nabla^s \mathbf{u}^h)} := P_H(\nabla^s \mathbf{u}^h), \qquad (\nabla^s \mathbf{u}^h)' = (\mathbb{I} - P_H)(\nabla^s \mathbf{u}^h).$$

Equation (11.19) thus simplifies, using orthogonality, to

$$\text{Bold Terms in (11.19)} = (\nu_T(\nabla^s \mathbf{u}^h)', (\nabla^s \mathbf{u}^h)').$$

Theorem 11.7. *The method (11.19) is equivalent to: find $\mathbf{u}^h : [0, T] \to \mathbf{X}^h$, and $p^h : (0, T] \to Q^h$ satisfying*

$$(\mathbf{u}_t^h, \mathbf{v}^h) + a(\mathbf{u}^h, \mathbf{v}^h) + b(\mathbf{u}^h, \mathbf{u}^h, \mathbf{v}^h) - (p^h, \nabla \cdot \mathbf{v}^h) + (q^h, \nabla \cdot \mathbf{u}^h)$$
$$(\nu_T(\nabla^s \mathbf{u}^h)', (\nabla^s \mathbf{v}^h)') = (\mathbf{f}, \mathbf{v}^h), \quad \forall \mathbf{v}^h \in \mathbf{X}^h, q^h \in Q^h. \qquad (11.20)$$

Thus, if we identify $\delta = H$, then the method (11.19), or equivalently (11.20), gives precisely the answer to finding an algorithmic realization of Richardson's idea of the cascade of energy through the cut-off length scale. Its stability (= kinetic energy balance) is equally easy and clear.

Theorem 11.8. *The solution $\mathbf{u}^h$ of (11.19), (11.20) satisfies $\forall t \in (0, T]$:*

$$\frac{1}{2}\|\mathbf{u}^h(\cdot, t)\|^2 + \int_0^t \left[\frac{2}{Re}\|\nabla^s \mathbf{u}^h\|^2 + \int_\Omega \nu_T|(\nabla^s \mathbf{u}^h)'|^2 \, d\mathbf{x}\right] dt'$$
$$= \frac{1}{2}\|\mathbf{u}^h(\cdot, 0)\|^2 + \int_0^t (\mathbf{f}, \mathbf{v}^h)(t') \, dt'.$$

Proof. Set $\mathbf{v}^h = \mathbf{u}^h$ and $q^h = p^h$ in (11.20) and repeat the proof of the NSE case. □

The numerical analysis of the method (11.19), (11.20) was begun in [204] for convection-dominated, convection diffusion problems (with error estimates that seem comparable to SUPG methods at a comparable stage of their development.) Recently, a complete error analysis of the method for the NSE was performed by S. Kaya [187], for the evolutionary convection diffusion problem by Heitmann [149] and a new approach to time stepping, exploiting the special structure of the discrete problem, by Anitescu, Layton, and Pahlevani [8]. There are many interesting possibilities for development and testing of this method. The two we want to summarize herein are (i) alternate formulations and (ii) the connection to VMMs [188].

Connection to Variational Multiscale Methods

The key idea is that a multiscale decomposition of the *deformation* induces a multiscale decomposition of the velocities [187, 188]. As usual, let $\mathbf{V}, \mathbf{V}^\mu$ denote the spaces of divergence-free functions and discretely divergence-free functions: For $\mu = h$ and H

$$\mathbf{V} := \left\{\mathbf{v} \in \mathbf{X} : (q, \nabla \cdot \mathbf{v}) = 0 \quad \forall q \in Q\right\},$$
$$\mathbf{V}^\mu := \left\{\mathbf{v}^\mu \in \mathbf{X}^\mu : (q^\mu, \nabla \cdot \mathbf{v}^\mu) = 0 \quad \forall q^\mu \in Q^\mu\right\}.$$

Definition 11.9 (Elliptic projection). *For $\mu = h, H$, $P_E^\mu : \mathbf{X} \to \mathbf{V}^\mu$ is the projection operator satisfying*

$$(\nabla^s[\mathbf{w} - P_E(\mathbf{w})], \nabla^s \mathbf{v}^\mu) = 0, \quad \forall \mathbf{v}^\mu \in \mathbf{V}^\mu.$$

If $\mathbf{w} \in \mathbf{V}$ (*i.e.* $\nabla \cdot \mathbf{w} = 0$), then $P_E \mathbf{w}$ is simply the discrete Stokes projection into $\mathbf{X}^\mu$.

Lemma 11.10. *For $\mu = h, H$, $P_E^\mu : \mathbf{X} \to \mathbf{V}^\mu \subset \mathbf{X}^\mu$ is a well-defined projection operator, with uniformly bounded norm in $\mathbf{X}$. Further, if $\mathbf{L}^H = \nabla^s \mathbf{X}^h$, and $P_H : \mathbf{L}^H \to \mathbf{L}$ is the L^2 projection, then, for any $\mathbf{v}^h \in \mathbf{V}^h$,*

$$P_H(\nabla^s \mathbf{v}^h) = \nabla^s(P_E^H \mathbf{v}^h). \tag{11.21}$$

Proof. That P_E^μ is well defined follows simply from the Lax–Milgram lemma, Poincaré inequality, and Körn's inequality. Equation (11.21) follows by untwisting the definitions of the L^2 and elliptic projectors.   □

Lemma 11.10 shows that the multiscale decomposition of deformations,

$$\nabla^s \mathbf{u}^h = (\overline{\nabla^s \mathbf{u}^h}) + (\nabla^s \mathbf{u}^h)',$$

is equivalent to one for discretely divergence-free velocities with $\overline{\mathbf{X}} := \mathbf{X}^H$ and

$$\mathbf{u}^h = \overline{\mathbf{u}}^h + (\mathbf{u}^h)', \qquad \overline{\mathbf{u}}^h = P_E^H \mathbf{u}^h, \qquad (\mathbf{u}^h)' = (\mathbb{I} - P_E^H)\mathbf{u}^h.$$

From this observation, it follows that (11.19) is a VMM.

Theorem 11.11. *The method* (11.19) *is a VMM. Specifically,* $\mathbf{u}^h = \overline{\mathbf{u}} + (\mathbf{u}^h)'$, *where*

$$\overline{\mathbf{u}} := P_E \mathbf{u}^h \in \mathbf{X}^H, \qquad (\mathbf{u}^h)' = (\mathbb{I} - P_E^H)\mathbf{u}^h.$$

The means $\overline{\mathbf{u}}$ *and the fluctuations* $(\mathbf{u}^h)'$ *satisfy the discrete VMM equations from Sect. 11.5.*

Of course, it is always interesting to establish a connection between "good" methods. The interest in this result goes beyond this connection however. For example, consider the case

$$\mathbf{X}^\mu := \{\mathcal{C}^0 \text{ piecewise linear on } \pi^\mu(\Omega)\}.$$

For a vertex N in the mesh $\pi^\mu(\Omega)$, let $\boldsymbol{\phi}_N(\mathbf{x})$ denote the usual piecewise linear finite element basis function associated with that vertex. Then,

$$\overline{\mathbf{X}} = \mathbf{X}^H = \text{span}\{\boldsymbol{\phi}_N(\mathbf{x}) : \text{ all vertices } N \in \pi^H(\Omega)\}^3,$$

while the discrete model of the fluctuations is

$$\mathbf{X}_b' := \text{span}\{\boldsymbol{\phi}_N(\mathbf{x}) : \text{ all vertices } N \in \pi^h(\Omega), N \notin \pi^H(\Omega)\}.$$

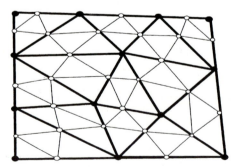

Fig. 11.1. Light nodes correspond to a velocity fluctuation model which is nonzero on element edges

In $2D$, this is illustrated in Fig. 11.1. It is clear that the model for the fluctuations allows them to be nonzero across mesh cells in $\pi^H(\Omega)$. Thus, fluctuations can move.

The secret to the computational feasibility of this choice is that the *deformation* of the means do *not* communicate across edges in $\pi^H(\Omega)$. The fluctuations' effects on means can be evaluated via:

$$(\nu_T \nabla^s(\mathbf{u}^h)', \nabla^s(\mathbf{v}^h)') = (\nu_T \nabla^s \mathbf{u}^h, \nabla^s \mathbf{v}^h) - (\nu_T P_H(\nabla^s \mathbf{u}^h), P_H(\nabla^s \mathbf{v}^h)).$$

Alternate Formulations

By the Helmholtz decomposition (see p. 41), stabilization of $\nabla \mathbf{u}^h$ is accomplished if we can stabilize $\nabla \cdot \mathbf{u}^h$ and $\nabla \times \mathbf{u}^h$. Now $\nabla \cdot \mathbf{u}^h$ is (approximately) zero and can be consistently stabilized by a least squares term

$$\alpha(\nabla \cdot \mathbf{u}^h, \nabla \cdot \mathbf{v}^h).$$

Furthermore, the other contribution to $\nabla \mathbf{u}^h$,

$$\boldsymbol{\omega}^h := \nabla \times \mathbf{u}^h$$

has one dimension less than $\nabla \mathbf{u}^h$. Thus, we can modify the stabilization to reduce the overall storage and computational effort as follows: choose a coarse mesh, discontinuous discrete vorticity space $\mathbf{L}^H$, scalar in $2D$ and vector in $3D$:

$$\mathbf{L}^H \subset L^2(\Omega) \text{ in 2D}, \qquad \mathbf{L}^H \subset [L^2(\Omega)]^3 \text{ in 3D}.$$

Adding multiscale stabilization of $\nabla \times \mathbf{u}^h$ and consistent least squares stabilization of $\nabla \cdot \mathbf{u}^h$ gives the method: find $\mathbf{u}^h : [0, T] \to \mathbf{X}^h$, $p^h : (0, T] \to Q^h$, and $\boldsymbol{\omega}^H : [0, T] \to \mathbf{L}^H$ satisfying

$$\begin{cases} (\mathbf{u}_t^h, \mathbf{v}^h) + \alpha_1(h)(\nabla \cdot \mathbf{u}_t^h, \nabla \cdot \mathbf{v}^h) + b(\mathbf{u}^h, \mathbf{u}^h, \mathbf{v}^h) + a(\mathbf{u}^h, \mathbf{v}^h) \\ + \alpha_2(h)(\nabla \cdot \mathbf{u}^h, \nabla \cdot \mathbf{v}^h) - (p^h, \nabla \cdot \mathbf{v}^h) + (q^h, \nabla \cdot \mathbf{u}^h) + (\nu_T \nabla \times \mathbf{u}^h, \nabla \times \mathbf{v}^h) \\ \qquad - (\nu_T \boldsymbol{\omega}^H, \nabla \times \mathbf{v}^h) = (\mathbf{f}, \mathbf{v}^h), \qquad \forall \mathbf{v}^h \in \mathbf{X}^h, q^h \in Q^h, \\ \qquad (\boldsymbol{\omega}^H - \nabla \times \mathbf{u}^h, \ell^H) = 0, \qquad \forall \ell^H \in \mathbf{L}^H. \end{cases}$$

11.7 Conclusions

Our goals in writing this chapter were two-fold: first, we briefly described the variational formulation and some of the corresponding numerical methods that are used in the numerical experiments described in Chap. 12. Second, we tried to give the reader a glimpse of the numerous challenges in the numerical analysis of LES, where the study of classic topics such as consistency, stability, and convergence of the LES discretization are still at an initial stage. Only the first few steps along these lines have been made, some of which are presented in the exquisite monograph of John [175]. Many open questions (and thus research opportunities!) still remain.

A rigorous numerical analysis for the LES discretization is urgently needed. This could help bring LES to a new level of robustness and universality. For example, the relationship between the filter radius δ and the mesh-size h is an important challenge. Currently, however, the "solution" to this challenge is based on heuristics coming from years of practical experience with LES discretizations. The most popular choice is a relationship of the form $\delta = C h$, where the usual value for the proportionality constant C is 2. The resulting LES discretizations, however, are *very sensitive* to the proportionality constant *and* the numerical method used. This is a clear indication that a rigorous numerical analysis to elucidate the relationship between δ and h is urgently needed!

We ended this chapter with two sections devoted to the Variational Multiscale Method (VMM) of Hughes and his collaborators [160, 161, 162] and one related approach of Layton [204]. This relatively new approach represents an exciting research area where numerical analysis can contribute. Indeed, because of the VMM's variational formulation, a thorough numerical analysis could yield new insight into classic LES challenges such as scale-separation and closure modeling.

Test Problems for LES

12.1 General Comments

Comparison of numerical simulations with physical experiments is an essential test for assessing the quality of an LES model. Numerical simulations are, however, demanding to perform and, once done, it is still nontrivial to extract useful information about an LES model or to make comparisons between LES models.

First, these simulations usually require a large amount of computer memory and time. Since each LES model should be tested on as many types of flows as possible, both memory and speed of execution become critical factors. Thus, it is preferable that the underlying code be parallel, and that we have access to a powerful parallel machine. We must also have a large enough storage capacity for the output files. These two practical issues are often bottlenecks in turbulent flow calculations, and should be considered carefully when starting LES model validation and testing.

Second, the numerical method underlying the code should be carefully assessed. Since we are testing subtle effects in the energy balance, it is easy for model effects to be masked by discretization errors. It is very important that we use a *stable* and *accurate* method which adds as little as possible (ideally none at all!) numerical dissipation and dispersion to the LES model. This apparently simple requirement is a very challenging task. A better understanding of the interplay between numerical discretizations and the LES models used is needed. This understanding is growing very slowly in LES, but it is growing!

Third, the numerical simulation should replicate (or be as close as possible to) an actual physical experiment. There are relatively few such clear cut experiments in turbulence and the data needed for a corresponding numerical simulation is often incomplete. The reason for this is two-fold: (i) physical experiments for turbulent flows are very challenging (LES, and numerical simulations in general, were designed as an alternative), and (ii) physical experiments in a simple enough setting (geometry, complexity), amenable to numerical simulations by LES are scarce and always include noise from many sources.

Fourth, we have to monitor meaningful quantities. For example, for the first steps in computational experiments, *statistics* of flow variables are preferred to the flow variables themselves. The goal of LES is to predict accurately pointwise values of the flow's large scales. However, these are very difficult to validate. *Statistics* of turbulent quantities are easier to evaluate and more stable to all the uncertainties in turbulent flow calculations. Clearly, a simulation being qualitatively correct (*i.e.* matching the correct statistics) is a necessary step to it being quantitatively accurate (matching point values as in, *e.g.* $\|\overline{\mathbf{u}} - \mathbf{w}^h\|$). In general, conclusions on the quality of the LES model based on what happens at a given time and location in the physical domain can be very elusive and uncertain.

We illustrate all these challenges and some possible answers by presenting two of the most popular test cases for the validation and testing of LES:

- turbulent channel flow;
- decay of free isotropic homogeneous turbulence.

We will center most of our discussion around turbulent channel flow, since this test problem involves one of the main challenges in turbulent flow simulations, *interaction with solid walls* (described in Part IV). For turbulent channel flow, we will carefully present many of the main challenges in the validation and testing of LES, such as experimental setting, essential flow parameters, initial conditions, numerical method, and statistics collected. We will also illustrate the entire discussion with LES runs for some of the LES models described in the book: the Smagorinsky model with Van Driest damping (12.11), the Gradient model (12.9), and the Rational LES model (12.10).

Although these two tests are the most popular test problems for the validation and testing of LES, there are many other interesting, challenging test cases, such as anisotropic homogeneous turbulence [14], the round jet [258], the plane mixing layer [258, 173, 175, 176, 243], the backward facing step [135], lid-driven cavity flow [319, 169], and the square-section cylinder [264]. An excellent presentation of some such test cases can be found in Chap. 11 of the exquisite monograph by Sagaut [267] or in Pope [258]. Each careful test and comparison increases our understanding of the relative strengths and (more importantly) weaknesses of LES models.

12.2 Turbulent Channel Flows

Many turbulent flows are (partly) bounded by one or more solid surfaces: flows through pipes and ducts, flows around aircraft and ships' hulls, and flows in the environment such as the atmospheric boundary layer and the flow of rivers. We present LES simulations for one of the simplest wall flows, turbulent channel flow. Most of the results presented in this section appeared in [165, 166, 106].

12.2.1 Computational Setting

3D channel flow (Fig. 12.1) is one of the most popular test problems for the investigation of wall bounded turbulent flows. It was pioneered as an LES test problem by Moin and Kim [240, 189].

The reason for its popularity is two-fold:

First, wall bounded turbulent flows are very challenging. The complex phenomena that take place in the vicinity of the solid surface are not fully understood, and their incorporation in the LES model is regarded as a central problem of LES. This is one of the reasons researchers in LES often consider the computational domain periodic in two directions and bounded just in the third direction. It should be pointed out that there is no physical support for this computational setting: designing a physical experiment with periodic boundaries is impossible. Great care has to be taken in specifying the dimensions of the computational domain in the two periodic directions: the dimensions of the channel in the x and z directions have to be chosen large enough to prevent these simple but artificial boundary conditions from seriously influencing the results. For a more detailed discussion of the choice of the computational domain based on the two-point correlation measurements of Comte-Bellot, the reader is referred to Sect. 3 of the pioneering paper by Moin and Kim [240].

The second reason for the popularity of periodic boundary conditions is their computational advantage: one can use a numerical scheme employing spectral or pseudo-spectral methods in the two periodic directions. This greatly reduces the computational time (one of the major bottlenecks in running LES simulations) and increases the accuracy and reliability of the simulation over lower order methods. Again, the reader is referred to Sect. 5 of the paper by Moin and Kim [240] for more details on such a numerical implementation.

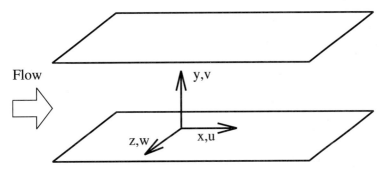

Fig. 12.1. Problem setup for the channel flow

Table 12.1. Dimensions of the computational domain

Nominal Re_τ	$L_x \times L_y \times L_z$
180	$4\pi \times 2 \times \frac{4}{3}\pi$
395	$2\pi \times 2 \times \pi$

12.2.2 Definition of Re_τ

As a benchmark for our LES simulations we used the fine DNS of Moser, Kim, and Mansour [242]. In comparing results of our numerical simulations, great care needs to be taken in simulating *the same* flows. For turbulent channel flow simulations, one important parameter is Re_τ, the Reynolds number based on the wall shear velocity, u_τ. In comparing results for two different simulations of the same flow, we need to make sure that Re_τ is the same in both flows. This is an important issue, and we need to explain it in more detail. Most of the discussion in this subsection is based on the exquisite presentation of turbulent channel flows in Sect. 7.1.3 of Pope [258].

In specifying a Reynolds number of a given flow, three parameters are needed: a characteristic length L, a characteristic velocity V, and the kinematic viscosity of the flow ν. Once these parameters are specified, the Reynolds number is computed via

$$Re = \frac{V L}{\nu}. \tag{12.1}$$

The parameters L and ν are specified in a straightforward manner: $L = H$, where H is half of the height of the channel, and ν is a parameter specific to the fluid in the channel.

The choice for the characteristic velocity V is not that straightforward and several are possible. In order to present the most popular choice for V in channel flow simulations, we need to introduce an important wall quantity, the *wall shear stress* τ_w. First, we note that the channel flow that we study is *fully developed, statistically steady* (we will explain these two concepts in more detail in the next two subsections), and *statistically one-dimensional*, with velocity statistics depending only on the vertical (wall-normal) coordinate y. Then, by using the mean continuity and the mean momentum equations, one can show (see Pope [258], pp. 266, 267) that the total shear stress

$$\tau(y) = \rho \nu \frac{d\langle u \rangle}{dy} - \rho \langle u'v' \rangle \tag{12.2}$$

satisfies

$$\tau(y) = \tau_w \left(1 - \frac{y}{H}\right), \tag{12.3}$$

where

$$\tau_w = \tau(0) \tag{12.4}$$

is the *wall shear stress*. In (12.2), ρ is the density of the fluid, and $\langle \cdot \rangle$ denotes *ensemble averaging*, defined below (see [258]):

Definition 12.1. *Let U denote a component of velocity at a given position and time in a repeatable turbulent flow experiment, and let $U^{(i)}$ denote U in the i-th repetition. Each repetition is performed under the same nominal conditions, and there is no dependence between different repetitions. Thus, the random variables $\{U^{(i)}\}_{i=1,N}$ are independent and identically distributed. The* ensemble average *(over N repetitions) is defined by*

$$\langle U \rangle := \frac{1}{N} \sum_{i=1}^{N} U^{(i)}.$$

Note that, since $\mathbf{u} = \mathbf{0}$ at the wall (no-slip boundary conditions), $\langle u'v' \rangle = 0$. Thus, (12.2) and (12.4) yield

$$\tau_w = \tau(0) = \rho\nu \left.\frac{d\langle u \rangle}{dy}\right|_{y=0} > 0.$$

For turbulent channel flow simulations, the usual choice in (12.1) is $V = u_\tau$, where u_τ is the wall shear velocity depending on the wall shear stress τ_w, and the density of the fluid, ρ.

Definition 12.2. *The wall shear velocity u_τ is defined as*

$$u_\tau = \sqrt{\frac{\tau_w}{\rho}}. \tag{12.5}$$

It is easy to check that u_τ has units of velocity and thus it is an acceptable choice of V in (12.1). The reason for choosing the wall shear velocity as characteristic velocity in (12.1) is that most of the turbulence in channel flow is due to the interaction of the flow with the solid boundary. In fact, choosing the characteristic velocity V the velocity of the flow *away* from the solid boundaries, can be *misleading*.

For example, in Fig. 12.2, the magnitude of the velocity in the center of the channel in the first flow (Fig. 12.2, top) is equal to that of the velocity near the wall in the second flow (Fig. 12.2, bottom). The Reynolds numbers (12.1) calculated with these two characteristic velocities would be the same. However, the two flows are *fundamentally different*! The first one is a laminar flow, whereas the second one is turbulent. Since we are studying *turbulence*, we should use a measure of V that will distinguish between these two cases, such as u_τ. In Table 12.2, for the same turbulent channel flow, we present the

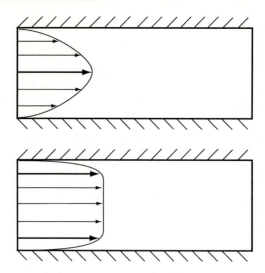

Fig. 12.2. Streamwise (x-) velocity profile: (*top*) laminar flow; (*bottom*) turbulent flow

Reynolds number based on the wall shear velocity u_τ and the corresponding Reynolds number based on the bulk velocity

$$u_m = \frac{1}{H} \int_0^H \langle u \rangle dy.$$

Table 12.2. Same channel flow, Reynolds number based on: wall shear velocity u_τ (left); bulk velocity u_m (right)

Re_τ	Re_m
180	5,600
395	13,750

12.2.3 Initial Conditions

Another essential issue in LES is the specification of initial conditions. The turbulent channel flow problem is a test of *fully developed turbulence*: in a physical experiment, near the entry of the channel ($x = 0$) there is a flow-development region. We investigate, however, the *fully developed* region (large x) in which velocity statistics no longer vary with x.

Thus, the initial flow for our LES simulations needs to be turbulent. There are a couple of approaches for obtaining turbulent initial flow fields.

The first and easiest approach is to take a flow field from a previous LES simulation for the same setting at a possible lower Re_τ (there are some databases with such flow fields). This flow field is then integrated over time at the actual Re_τ until a *statistically steady state* is reached, that is, the statistics of the flow variables do not vary over time. We will explain in more detail the way we collect statistics and how we identify the statistically steady state in the next subsection.

The second approach is to start with a laminar flow, impose a certain set of disturbances on it, and integrate it over time until it transitions to a turbulent flow. This approach, however, is usually computationally intensive: the time to transition to turbulence is very long.

In the results for the LES simulations presented in this chapter, we used both methods. First, the initial conditions for the $Re_\tau = 180$ simulations were obtained by superimposing some perturbations (a 2D Tollmien–Schlichting (TS) mode of 2% amplitude and a 3D TS mode of 1% amplitude, see again Moin and Kim [240, 189] for details) on a parabolic mean flow (Poiseuille flow). We then integrated the flow for a long time (approximately 200 H/u_τ) on a finer mesh. The final field file was further integrated on the actual coarse LES mesh for approximately 50 H/u_τ to obtain the initial condition for *all* $Re_\tau = 180$ simulations.

The initial condition for the $Re_\tau = 395$ case was obtained as follows: we started with a flow field corresponding to an $Re_\tau = 180$ simulation, and we integrated it on a finer mesh for a long time (approximately 50 H/u_τ). Then, we integrated the resulting flow on the actual coarser LES mesh and for $Re_\tau = 395$ for another 40 H/u_τ, and the final flow field was used as initial condition for *all* simulations.

12.2.4 Statistics

Since there is no interaction between the flow and the exterior, by keeping a constant mass flux through the channel, the flow will evolve to a state in which the *statistics* for all quantities of interest will be constant. It should be emphasized that the variables of interest (like the velocity of the flow) will vary in time – the statistics of these variables, however, will remain constant. Thus, while still considering a turbulent flow, we can monitor statistics of the flow variables. These statistics are considered more reliable than instantaneous values of the flow variables.

In the numerical results presented in this chapter, for all the LES models we tested and for both $Re_\tau = 180$ and $Re_\tau = 395$, the flow was integrated further over time until the statistically steady state was reached (for approximately 15 H/u_τ). The statistically steady state was identified by a linear total shear stress profile (see Fig. 12.3). For details on why the linear shear stress profile is an indicator of statistically steady state, the reader is again referred to Pope [258]. The statistics were then collected over another 5 H/u_τ and contained samples taken after each time step ($\Delta t = 0.0002$ for $Re_\tau = 180$ and

$\Delta t = 0.00025$ for $Re_\tau = 395$). We also averaged over the two halves of the channel, to increase the reliability of the statistical sample.

The numerical results include plots of the following time- and plane-averaged (denoted by $\langle \cdot \rangle$) quantities normalized by the *computed* u_τ: the mean streamwise (x-) velocity, the x, y-component of the Reynolds stress, and the root mean square (rms) values of the streamwise (x-), wall-normal (y-), and spanwise (z-) velocity fluctuations. We computed these statistics following the approach in [315], where it was proved that the best way to reconstruct the Reynolds stresses from LES is

$$R_{ij}^{DNS} \approx R_{ij}^{LES} + \langle \overline{A}_{ij}^M \rangle, \qquad (12.6)$$

where $R_{ij}^{DNS} \equiv \langle u_i u_j \rangle - \langle u_i \rangle \langle u_j \rangle$ are the Reynolds stresses from the fine DNS in [242], $R_{ij}^{LES} \equiv \langle \overline{u}_i \overline{u}_j \rangle - \langle \overline{u}_i \rangle \langle \overline{u}_j \rangle$ are the Reynolds stresses coresponding to the dynamics of the LES field, and $\langle \overline{A}_{ij}^M \rangle$ are the averaged values of the modeled subfilter-scale stresses $\langle \tau_{ij} \rangle = \langle \overline{u_i u_j} - \overline{u}_i \overline{u}_j \rangle$.

As pointed out in [315], the Reynolds stresses from an LES can only be compared with those from a DNS by also taking into account the significant contribution from the averaged subfilter-scale stresses. Since we include results for the Smagorinsky model with Van Driest damping, we need to be careful

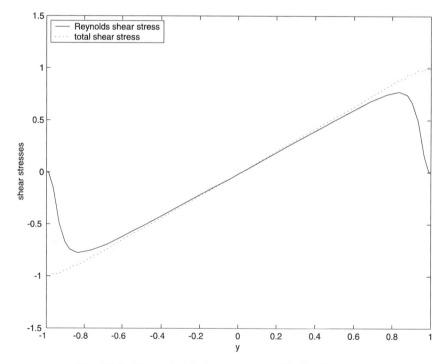

Fig. 12.3. Linear total shear stress profile for $Re_\tau = 180$

with the reconstruction of the diagonal Reynolds stresses (the rms turbulence intensities). Specifically, for this eddy viscosity model, only the *anisotropic* part of R_{ij}^{DNS} can be reconstructed (and thus compared with DNS):

$$R_{ii}^{*DNS} \approx R_{ii}^{*LES} + \langle \overline{A}_{ii}^{*M} \rangle, \tag{12.7}$$

where

$$R_{ii}^{*DNS} \equiv \langle u_i' u_i' \rangle - \frac{1}{3} \sum_{k=1}^{3} \langle u_k' u_k' \rangle = R_{ii}^{DNS} - \frac{1}{3} \sum_{k=1}^{3} R_{kk}^{DNS}$$

$$R_{ii}^{*LES} \equiv R_{ii}^{LES} - \frac{1}{3} \sum_{k=1}^{3} R_{kk}^{LES},$$

$$\overline{A}_{ii}^{*} \equiv \overline{A}_{ii} - \frac{1}{3} \sum_{k=1}^{3} \overline{A}_{kk},$$

and $\overline{A}_{ii}^{*M}$ is modeling $\overline{A}_{ii}^{*}$.

The reconstruction of the off-diagonal stresses R_{xy} is straightforward:

$$R_{xy}^{DNS} \approx R_{xy}^{LES} + \langle \overline{A}_{xy}^{*M} \rangle, \tag{12.8}$$

since $\langle \overline{A}_{xy}^{*M} \rangle = \langle \overline{A}_{xy}^{M} \rangle$.

In computing R_{xy}, u_{rms}^*, v_{rms}^*, and w_{rms}^*, for the three LES models, we used formulas (12.6)–(12.8). These results were then compared with the corresponding ones in [242].

12.2.5 LES Models Tested

In Chap. 7, we presented the Gradient model

$$\boldsymbol{\tau} = \overline{\mathbf{u}\,\mathbf{u}^T} - \overline{\mathbf{u}}\,\overline{\mathbf{u}}^T \approx \frac{\delta^2}{2\gamma} \nabla \overline{\mathbf{u}} \, \nabla \overline{\mathbf{u}}^T, \tag{12.9}$$

where δ is the filter radius, γ is a shape parameter in the definition of the Gaussian filter, and

$$(\nabla \overline{\mathbf{u}} \, \nabla \overline{\mathbf{u}}^T)_{i,j} = \sum_{l=1}^{d} \frac{\partial \overline{\mathbf{u}}_i}{\partial \mathbf{x}_l} \frac{\partial \overline{\mathbf{u}}_j}{\partial \mathbf{x}_l},$$

and the Rational LES model

$$\boldsymbol{\tau} = \left[\left(-\frac{\delta^2}{4\gamma} \Delta + \mathbb{I} \right)^{-1} \left(\frac{\delta^2}{2\gamma} \nabla \overline{\mathbf{u}} \, \nabla \overline{\mathbf{u}}^T \right) \right]. \tag{12.10}$$

These two models are subfilter-scale (SFS) models: they aim at computing an improved approximation of the stress tensor $\tau = \overline{\mathbf{u}\,\mathbf{u}^T} - \overline{\mathbf{u}}\,\overline{\mathbf{u}}^T$ by replacing the unknown unfiltered variables with approximately deconvolved filtered variables [276, 130, 194, 93, 92, 285, 289, 290].

The gradient (also called nonlinear, or tensor-diffusivity) model, has been used in numerous studies [212, 65, 55, 41, 316, 309, 223, 7, 182]. In *all* these numerical tests, the Gradient model (12.9) was found to be *very unstable*. To stabilize the Gradient model, Clark, Ferziger, and Reynolds [65] combined it with a Smagorinsky term, but the resulting mixed model inherited the excessive dissipation of the Smagorinsky model. A different approach was proposed by Liu *et al.* [223], who supplied the Gradient model with a "limiter"; this clipping procedure ensures that the model dissipates energy from large to small scales. This approach was also used in [85, 78, 77].

In this section, we present a comparison of

- the RLES model (12.10);
- the Gradient model (12.9);
- the Smagorinsky model with Van Driest damping (12.11);

in the numerical simulation of 3D turbulent channel flows at Reynolds numbers based on the wall shear velocity $Re_\tau = 180$ and $Re_\tau = 395$.

To give a measure of the success of the first two SFS models, we compare them with a classical eddy viscosity model, the Smagorinsky model with Van Driest damping [302] (see Chap. 3)

$$\tau = -(C_S\,\delta\,(1 - \exp(-y^+/A))^2\,\|\nabla^s\overline{\mathbf{u}}\|_F\,\nabla^s\overline{\mathbf{u}}, \tag{12.11}$$

where $\nabla^s\overline{\mathbf{u}} := \frac{1}{2}(\nabla\overline{\mathbf{u}} + \nabla\overline{\mathbf{u}}^T)$ is the deformation tensor of the filtered field, $\|\cdot\|_F$ is the Frobenius norm, $C_S \approx 0.17$ is the Smagorinsky constant, y^+ is the nondimensional distance from the wall, $H = 1$ is the channel half-width, u_τ is the wall shear velocity, and $A = 25$ is the Van Driest constant [302].

In (12.11), we encountered the variable y with a superscript $^+$. Since this is an important quantity in channel flow simulations, we define it below.

Definition 12.3. *The distance from the wall measured in wall units is defined by*

$$y^+ = \frac{u_\tau\,(H - |y|)}{\nu}, \tag{12.12}$$

and determines the relative importance of viscous and turbulent phenomena.

The numerical results of our comparison were presented in [106, 165], and they shed light on two important issues:

- a comparison between the Gradient model (12.9) and the RLES model (12.10) as *subfilter-scale (SFS) models;*
- a comparison of these two SFS models with a classical eddy viscosity model, the Smagorinsky model with Van Driest damping (12.11).

12.2.6 Numerical Method and Numerical Setting

The numerical simulations were performed by using a spectral element code based on the $\mathbb{P}_N - \mathbb{P}_{N-2}$ velocity and pressure spaces introduced by Maday and Patera [224].

The domain was decomposed into spectral elements, as shown in Fig. 12.4. In an attempt to keep the numerical setting as close as possible to that used for our benchmark results (the fine DNS in [242]), the mesh spacing in the wall-normal direction (y) was chosen to be roughly equivalent to a Chebychev distribution having the same number of points.

The velocity is continuous across element interfaces and is represented by Nth-order tensor-product Lagrange polynomials based on the Gauss–Lobatto–Legendre (GLL) points. The pressure is discontinuous and is represented by tensor-product polynomials of degree N–2. Time-stepping is based on operator-splitting of the discrete system, which leads to separate convective, viscous, and pressure subproblems without the need for *ad hoc* pressure boundary conditions. A filter, which removes 2%–5% of the highest velocity mode, is used to stabilize the Galerkin formulation [108]; the filter does not compromise the spectral accuracy. Details of the discretization and solution algorithm are given in [105, 107].

As we have seen in Sect. 12.2.1, in comparing results for the numerical simulation of turbulent flows, one needs to ensure that the flow parameters are *the same*. In particular, for channel flow simulation, one must evolve the flow so that the wall shear velocity u_τ (and thus, the corresponding Re_τ) be kept close to the desired value. The most popular approaches in ensuring this in channel flow simulations are

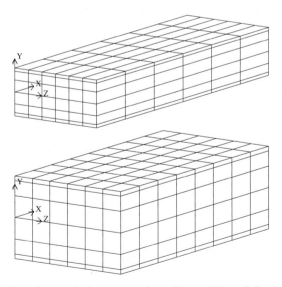

Fig. 12.4. Spectral element meshes: $Re_\tau = 180$ and $Re_\tau = 395$

- by enforcing a constant mass flux through the channel;
- by enforcing a constant pressure gradient through the channel.

These approaches adjust *dynamically* the forcing term so that the mass flux and the pressure gradient, respectively, are constant.

In our numerical simulations we chose the former: the forcing term that drives the flow was adjusted dynamically to maintain a constant mass flux through the channel. Thus, in our simulations the bulk velocity u_m was fixed to match the corresponding one in [242] (see Table 12.3), and the wall shear velocity u_τ was a result of the simulations. Table 12.3 presents the *actual* values of Re_τ corresponding to the wall shear velocity u_τ computed for all three numerical tests and two nominal Reynolds numbers. We note that the friction velocity u_τ is within 1%–2% of the nominal value, and, as a result, so is the actual Re_τ.

Table 12.3. Computed u_τ and Re_τ

Fixed U_m	Nominal Re_τ	Case	Computed u_τ	Computed Re_τ
15.63	180	RLES	0.9879448	177.8352
		gradient	0.9890118	178.0222
		Smagorinsky	0.9917144	178.5120
		with Van Driest damping		
17.54	395	RLES	1.001025319	395.4071960
		gradient	1.005021334	396.9859924
		Smagorinsky	0.9974176884	393.9718933
		with Van Driest damping		

The filter width δ was computed by using the most popular formula $\delta = \sqrt[3]{\Delta_x \, \Delta_z \, \Delta_y(y)}$, where Δ_x and Δ_z are the largest spaces between the Gauss–Lobatto–Legendre (GLL) points in the spectral element in the x and z directions, respectively, and $\Delta_y(y)$ is inhomogeneous and is computed as an interpolation function that is zero at the wall and is twice the normal mesh size for the elements in the center of the channel. Note that, since we filter in all three directions, the filter width δ never vanishes away from the wall. This, however, could be a serious problem for tests in which one filtering direction is discarded; in this case, the LES model vanishes although the other two directions are poorly resolved. To avoid this difficulty, one should instead use the anisotropic version of the RLES model (12.10), in which δ_x, δ_y, and δ_z are all different. The derivation of this anisotropic form of the RLES model is straightforward and the resulting model remains easy to implement.

We used as a first step the RLES model (12.10) with the inverse operator equipped with Neumann boundary conditions. This is clearly not the best choice, since the subfilter-scale stresses $\boldsymbol{\tau} = \overline{\mathbf{u}\,\mathbf{u}} - \overline{\mathbf{u}}\,\overline{\mathbf{u}}$ modeled by the RLES model vanish on the boundary if $\delta = 0$ at the wall (which is our case). We plan to investigate the RLES model with the inverse operator equipped with homogeneous Dirichlet boundary conditions instead of Neumann boundary conditions as in the present simulations. These new boundary conditions could yield better behavior near the wall for the RLES model (12.10).

In our numerical experiments, we considered, as a first step, homogeneous boundary conditions for all LES models tested. As argued in Part IV, this is clearly not the right approach, because of its high computational cost and the commutation error introduced by the fact that differentiation and convolution might not commute for variable filter radius δ (which is our case). We chose this popular approach, however, because of its simplicity and because we wanted to focus on the comparison of the RLES and gradient models as subfilter-scale models. In other words, the boundary conditions might be inappropriate, but they are the same for both LES models. Obviously, we plan to investigate better boundary conditions, such as those indicated in Part IV.

12.2.7 *A Posteriori* Tests for $Re_\tau = 180$

A few words on the terminology are necessary. An *a posteriori* test is a numerical simulation which employs an actual LES model. In other words, one first needs to run the simulation with the LES model included in the numerical method, and only then collect the results. This is in contrast with the *a priori* tests, where one uses a data set from a previous DNS calculation, and then computes the corresponding *filtered* quantities, without effectively using any LES model in the numerical simulation.

It should be emphasized that the two approaches yield *different* results. The reason is that the flows in the two simulations evolve differently: the former includes the contribution of the LES model, while the latter is a DNS. This difference has been noticed time and again in the validation of LES models. For example, the scale-similarity model of Bardina [13] presented in Chap. 8 yielded very good results in *a priori tests*, but performed poorly in *a posteriori* tests. Thus, it is imperative to test an LES model in *a posteriori* tests in order to assess its performance.

In [165], we ran *a posteriori* tests for the RLES model (12.10), the Gradient model (12.9), and the Smagorinsky model with Van Driest damping (12.11). We compared the corresponding results with the fine DNS simulation of Moser, Kim, and Mansour [242]. Having an extensive database such as that in [242] makes our task much easier. Without such a database, we would have had to run extremely long fine DNS tests and collect statistics. Thus, it is preferable to start with a test problem for which such extensive databases exist.

Figure 12.5 shows the normalized mean streamwise velocity u^+, where a "+" superscript denotes the variable in wall-units; note the almost perfect overlapping of the results corresponding to the models tested. We interpret this behavior as a measure of our success in enforcing a constant mass flux through the channel. Since we have only two mesh points with $y^+ \leq 10$ away from the wall, the plotting by linear interpolation between these two points produces inadequate results. The mean streamwise velocity u^+ at these points is, however, very close to that in the fine DNS.

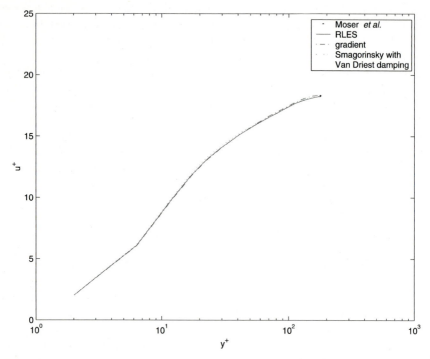

Fig. 12.5. Mean streamwise velocity, $Re_\tau = 180$. We compared the RLES model (12.10), the Gradient model (12.9), and the Smagorinsky model with Van Driest damping (12.11) with the fine DNS of Moser, Kim, and Mansour [242]

Figure 12.6 presents the normalized x, y-component of the Reynolds stress, R_{xy}, computed by using (12.8). Note that R_{xy} includes contributions from the subgrid-scale stresses, which, in turn, include terms containing the gradient of the computed velocity. Since this gradient is not continuous across the spectral elements, we obtain spikes in the Gradient (12.9) and Smagorinsky with Van Driest damping models. The inverse operator in the RLES model (12.10) has a smoothing effect on the subgrid-scale stress tensor and attenuates these spikes. This behavior is apparent in all the other plots for the Reynolds stresses. The R_{xy} for the RLES model (12.10) is better than that for the Gradient model (12.9) (there are no spikes), with the exception of the near-wall region; here, the inverse (smoothing) operator equipped with Neumann boundary conditions introduces a nonzero R_{xy} for the RLES model (12.10).

Nevertheless, both the RLES (12.10) and the Gradient (12.9) model yield much better results for R_{xy} than the Smagorinsky model with Van Driest damping; the latter performs poorly.

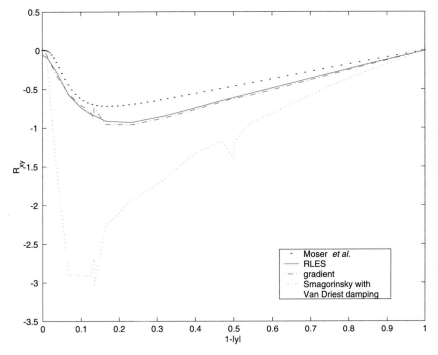

Fig. 12.6. The x, y-component of the Reynolds stress, $Re_\tau = 180$. We compared the RLES model (12.10), the Gradient model (12.9), and the Smagorinsky model with Van Driest damping (12.11) with the fine DNS of Moser, Kim, and Mansour [242]

The situation is completely different for the diagonal stresses (rms turbulence intensities) in Figs. 12.7–12.9. Here, the Smagorinsky model with Van Driest damping performs significantly better than both the RLES (12.10) and the gradient (12.9) models. As for the R_{xy}, the inverse operator in the RLES model has a smoothing effect and attenuates the spikes in the diagonal Reynolds stresses of the Gradient model (12.9), yielding improved results, with the exception of the near-wall region where it introduces a nonzero diagonal Reynolds stress. We also note that the first spike in the rms turbulence intensities for Gradient model (12.9) away from the wall is *not* at the spectral element interface. Nevertheless, the smoothing operator in the RLES model (12.10) attenuates it significantly. The inverse operator is also responsible for the much increased numerical stability of the RLES model (12.10) over the Gradient model (12.9). In order to prevent numerical simulations with the Gradient model from blowing up, we had to use a very small time-step; the

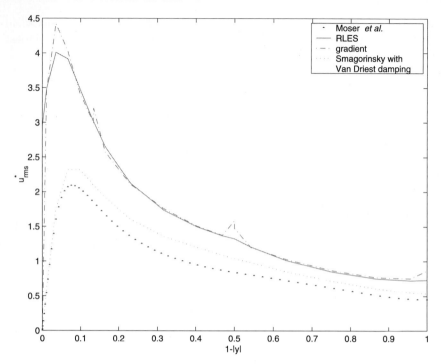

Fig. 12.7. Rms values of streamwise velocity fluctuations, $Re_\tau = 180$. We compared the RLES model (12.10), the Gradient model (12.9), and the Smagorinsky model with Van Driest damping (12.11) with the fine DNS of Moser, Kim, and Mansour [242]

simulations with the RLES model (12.10) ran with much larger time-steps. (To collect statistics, however, we ran the two LES models with the same time-step.)

12.2.8 *A Posteriori* Tests for $Re_\tau = 395$

In [165], we ran simulations with all three LES models for $Re_\tau = 395$, and compared our results with the fine DNS in [242]. Again, as in the $Re_\tau = 180$ case, the normalized mean streamwise velocity profiles in Fig. 12.10 are practically identical; this time, however, they do not overlap the profiles for the fine DNS. Nevertheless, the mean flows are the same, and this is supported by the fact that the models underpredict the correct value near the wall but overpredict it away from the wall. The inadequate behavior near the wall is due to the plotting, as in the $Re_\tau = 180$ case (we used linear interpolation for the two mesh points with $y^+ \leq 10$ away from the wall). In fact, u^+ at these two mesh points compares very well with the fine DNS results in [242]. In the buffer and log layers the three LES models deviate from the correct DNS results, but they perform well at the center of the channel.

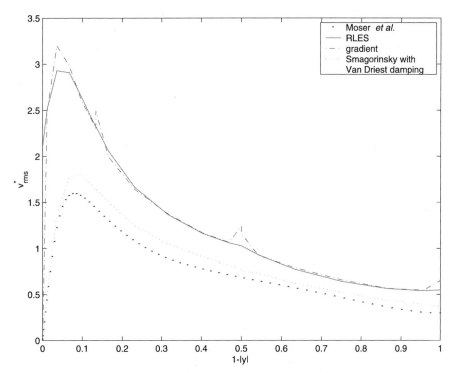

Fig. 12.8. Rms values of wall-normal velocity fluctuations, $Re_\tau = 180$. We compared the RLES model (12.10), the Gradient model (12.9), and the Smagorinsky model with Van Driest damping (12.11) with the fine DNS of Moser, Kim, and Mansour [242]

The results for the normalized Reynolds stresses in Figs. 12.11–12.14 parallel the corresponding ones for the $Re_\tau = 180$ case. The RLES model (12.10) performs better than the Gradient model (12.9) (the smoothing operator eliminates the spikes), with the exception of the near-wall region, where the smoothing operator introduces a nonzero value.

Both the RLES (12.10) and the Gradient (12.9) models yield much better results for the off-diagonal Reynolds stress tensor R_{xy} than the Smagorinsky model with Van Driest damping (Fig. 12.11).

However, the Smagorinsky model with Van Driest damping performs much better than both the RLES (12.10) and the Gradient (12.9) models in predicting the diagonal stresses (Figs. 12.12–12.14), with the exception of R_{zz} in Fig. 12.14, where the improvement is not that dramatic.

Again, as in the $Re_\tau = 180$ case, the RLES model (12.10) is much more stable numerically than the Gradient model.

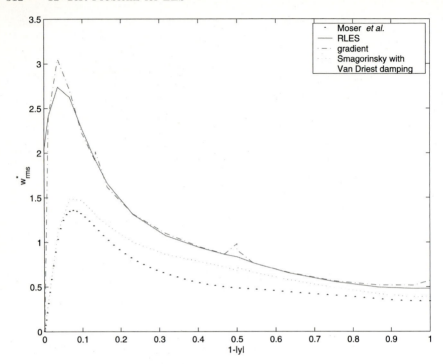

Fig. 12.9. Rms values of spanwise velocity fluctuations, $Re_\tau = 180$. We compared the RLES model (12.10), the Gradient model (12.9), and the Smagorinsky model with Van Driest damping (12.11) with the fine DNS of Moser, Kim, and Mansour [242]

12.2.9 Backscatter in the Rational LES Model

We close this section on the numerical simulation of turbulent channel flow with a very interesting and important phenomenon, *backscatter*. A detailed description of backscatter and its relationship to the concept of energy cascade is given in Sect. 3.5. Here we present numerical results for backscatter in turbulent channel flow simulations. The results in this subsection were published in [166].

Based on the concept of energy cascade (see Fig. 3.1), most of the commonly used LES models assume that the essential function of the unresolved (modeled) scales is to remove energy from the large scales and dissipate it through the action of viscous forces. While, on average, energy is transferred from the large to the small scales ("forward scatter"), it has been recognized that the inverse transfer of energy from small to large scales ("backscatter") may be quite significant (see Fig. 3.3) and should be included in the LES model. Indeed, Piomelli *et al.* [254] performed DNS of transitional and turbulent channel flow and compressible isotropic turbulence. In all flows considered, approximately 50% of the grid points experienced backscatter.

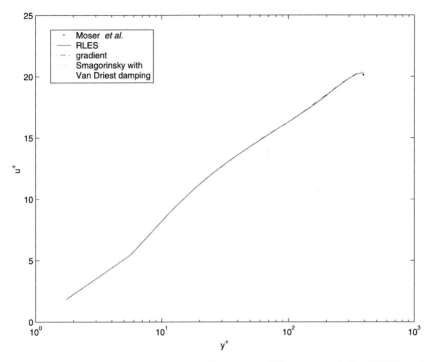

Fig. 12.10. Mean streamwise velocity, $Re_\tau = 395$. We compared the RLES model (12.10), the Gradient model (12.9), and the Smagorinsky model with Van Driest damping (12.11) with the fine DNS of Moser, Kim, and Mansour [242]

To illustrate the importance of including backscatter in the LES model, note that the Smagorinsky model [277], the most popular eddy viscosity model, is purely dissipative and cannot predict backscatter. To include backscatter, the Smagorinsky model is usually used in the dynamical framework of Germano *et al.* [129]. This approach may, however, lead to numerical instabilities. The reason could be the fact that backscatter is not introduced in a *natural* way: we start with a purely dissipative model (the Smagorinsky model), and through some clever manipulations, we get a model that could yield backscatter (the dynamic subgrid-scale model).

A few LES models introduce backscatter in a *natural* way. We present numerical investigation of backscatter in two such LES models, the RLES (12.10) and the Gradient (12.9) LES models applied to the numerical simulation of turbulent channel flows at $Re_\tau = 180$ and $Re_\tau = 395$.

We collected statistics for SGS dissipation, forward scatter and backscatter. (We define these quantities below.) We started with field files corresponding to LES simulations in [165], which had already reached a statistically steady state. We then integrated the flow further over time and collected statistics for the above three quantities, which were averaged over time and

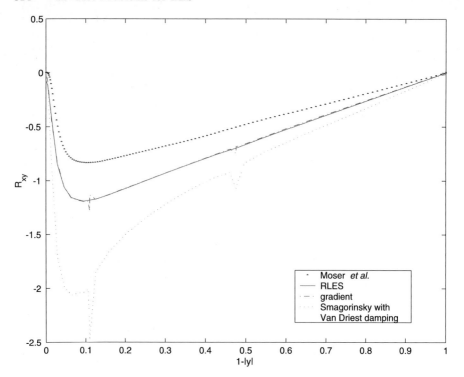

Fig. 12.11. The x, y-component of the Reynolds stress, $Re_\tau = 395$. We compared the RLES model (12.10), the Gradient model (12.9), and the Smagorinsky model with Van Driest damping (12.11) with the fine DNS of Moser, Kim, and Mansour [242]

homogeneous directions (streamwise and spanwise). All three statistics were normalized by u_τ^3, where u_τ is the computed wall-shear velocity, which was found to be within 1%–2% of the nominal value.

The *model subgrid-scale dissipation* was computed as

$$\varepsilon_{SGS} := \boldsymbol{\tau}_{ij} \left(\nabla^s \overline{\mathbf{u}} \right)_{ij}, \qquad (12.13)$$

where $(\nabla^s \overline{\mathbf{u}})_{ij} = \frac{1}{2} \left(\frac{\partial \overline{u}_i}{\partial \mathbf{x}_j} + \frac{\partial \overline{u}_j}{\partial \mathbf{x}_i} \right)$ represents the large-scale strain-rate tensor, and $\boldsymbol{\tau} = \overline{\mathbf{u}\,\mathbf{u}} - \overline{\mathbf{u}}\,\overline{\mathbf{u}}$ is the subfilter-scale stress tensor. To collect statistics of ε_{SGS}, at each coordinate y in the computational domain, we averaged over the horizontal directions of homogeneity and in time.

The model subgrid-scale dissipation ϵ_{SGS} represents the energy transfer between the resolved and the unresolved (subfilter-scale) scales. If ε_{SGS} is negative, energy is transferred from large scales to small scales (forward scatter); if ε_{SGS} is positive, energy is transferred from small scales to large scales (backscatter). We denote the forward scatter by $\varepsilon_+ = \frac{1}{2}(\varepsilon_{SGS} + |\varepsilon_{SGS}|)$ and the backscatter by $\varepsilon_- = \frac{1}{2}(\varepsilon_{SGS} - |\varepsilon_{SGS}|)$.

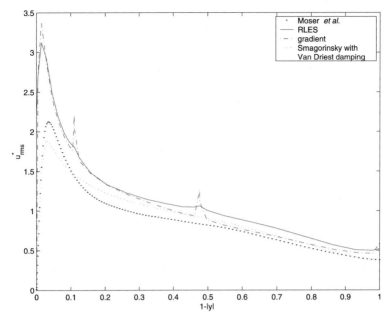

Fig. 12.12. Rms values of streamwise velocity fluctuations, $Re_\tau = 395$. We compared the RLES model (12.10), the Gradient model (12.9), and the Smagorinsky model with Van Driest damping (12.11) with the fine DNS of Moser, Kim, and Mansour [242]

12.2.10 Numerical Results

In the $Re_\tau = 180$ case, the model subgrid-scale dissipation ε_{SGS} in Fig. 12.15 shows the correct behavior for the RLES model (12.10): the forward scatter is dominant throughout the channel, with a peak near the wall. This behavior can be noticed in the DNS results in [93] (Fig. 8a, p. 2159).

The correct ε_{SGS} is quite challenging to capture in LES: the velocity estimation model in [93] (Fig. 8a, p. 2159) underpredicts the correct peak value of ε_{SGS}. The variational multiscale approach in [162] underpredicts significantly the correct peak value for ε_{SGS} (Fig. 14, p. 1791). The ε_{SGS} corresponding to the RLES model in Fig. 12.15 performs better than both previous methods; the RLES model actually performs similarly to the classical eddy viscosity models (the Smagorinsky model in [93] and the Smagorinsky model with Van Driest damping in [162]). This is quite remarkable for a *non* eddy viscosity model such as the RLES model, which introduces a significant amount of backscatter.

The Gradient model (12.9) has an incorrect behavior: it starts with a huge amount of backscatter near the wall and then reaches the peak value of forward scatter away from the correct location [93].

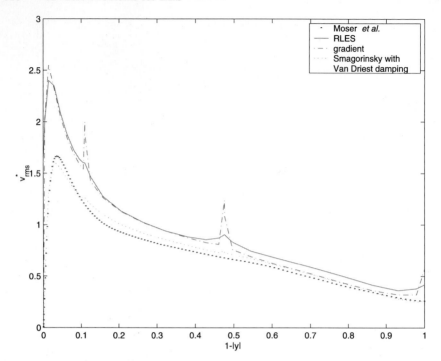

Fig. 12.13. Rms values of wall-normal velocity fluctuations, $Re_\tau = 395$. We compared the RLES model (12.10), the Gradient model (12.9), and the Smagorinsky model with Van Driest damping (12.11) with the fine DNS of Moser, Kim, and Mansour [242]

The forward scatter and backscatter in Fig. 12.15 illustrate the smoothing character of the inverse filtering in the RLES model (12.10): the "spikes" seen in the Gradient model are damped in the RLES model. This process has a positive effect on the numerical stability of the RLES model. The huge amount of forward scatter and backscatter introduced by the gradient model in the near-wall region is responsible for the unstable behavior in wall-bounded flow simulations [316].

For both LES models, the backscatter and the forward scatter contributions to the SGS dissipation were comparable, and each was much larger than the total SGS dissipation. This behavior was also noticed in [254].

In the $Re_\tau = 395$ case, the SGS dissipation corresponding to the RLES model (12.10) in Fig. 12.16 is much less than that for the Gradient model (12.9); the latter seems exaggerated for this Reynolds number. The forward and backscatter for the RLES model are, however, larger than those for the Gradient model. This fact does not contradict the observation about the SGS dissipation, since ε_{SGS} is the sum of the forward and backscatter. We also need to keep in mind that, although both LES models are started from the same initial conditions, the corresponding flows evolve in time *differently*. Thus, in

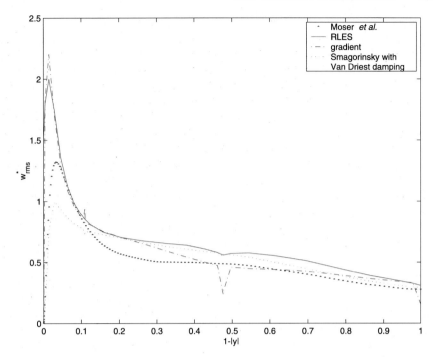

Fig. 12.14. Rms values of spanwise velocity fluctuations, $Re_\tau = 395$. We compared the RLES model (12.10), the Gradient model (12.9), and the Smagorinsky model with Van Driest damping (12.11) with the fine DNS of Moser, Kim, and Mansour [242]

the numerical simulations, the SFS stress tensor τ in the RLES model is *not* simply the inverse operator in (12.10) applied to the SFS stress tensor τ in the Gradient model.

As in the $Re_\tau = 180$ case, for both LES models the backscatter and the forward scatter contributions were comparable, and each was much larger than the total SGS dissipation [254].

We note the nonphysical spikes corresponding to the Gradient model (12.9) in all three quantities monitored: ε_{SGS}, ε_-, and ε_+. These spikes are located exactly at the interfaces between adjacent spectral elements. This behavior is natural, since the SGS tensor τ for the Gradient model (12.9) contains products of gradients of the computed velocity (see (12.9)). The RLES model, on the other hand, smooths out these spikes through its inverse operator; this smoothing makes the RLES model more stable numerically. Further investigation of these issues is necessary.

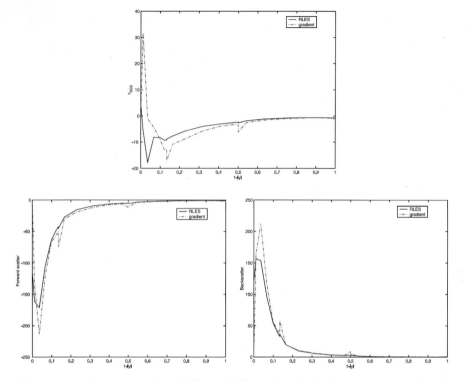

Fig. 12.15. $Re_\tau = 180$, the RLES model (12.10) and the Gradient model (12.9): SGS dissipation (*top*); forward scatter (*bottom, left*); backscatter (*bottom, right*)

12.2.11 Summary of Results

The RLES model (12.10) yielded better results than the Gradient model (12.9) for both $Re_\tau = 180$ and $Re_\tau = 395$, and for all Reynolds stresses. This was due to the inverse operator in the RLES model, which had a smoothing effect over the modeled subfilter-scale stress tensor and eliminated (or attenuated) the spikes in the Gradient model. The inverse operator, however, introduced nonzero Reynolds stresses in the near-wall region. The Neumann boundary conditions need to be replaced by homogeneous boundary conditions, as argued before.

But the most significant improvement of the RLES model over the Gradient model is the much increased numerical stability, which is also due to the smoothing effect of the inverse operator.

The Smagorinsky model with Van Driest damping (12.11) performed worse than both the RLES and the Gradient models in predicting the off-diagonal Reynolds stresses, but predicted very accurately the diagonal ones.

We believe that these results for the RLES model are encouraging. They also support our initial thoughts: the RLES model is an improvement over the

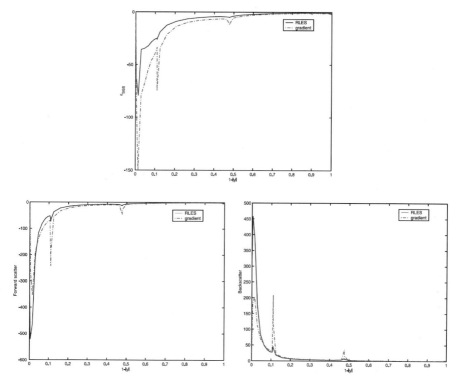

Fig. 12.16. $Re_\tau = 395$, the RLES model (12.10) and the Gradient model (12.9): SGS dissipation (*top*); forward scatter (*bottom, left*); backscatter (*bottom, right*)

gradient model as a *subfilter-scale* model. The RLES model is also more stable numerically because of the additional smoothing operator, and this feature is manifest for both low ($Re_\tau = 180$) and moderate ($Re_\tau = 395$) Reynolds number flows.

However, the RLES model accounts just for the subfilter-scale part of the stress reconstruction. The information lost at the subgrid-scale level must be accounted for in a different way, as advocated by Carati *et al.* [55]. This was illustrated by the dramatic improvement for the diagonal Reynolds stresses, for both $Re_\tau = 180$ and $Re_\tau = 395$, yielded by the Smagorinsky model with Van Driest damping, a classical eddy viscosity model.

It seems that the RLES model (12.10), although an improvement over the Gradient model (12.9), should probably be supplemented by an eddy viscosity mechanism (a mixed model) to be competitive in challenging wall-bounded turbulent flow simulations. One should investigate this mixed model in more challenging simulations and compare the results with state-of-the-art LES models such as the dynamic Smagorinsky model [129] and the variational multiscale method of Hughes *et al.* [160, 161, 162].

Other possible research directions include the study of improved boundary conditions, the commutation error [136, 133], and the relationship between the filter radius and the mesh size in a spectral element discretization.

We also gathered statistics for the model SGS dissipation, the forward scatter, and the backscatter. In the $Re_\tau = 180$ case, the RLES model (12.10) yielded much improved results, closer to the DNS results in [93]. The Gradient model introduced an unphysical amount of backscatter near the wall, which made the computations more unstable. In the $Re_\tau = 395$ case, the RLES model's SGS dissipation was closer to a realistic value. The SGS dissipation for the Gradient model seemed unrealistically high. The amount of forward and backscatter was, however, higher for the RLES model. Despite this, the Gradient model (12.9) was more unstable numerically, as reported in [316]. This issue deserves further investigation.

Both the RLES and the Gradient models introduce backscatter in a *natural* way. The Gradient model is unstable in numerical simulations. On the contrary, the RLES model, through the action of its smoothing filter, makes the computations much more stable; it can run for thousands of time-steps without additional numerical stabilization procedures.

As mentioned at the beginning of this chapter, we will not go into great detail in the presentation of the next test case for LES. We will just point out the main features and challenges, and direct the interested reader to other more detailed references.

12.3 A Few Remarks on Isotropic Homogeneous Turbulence

Isotropic homogeneous turbulence is the simplest turbulence flow on which LES models can be validated. It has two main advantages: first, the computational domain is equipped with periodic boundary conditions in all three directions, and thus the challenge of filtering in the presence of solid boundaries (see Part IV) is completely eliminated. Second, one can use pseudo-spectral methods in all three dimensions, this speeding up considerably the calculations.

It should be mentioned, however, that success in LES of isotropic homogeneous turbulence does not automatically imply success in LES of wall-bounded flows. This remark is in the same spirit as those made in the Introduction: to assess the quality of an LES model, one should test the model in as many different test problems as possible.

The isotropic homogeneous turbulence can be of two types, see Sagaut [267].

- Decay of free isotropic homogeneous turbulence, in which energy is initially located in a low spectral band, after which the energy cascade sets in (energy is transferred to smaller and smaller scales until eventually is dissipated through viscous effects). While the energy cascade sets in, the kinetic energy remains constant. After that, the kinetic energy decreases.

- Sustained isotropic homogeneous turbulence, in which the total dissipation of the kinetic energy is prevented by injecting energy at each time-step. After a transitory phase, an equilibrium solution (including an inertial range) sets in.

In this section, we will focus on the decay of free isotropic homogeneous turbulence. In our presentation, we will use Chap. 11 in the unique monograph of Sagaut [267], Sect. 9.1 in Pope [258], and Scott Collis' notes [69].

For this test problem, we will not reach the level of detail in the previous description of turbulent channel flow. Instead, we will try to outline the features that distinguish this test problem from turbulent channel flow.

12.3.1 Computational Setting

For the numerical simulation of the decay of free isotropic homogeneous turbulence, pseudo-spectral methods are the most popular. For more details on spectral methods, the reader is referred to the excellent introduction in the books of Canuto, Hussaini, Quarteroni, and Zang [54], and Peyret [252].

The computational domain is a cubic box of dimension L, where L is large compared to the integral scales of the turbulence contained in the box. Thus, we can treat the velocity and pressure as periodic functions and expand them as truncated three-dimensional Fourier series:

$$u_l(\mathbf{x}) = \sum_{k_1=-N/2}^{N/2-1} \sum_{k_2=-N/2}^{N/2-1} \sum_{k_3=-N/2}^{N/2-1} \widehat{u}_l(\mathbf{k}) \exp\left[i\,\frac{2\pi}{L}\mathbf{k}\cdot\mathbf{x}\right], \quad \text{for } l = 1, 2, 3,$$

$$p(\mathbf{x}) = \sum_{k_1=-N/2}^{N/2-1} \sum_{k_2=-N/2}^{N/2-1} \sum_{k_3=-N/2}^{N/2-1} \widehat{p}(\mathbf{k}) \exp\left[i\,\frac{2\pi}{L}\mathbf{k}\cdot\mathbf{x}\right].$$

The efficient way of calculating the Fourier coefficients is by using the Fast Fourier Transform (FFT), whose computational cost is $O(N \log_2(N))$ (for details see [54]).

Inserting these truncated Fourier series into the NSE, we obtain the following Galerkin approximation to the NSE:

$$\begin{cases} \left(\dfrac{d}{dt} + \dfrac{1}{Re}|\mathbf{k}|^2\right)\widehat{u}_l(\mathbf{k}) = -i\,k_l\,\widehat{p}(\mathbf{k}) - \displaystyle\sum_{j=1}^{3}[\widehat{u_j\,u_{l,j}}](\mathbf{k}) + \widehat{f}_l(\mathbf{k}), \ l = 1, 2, 3, \\ i\,\displaystyle\sum_{j=1}^{3} k_j\,\widehat{u}_j(\mathbf{k}) = 0, \end{cases} \tag{12.14}$$

for $N/2 \leq k_j \leq N/2 - 1$, $j = 1, 2, 3$. We mention that the pressure can be eliminated from (12.14) by multiplying the first equation in (12.14) by $i\,k_l$

and summing over l (the equivalent of taking the divergence of the NSE in physical space). This yields

$$\widehat{p}(\mathbf{k}) = -i \sum_{j=1}^{3} \frac{k_j \, \widehat{f_j}(\mathbf{k})}{|\mathbf{k}|^2},$$

which is exactly the solution of the Poisson equation for pressure in Fourier space. Thus, the conservation of mass can be incorporated directly into the conservation of momentum by eliminating pressure:

$$\frac{d\widehat{u_l}}{dt}(\mathbf{k}) = \widehat{f_l}(\mathbf{k}) - k_l \sum_{j=1}^{3} \frac{k_j \, \widehat{f_j}(\mathbf{k})}{|\mathbf{k}|^2} - \frac{1}{Re} |\mathbf{k}|^2 \, \widehat{u_l}(\mathbf{k}) - \sum_{j=1}^{3} [\widehat{u_j \, u_{l,j}}](\mathbf{k}),$$

for $l = 1, 2, 3$. $\hspace{8cm}$ (12.15)

Therefore, we need to explicitly store and solve only for the velocity field.

Remark 12.4. We note that the numerical treatment of the quadratic nonlinearity due to the convective term needs care. This is the source of the well-known *aliasing error*: energy from outside the truncated range of wavenumbers is mapped back onto the truncated range. The most common remedy for the aliasing error is the "3/2 rule" (see [54] for alternative techniques).

With this rule, the above equations become a system of ordinary equations in time, which needs to be solved for each Fourier coefficient.

12.3.2 Initial Conditions

In general, the initial velocity field for isotropic turbulence must satisfy at least three conditions:

- conservation of mass;
- real function of space and time;
- realistic energy spectrum.

The first two conditions are required in order to obtain a stable numerical solution. The third condition is required to reduce the initial transient of the flow to realistic isotropic turbulence. Actually, one also needs to specify the relative phases of the modes. Since this information is not available in practice, the initial velocity fields are usually constructed with random phases, while having a prescribed initial energy spectrum. Since these random phases do not resemble those encountered in isotropic turbulence, this initial condition will lead to a transient in which the phases adjust themselves to appropriate values.

12.3.3 Experimental Results

Results on the decay of isotropic turbulence were provided by the experiment of Comte-Bellot and Corrsin [70]. They simulated isotropic turbulence in a wind tunnel by passing the flow through a grid of square rods (see Barenblatt [15] for an explanation). Conceptually, isotropic homogeneous turbulence decays in time. However, in the wind tunnel of this experiment, the turbulence decays in space as it evolves downstream. In order to convert to a temporal decay, the authors make use of Taylor's hypothesis (or the frozen turbulence approximation, (6.203) in Pope [258]), in which the decay of turbulence in space is related to the decay in time of a fictitious box of homogeneous turbulence. The accuracy of Taylor's hypothesis depends both on the properties of the flow and on the statistic being measured. In grid turbulence with $\mathbf{u}' \ll \langle \overline{\mathbf{u}} \rangle$, it is quite accurate. In free shear flows, however, many experiments have shown Taylor's hypothesis to fail. For more details, the reader is referred to pp. 223, 224 in Pope [258] and the references therein.

The experiment of Comte-Bellot and Corrsin clearly displays a $k^{-5/3}$ decay in the *inertial range* as well as the dissipating range of scales at very large wavenumbers. The inertial range is the region displaying the *energy cascade*, described in Chap. 3: energy is transferred *in the average* from large scales to smaller and smaller scales. We should notice that, as we have seen in Sect. 12.2.9, the local inverse phenomenon of transferring energy from small to large scales (backscatter) can be significant.

If one wishes to simulate the experiment of Comte-Bellot and Corrsin, then one needs, at the very least, to construct an initial condition that has the same spectrum.

12.3.4 Computational Cost

The computational cost of a simulation is largely determined by the resolution requirements. The box size must be large enough to represent the energy-containing motions. The grid size must be small enough to represent the dissipative scales. Moreover, the time-step used to advance the solution is limited by considerations of numerical accuracy.

Based on the above requirements, Pope [258] presented an analysis of the computational cost for the DNS of isotropic homogeneous turbulence. The number of floating-point operations required to perform a simulation is proportional to the product of the number of modes and the number of steps, which was estimated as $160\,Re^3$ (Pope [258], pp. 348, 349). Assuming that 1 000 floating point operations are needed per mode per time-step, the time in days, T_G, needed to perform a simulation at a computing rate of 1 gigaflop is given by

$$T_G \approx \left(\frac{Re}{800} \right)^3 . \tag{12.16}$$

This estimate matches the practical findings (Fig. 9.3 in Pope [258]).

The obvious conclusion from this estimate is that the computational cost increases so steeply with Re that it is impractical to go much higher than $Re \approx 1\,500$ with gigaflop computers. This estimate also gives an estimate for the factor of improvement needed to perform a DNS for $Re = 1.5 \cdot 10^5$ in one day: one million fold.

With this analysis, we can conclude that DNS of even simple, isotropic turbulence such as the Comte-Bellot and Corrsin experiment can quickly become impractical. It also means that DNS of practical, engineering or geophysical flows is entirely impractical and will likely remain so for the foreseeable future.

However, there is hope: the analysis on p.349 in Pope [258] shows that, in a well-resolved simulation, less than 0.02% of the modes represent motions in the energy-containing or in the inertial subrange.

12.3.5 LES of the Comte-Bellot Corrsin Experiment

The initial condition for the LES is constructed by using the energy spectrum in the Comte-Bellot Corrsin experiment. An important fact when constructing the initial condition is that the turbulent kinetic energy computed from the LES field will not equal that of the experiment. The reason is that only the portion of the spectrum from $k = 0$ to $k = N/2$ is available in the simulation. Thus, in comparing the simulation and the experiment, we must also filter the experimental results. Numerical results for the LES of the Comte-Bellot and Corrsin experiment are presented in [69].

The first LES for the decay of free isotropic homogeneous turbulence were performed more than twenty years ago [62] with coarse resolutions (16^3 and 32^3 grid points) with satisfactory results [14]. Higher resolution simulations (128^3 grid points) have been performed recently, yielding improved results [215, 235].

Although the decay of free isotropic homogeneous turbulence is the simplest LES test case, it has complex dynamics resulting from the interaction of many elongated vortex structures called "worms" (Fig. 11.1 in Sagaut [267]). Thus, a good LES model should reflect the correct dynamics of these structures.

12.4 Final Remarks

In this chapter, we have presented two of the most popular test problems for the validation and testing of LES models: turbulent channel flow and the decay of free isotropic homogeneous turbulence. Once a new LES model is created, it should be tried first in these two settings.

These two tests, however, are by no means the ultimate criteria for the success of an LES model. In fact, an LES model could perform well for one test case, and poorly for the other. For example, a classical eddy viscosity LES model, such as Smagorinsky [277] could yield very good results for the decay of free isotropic homogeneous turbulence, but it could yield poor results in the numerical simulation of turbulent channel flows.

Then how do we decide whether an LES model is good or not? The answer to this question is that, most probably, there is no universally best LES model. For each application (or class of applications) of interest there is usually an LES model that outperforms the others, although the same model could fail to produce the desired results in other applications. By testing the LES model on as many test problems as possible, one can gain better insight into the qualities and drawbacks of the model, and possibly devise better, improved LES models.

References

1. M. Abramowitz and I.A. Stegun. *Handbook of mathematical functions with formulas, graphs, and mathematical tables*, volume 55 of *National Bureau of Standards Applied Mathematics Series*. For sale by the Superintendent of Documents, U.S. Government Printing Office, Washington, D.C., 1964.
2. N.A. Adams and S. Stolz. Deconvolution methods for subgrid-scale approximation in large-eddy simulation. In B. Geurts, editor, *Modern Simulation Strategies for Turbulent Flow*, pages 21–41. R.T. Edwards, 2001.
3. N.A. Adams and S. Stolz. A subgrid-scale deconvolution approach for shock capturing. *J. Comput. Phys.*, 178(2):391–426, 2002.
4. R.A. Adams. *Sobolev spaces*. Academic Press, New York-London, 1975. Pure and Applied Mathematics, Vol. 65.
5. S. Agmon, A. Douglis, and L. Nirenberg. Estimates near the boundary for solutions of elliptic partial differential equations satisfying general boundary conditions. I. *Comm. Pure Appl. Math.*, 12:623–727, 1959.
6. S. Agmon, A. Douglis, and L. Nirenberg. Estimates near the boundary for solutions of elliptic partial differential equations satisfying general boundary conditions. II. *Comm. Pure Appl. Math.*, 17:35–92, 1964.
7. A.A. Aldama. Filtering techniques for turbulent flow simulation. In *Springer Lecture Notes in Eng.*, volume 56. Springer, Berlin, 1990.
8. M. Anitescu, W.J. Layton, and F. Pahlevani. Implicit for local effects and explicit for nonlocal effects is unconditionally stable. *Electron. Trans. Numer. Anal.*, 18:174–187 (electronic), 2004.
9. V. Arnol'd. Sur la topologie des écoulements stationnaires des fluides parfaits. *C. R. Acad. Sci. Paris*, 261:17–20, 1965.
10. T.G. Bagwell. *Stochastic estimation of near wall closure in turbulence models*. PhD thesis, University of Illinois at Urbana-Champaign, 1994.
11. T.G. Bagwell, R.J. Adrian, R.D. Moser, and J. Kim. Improved approximation of wall shear stress boundary conditions for large eddy simulation. In R.M.C. So, C.G. Speziale, and B.E. Launder, editors, *Near-Wall Turbulent Flows*, New York, 1993. Elsevier Science.
12. D. Barbato, L.C. Berselli, and C.R. Grisanti. Analytical and numerical results for the rational large eddy simulation model. Technical report, 2005. Accepted for publication.

13. J. Bardina, J.H. Ferziger, and W.C. Reynolds. Improved subgrid scale models for large eddy simulation. *AIAA paper*, 80:1357–1366, 1980.

14. J. Bardina, J.H. Ferziger, and W.C. Reynolds. Improved turbulence models based on large eddy simulation of homogeneous, incompressible, turbulent flows. Technical Report TF-19, Thermosciences Division, Dept. Mech. Engineering, Stanford University, 1983.

15. G.I. Barenblatt. *Scaling, self-similarity, and intermediate asymptotics*. Cambridge University Press, Cambridge, 1996. With a foreword by Ya. B. Zeldovich.

16. G.I. Barenblatt and A.J. Chorin. New Perspectives in Turbulence Scaling laws, asymptotics, and intermittency. *SIAM Review*, 40(2):265–291, 1998.

17. G.K. Batchelor. *An introduction to fluid dynamics*. Cambridge Mathematical Libr ary. Cambridge University Press, Cambridge, paperback edition, 1999.

18. J.T. Beale, T. Kato, and A. Majda. Remarks on the breakdown of smooth solutions for the 3-D Euler equations. *Comm. Math. Phys.*, 94(1):61–66, 1984.

19. H. Beirão da Veiga. A new regularity class for the Navier-Stokes equations in $\mathbf{R}^n$. *Chinese Ann. Math. Ser. B*, 16(4):407–412, 1995.

20. H. Beirão da Veiga. Regularity for Stokes and generalized Stokes systems under non homogeneous slip type boundary conditions. *Adv. Differential Equations*, 9(9/10):1079–1114, 2004.

21. H. Beirão da Veiga. A new approach to the L^2-regularity theorems for linear stationary nonhomogeneous Stokes systems. *Portugal. Math.*, 54(3):271–286, 1997.

22. H. Beirão da Veiga and L.C. Berselli. On the regularizing effect of the vorticity direction in incompressible viscous flows. *Differential Integral Equations*, 15(3):345–356, 2002.

23. E. Beltrami. Sui principii fondamentali dell'idrodinamica razionale. *Mem. dell'Accad. Scienze Bologna*, page 394, 1873.

24. A. Bensoussan and R. Temam. Équations stochastiques du type Navier-Stokes. *J. Functional Analysis*, 13:195–222, 1973.

25. J. Bergh and J. Löfström. *Interpolation spaces. An introduction*. Springer-Verlag, Berlin, 1976. Grundlehren der Mathematischen Wissenschaften, No. 223.

26. R. Berker. Intégration des équations du mouvement d'un fluide visqueux incompressible. In *Handbuch der Physik, Bd. VIII/2*, pages 1–384,. Springer, Berlin, 1963.

27. L.C. Berselli. On a regularity criterion for the solutions to the 3D Navier-Stokes equations. *Differential Integral Equations*, 15(9):1129–1137, 2002.

28. L.C. Berselli. On the large eddy simulation of the Taylor-Green vortex. *J. Math. Fluid Mech.*, 7:S164–S191, 2005.

29. L.C. Berselli, G.P. Galdi, T. Iliescu, and W.J. Layton. Mathematical analysis for the rational large eddy simulation model. *Math. Models Methods Appl. Sci.*, 12:1131–1152, 2002.

30. L.C. Berselli and C.R. Grisanti. On the mathematical theory of the rational LES model. In preparation.

31. L.C. Berselli and C.R. Grisanti. On the consistency of the Rational large eddy simulation model. *Comput. Vis. Sci.*, 6:75–82, 2004.

32. L.C. Berselli, C.R. Grisanti, and V. John. On commutation errors in the derivation of the space averaged navier-stokes equations. Technical Report 12, Dept. Appl. Math., Pisa University, 2004. Downloadable at http://www.ing.unipi.it/~d9378.

33. L.C. Berselli and T. Iliescu. A higher-order subfilter-scale model for large eddy simulation. *J. Comput. Appl. Math.*, 159(2):411–430, 2003.

34. L.C. Berselli and V. John. On the comparison of a commutation error and the Reynolds stress tensor for flows obeying a wall law. Technical Report 18, Dept. Appl. Math., Pisa University, 2004. Downloadable at http://www.ing.unipi.it/~d9378.

35. L.C. Berselli and M. Romito. On the existence and uniqueness of weak solutions for a vorticity seeding model. *SIAM J. Math. Anal.*, 2005. Accepted for publication.

36. M. Bertero and P. Boccacci. *Introduction to inverse problems in imaging*. Institute of Physics Publishing, Bristol, 1998.

37. J.T. Borggaard, J. Burkardt, and M. Gunzburger, editors. *Optimal design and control*, volume 19 of *Progress in Systems and Control Theory*, Boston, MA, 1995. Birkhäuser Boston Inc.

38. J.T. Borggaard, P. Hovland, T. Iliescu, and A. Miroshnikov. Sensitivity computations for large eddy simulation of turbulent flows. Technical report, Virginia Tech, 2005.

39. J.T. Borggaard and T. Iliescu. Approximate deconvolution boundary conditions for large eddy simulation. *Appl. Math. Lett.*, 2005. accepted for publication.

40. J.T. Borggaard, T. Iliescu, and A. Miroshnikov. Large eddy simulation of turbulent channel flows by a scale-similarity eddy-viscosity model. Technical report, Virginia Tech, 2004.

41. V. Borue and S.A. Orszag. Local energy flux and subgrid-scale statistics in three-dimensional turbulence. *J. Fluid Mech.*, 366:1, 1998.

42. J.P. Bourguignon and H. Brezis. Remarks on the Euler equation. *J. Functional Analysis*, 15:341–363, 1974.

43. J. Boussinesq. Essai sur la théorie des eaux courantes,. *Mém. prés par div. savants à la Acad. Sci.*, 23:1–680, 1877.

44. M.E. Brachet, D. Meiron, S. Orszag, B. Nickel, R. Morf, and U. Frisch. Small-scale structure of the Taylor-Green vortex. *J. Fluid Mech.*, 130:411–452, 1983.

45. H. Brezis. *Analyse fonctionnelle*. Collection Mathématiques Appliquées pour la Maîtrise. [Collection of Applied Mathematics for the Master's Degree.]. Masson, Paris, 1983. Théorie et applications. [Theory and applications.].

46. F.E. Browder. Nonlinear elliptic boundary value problems. *Bull. Amer. Math. Soc.*, 69:862–874, 1963.

47. R.M. Brown, P. Perry, and Z. Shen. The additive turbulent decomposition for the two-dimensional incompressible Navier-Stokes equations: convergence theorems and error estimates. *SIAM J. Appl. Math.*, 59(1):139–155 (electronic), 1999.

48. W.H. Cabot. Large-eddy simulations with wall models. In *Annual Research Briefs*, pages 41–58. Center for Turbulence Research, Stanford, 1995.

49. W.H. Cabot. Near-wall models in large eddy simulations of flow behind a backward-facing step. In *Annual Research Briefs*, pages 199–210. Center for Turbulence Research, Stanford, 1996.

50. A. Çağlar. Weak imposition of boundary conditions for the Navier-Stokes equations by a penalty-Lagrange multiplier method. Technical report, University of Pittsburgh, 2001.

51. A. Çağlar. Weak imposition of boundary conditions for the Navier-Stokes equations by a penalty-penalty method. Technical report, University of Pittsburgh, 2002.

52. A. Çağlar. Weak imposition of boundary conditions for the Navier-Stokes equations. *Appl. Math. Comput.*, 149(1):119–145, 2004.

53. R. Camassa and D.D. Holm. An integrable shallow water equation with peaked solitons. *Phys. Rev. Lett.*, 71(11):1661–1664, 1993.

54. C. Canuto, M.Y. Hussaini, A. Quarteroni, and T.A. Zang. *Spectral methods in fluid dynamics.* Springer Series in Computational Physics. Springer-Verlag, New York, 1988.

55. D. Carati, G.S. Winckelmans, and H. Jeanmart. On the modeling of the subgrid-scale and filtered-scale stress tensors in large-eddy simulation. *J. Fluid Mech*, 441:119–138, 2001.

56. L. Cattabriga. Su un problema al contorno relativo al sistema di equazioni di Stokes. *Rend. Sem. Mat. Univ. Padova*, 31:308–340, 1961.

57. F. Catté, P.-L. Lions, J.M. Morel, and T. Coll. Image selective smoothing and edge detection by nonlinear diffusion. *SIAM J. Numer. Anal.*, 129:182–193, 1992.

58. D.R. Chapman. Computational aerodynamics development and outlook. *AIAA Journal*, 17:1293–1313, 1979.

59. A. Cheskidov, E. Olson D.D. Holm, and E.S. Titi. On a Leray-α model of turbulence. *Proc. R. Soc. Lond. Ser. A Math. Phys. Eng. Sci.*, 146:1–21, 2005.

60. S. Childress. New solutions of the kinematics dynamo problem. *J. Math. Phys.*, 11(10):3063–3076, 1970.

61. S. Childress, R.R. Kerswell, and A.D. Gilbert. Bounds on dissipation for Navier-Stokes flow with Kolmogorov forcing. *Phys. D*, 158(1-4):105–128, 2001.

62. J.P. Chollet and M. Lesieur. Parametrization of small scales of three-dimensional isotropic turbulence utilizing spectral closures. *J. Atmos. Sci.*, 38:2747–2757, 1981.

63. A.J. Chorin. *Vorticity and turbulence.* Springer-Verlag, New York, 1994.

64. A.J. Chorin and J.E. Marsden. *A mathematical introduction to fluid mechanics.* Springer-Verlag, New York, third edition, 1993.

65. R.A. Clark, J.H. Ferziger, and W.C. Reynolds. Evaluation of subgrid scale models using an accurately simulated turbulent flow. *J. Fluid Mech.*, 91:1–16, 1979.

66. P. Coletti. A global existence theorem for large eddy simulation turbulence model. *Math. Models Methods Appl. Sci.*, 7(5):579–591, 1997.

67. P. Coletti. *Analytical and Numerical Results for k-ϵ and large eddy simulation turbulence models.* PhD thesis, Università di Trento, 1998.

68. S.S. Collis. Monitoring unresolved scales in multiscale turbulence modeling. *Phys. Fluids*, 13(6):1800–1806, 2001.

69. S.S. Collis. Decay of free isotropic homogeneous turbulence. Unpublished, (2004).

70. Comte-Bellot and Corrsin. Simple Eulerian time correlation of full- and narrow-band velocity signals in grid-generated, "isotropic" turbulence. *J. Fluid Mech.*, 48:273–337, 1971.

71. L. Consiglieri. A nonlocal friction problem for a class of non-Newtonian flows. *Port. Math. (N.S.)*, 60(2):237–252, 2003.

72. P. Constantin. Geometric statistics in turbulence. *SIAM Rev.*, 36(1):73–98, 1994.

73. P. Constantin and C. Fefferman. Direction of vorticity and the problem of global regularity for the Navier-Stokes equations. *Indiana Univ. Math. J.*, 42(3):775–789, 1993.

74. P. Constantin and C. Foiaş. *Navier-Stokes equations*. University of Chicago Press, Chicago, IL, 1988.

75. G.-H. Cottet. Anisotropic subgrid-scale numerical schemes for large eddy simulations of turbulent flows. Unpublished, downlodable at `http://www-lmc.imag.fr/lmc-edp/Georges-Henri.Cottet`, (1997).

76. G.-H. Cottet, D. Jiroveanu, and B. Michaux. Vorticity dynamics and turbulence models for large eddy simulations. *M2AN Math. Model. Numer. Anal*, 37:187–207, 2003.

77. G.-H. Cottet and O.V. Vasilyev. Comparison of dynamic smagorinsky and anisotropic subgrid-scale models. In *Proceedings of the Summer Program, Center for Turbulence Research*, pages 367–388. Stanford University and NASA Ames, 1998.

78. G.-H. Cottet and A.A. Wray. Anisotropic grid-based formulas for subgrid-scale models. In *Annual Research Briefs, Center for Turbulence Research*, pages 113–122. Stanford University and NASA Ames, 1997.

79. J. Cousteix. Turbulence et couche limite. CEPADUES, 1989.

80. C. Cuvelier, A. Segal, and A.A. van Steenhoven. *Finite element methods and Navier-Stokes equations*, volume 22 of *Mathematics and its Applications*. D. Reidel Publishing Co., Dordrecht, 1986.

81. G. Da Prato and J. Zabczyk. *Stochastic equations in infinite dimensions*. Cambridge University Press, Cambridge, 1992.

82. G. Da Prato and J. Zabczyk. *Ergodicity for infinite-dimensional systems*. Cambridge University Press, Cambridge, 1996.

83. A. Das and R. Moser. Filtering boundary conditions for LES and embedded boundary simulations. In C. Liu, L. Sakell, and T. Beutner, editors, *DNS/LES progress and challenges*, pages 389–396, Columbus, 2001. Greyden Press.

84. R. Dautray and J.-L. Lions. *Mathematical analysis and numerical methods for science and technology. Vol. 1–9*. Springer-Verlag, Berlin, 1990.

85. E. David. *Modélisation des écoulements compressible et hypersoniques: une approche instationnaire*. PhD thesis, INPG-LEGI, Grenoble, 1993.

86. G. De Stefano, F.M. Denaro, and G. Riccardi. High-order filtering for control volume flow simulation. *Int. J. Numer. Meth. Fluids*, 37(7):797–835, 2001.

87. J.W. Deardorff. A numerical study of three-dimensional turbulent channel flow at large Reynolds numbers. *J. Fluid Mech.*, 41(453–480), 1970.

88. P. Deuring and W. von Wahl. Strong solutions of the Navier-Stokes system in Lipschitz bounded domains. *Math. Nachr.*, 171:111–148, 1995.

89. M.O. Deville, P.F. Fischer, and E.H. Mund. *High-order methods for incompressible fluid flow*, volume 9 of *Cambridge Monographs on Applied and Computational Mathematics*. Cambridge University Press, Cambridge, 2002.

90. C.R. Doering and P. Constantin. Energy-dissipation in shear driven turbulence. *Phys. Rev. Letters*, 69(11):1648–1651, 1992.

91. C.R. Doering and C. Foiaş. Energy dissipation in body-forced turbulence. *J. Fluid Mech.*, 467:289–306, 2002.

92. J.A. Domaradzki and K.-C. Loh. The subgrid-scale estimation model in the physical space representation. *Phys. Fluids*, 11(8):2330–2342, 1999. The International Conference on Turbulence (Los Alamos, NM 1998).

93. J.A. Domaradzki and E.M. Saiki. A subgrid-scale model based on the estimation of unresolved scales of turbulence. *Phys. Fluids*, 9:2148–2164, 1997.

94. Q. Du and M.D. Gunzburger. Finite-element approximations of a Ladyzhenskaya model for stationary incompressible viscous flow. *SIAM J. Numer. Anal.*, 27:1–19, 1990.

95. Q. Du and M.D. Gunzburger. Analysis of a Ladyzhenskaya model for incompressible viscous flow. *J. Math. Anal. Appl.*, 155:21–45, 1991.

96. T. Dubois, F. Jauberteau, and R. Temam. *Dynamic multilevel methods and the numerical simulation of turbulence.* Cambridge University Press, Cambridge, 1999.

97. J. Duchon and R. Robert. Inertial energy dissipation for weak solutions of incompressible Euler and Navier-Stokes equations. *Nonlinearity*, 13:249–255, 2000.

98. A. Dunca and E. Epshteyn. On the Stolz-Adams deconvolution models for LES. *SIAM J. Math. Anal.*, 2005. Accepted for publication.

99. A. Dunca and V. John. Finite element error analysis of space averaged flow fields defined by a differential filter. *Math. Models Methods Appl. Sci.*, 14(4):603–618, 2004.

100. A. Dunca, V. John, and W.J Layton. Approximating local averages of fluid velocities: The equilibrium Navier-Stokes equations. *Appl. Numer. Math.*, 49(2):187–205, 2004.

101. A. Dunca, V. John, and W.J. Layton. The commutation error of the space averaged Navier-Stokes equations on a bounded domain. In *Contributions to current challenges in mathematical fluid mechanics*, Adv. Math. Fluid Mech., pages 53–78. Birkhäuser, Basel, 2004.

102. A. Dunca, V. John, W.J. Layton, and N. Sahin. Numerical analysis of large eddy simulation. In C. Liu, L. Sakell, and T. Beutner, editors, *DNS/LES progress and challenges*, pages 359–364, Columbus, 2001. Greyden Press.

103. D.G. Ebin and J. Marsden. Groups of diffeomorphisms and the notion of an incompressible fluid. *Ann. of Math. (2)*, 92:102–163, 1970.

104. S. Faedo. Un nuovo metodo per l'analisi esistenziale e quantitativa dei problemi di propagazione. *Ann. Scuola Norm Super. Pisa (3)*, 1:1–41 (1949), 1947.

105. P.F. Fischer. An overlapping Schwarz method for spectral element solution of the incompressible Navier-Stokes equations. *J. Comp. Phys.*, 133:84–101, 1997.

106. P.F. Fischer and T. Iliescu. A 3d channel flow simulation at $re_\tau = 180$ using a rational LES model. In C. Liu, L. Sakell, and T. Beutner, editors, *DNS/LES progress and challenges*, pages 283–290, Columbus, 2001. Greyden Press.

107. P.F. Fischer, N.I. Miller, and H.M. Tufo. An overlapping Schwarz method for spectral element simulation of three-dimensional incompressible flows. In P. Björstad and M. Luskin, editors, *Parallel Solution of Partial Differential Equations*, pages 159–181. Springer, 2000.

108. P.F. Fischer and J. Mullen. Filter-based stabilization of spectral element methods. *C. R. Acad. Sci. Paris Sér. I Math.*, 332(3):265–270, 2001.

109. C. Foiaş. What do the Navier-Stokes equations tell us about turbulence? In *Harmonic analysis and nonlinear differential equations (Riverside, CA, 1995)*,

volume 208 of *Contemp. Math.*, pages 151–180. Amer. Math. Soc., Providence, RI, 1997.

110. C. Foiaş, D.D. Holm, and E.S. Titi. The Navier-Stokes-alpha model of fluid turbulence. *Phys. D*, 152/153:505–519, 2001. Advances in nonlinear mathematics and science.

111. C. Foiaş, D.D. Holm, and E.S. Titi. The three dimensional viscous Camassa-Holm equations, and their relation to the Navier-Stokes equations and turbulence theory. *J. Dynam. Differential Equations*, 14(1):1–35, 2002.

112. C. Foiaş, O. Manley, R. Rosa, and R. Temam. *Navier Stokes Equations and Turbulence*. Cambridge University Press, 2001.

113. M. Fortin, R. Guénette, and R. Pierre. Numerical analysis of the modified EVSS method. *Comput. Methods Appl. Mech. Engrg.*, 143(1-2):79–95, 1997.

114. L.P. Franca and C. Farhat. Bubble functions prompt unusual stabilized finite element methods. *Comput. Methods Appl. Mech. Engrg.*, 123(1-4):299–308, 1995.

115. L.P. Franca, A. Nesliturk, and M. Stynes. On the stability of residual-free bubbles for convection-diffusion problems and their approximation by a two-level finite element method. *Comput. Methods Appl. Mech. Engrg.*, 166(1-2):35–49, 1998.

116. J. Frehse and J. Málek. Problems due to the no-slip boundary in incompressible fluid dynamics. In S. Hildebrandt and H. Karcher, editors, *Geometric Analysis and Nonlinear PArtial Differential Equations*, pages 559–571. Springer-Verlag, Berlin, Heidelberg, New York, 2003.

117. U. Frisch. *Turbulence, The Legacy of A.N. Kolmogorov*. Cambridge University Press, Cambridge, 1995.

118. H. Fujita. A coherent analysis of Stokes flows under boundary conditions of friction type. *J. Comput. Appl. Math.*, 149(1):57–69, 2002. Scientific and engineering computations for the 21st century—methodologies and applications (Shizuoka, 2001).

119. C. Fureby and G. Tabor. Mathematical and physical constraints on large-eddy simulations. *Theoret. Comput. Fluid Dynamics*, 9:85–102, 1997.

120. G.P. Galdi. *An introduction to the mathematical theory of the Navier-Stokes equations. Vol. I*, volume 38 of *Springer Tracts in Natural Philosophy*. Springer-Verlag, New York, 1994. Linearized steady problems.

121. G.P. Galdi. An introduction to the Navier-Stokes initial-boundary value problem. In *Fundamental directions in mathematical fluid mechanics*, Adv. Math. Fluid Mech., pages 1–70. Birkhäuser, Basel, 2000.

122. G.P. Galdi and W.J. Layton. Approximation of the larger eddies in fluid motions. II. A model for space-filtered flow. *Math. Models Methods Appl. Sci.*, 10(3):343–350, 2000.

123. B.G. Galerkin. Rods and plates. Series on some problems of elastic equilibrium of rods and plates (in Russian). *Vest. Inzh. Tech.*, 19:897–908, 1915.

124. G. Gallavotti. *Foundations of fluid dynamics*. Texts and Monographs in Physics. Springer-Verlag, Berlin, 2002. Translated from the Italian.

125. T. Gallouët, J. Lederer, R. Lewandowski, F. Murat, and L. Tartar. On a turbulent system with unbounded eddy viscosities. *Nonlinear Anal.*, 52(4):1051–1068, 2003.

126. M. Germano. Differential filters for the large eddy numerical simulation of turbulent flows. *Phys. Fluids*, 29(6):1755–1757, 1986.

127. M. Germano. Differential filters of elliptic type. *Phys. Fluids*, 29(6):1757–1758, 1986.

128. M. Germano. Fundamentals of large eddy simulation. In *Advanced turbulent flow computations (Udine, 1998)*, volume 395 of *CISM Courses and Lectures*, pages 81–130. Springer, Vienna, 2000.

129. M. Germano, U. Piomelli, P. Moin, and W.H. Cabot. A dynamic subgrid-scale eddy viscosity model. *Phys. Fluids A*, 3:1760–1765, 1991.

130. B.J. Geurts. Inverse modeling for large-eddy simulation. *Phys. Fluids*, 9:3585–3587, 1997.

131. B.J. Geurts. *Elements of Direct and Large-Edddy Simulation*. R.T. Edwards, Inc, 2003.

132. B.J. Geurts and D.D. Holm. Alpha-modeling strategy for LES of turbulent mixing. In *Turbulent flow computation*, volume 66 of *Fluid Mech. Appl.*, pages 237–278. Kluwer Acad. Publ., Dordrecht, 2002.

133. S. Ghosal. An analysis of numerical errors in large-eddy simulations of turbulence. *J. Comput. Phys.*, 125(1):187–206, 1996.

134. S. Ghosal. Mathematical and physical constraints on large eddy simulation of turbulence. *AIAA J.*, 37:425–433, 1999.

135. S. Ghosal, T. S. Lund, P. Moin, and K. Akselvoll. A dynamic localization model for large-eddy simulation of turbulent flows. *J. Fluid Mech.*, 286:229–255, 1995.

136. S. Ghosal and P. Moin. The basic equations for the large-eddy simulation of turbulent flows in complex geometry. *J. Comput. Phys.*, 118:24–37, 1995.

137. V. Girault and P.-A. Raviart. *Finite element methods for Navier-Stokes equations*, volume 5 of *Springer Series in Computational Mathematics*. Springer-Verlag, Berlin, 1986. Theory and algorithms.

138. A.E. Green and G.I. Taylor. Mechanism of the production of small eddies from larger ones. *Proc. Royal Soc. A.*, 158:499–521, 1937.

139. P.M. Gresho and R.L. Sani. *Incompressible Flow and the Finite Element Method, Advection-Diffusion and Isothermal Laminar Flow*, volume 1. John Wiley & Sons, 2000. Paperback.

140. P.M. Gresho and R.L. Sani. *Incompressible Flow and the Finite Element Method, Isothermal Laminar Flow*, volume 2. John Wiley & Sons, 2000. Paperback.

141. G. Grötzbach. Direct numerical and large eddy simulation of turbulent channel flows. In N.P. Cheremisinoff, editor, *Encyclopedia of Fluid Mechanics*. Gulf, West Orange, NJ, 1987.

142. J.-L. Guermond. Stabilization of Galerkin approximations of transport equations by subgrid modeling. *M2AN Math. Model. Numer. Anal.*, 33(6):1293–1316, 1999.

143. J.-L. Guermond. Stablization par viscosité de sous-maille pour l'approximation de Galerkin des opérateurs linéaires monotones. *C.R.A.S.*, 328:617–622, 1999.

144. J.-L. Guermond, J. T. Oden, and S. Prudhomme. Mathematical perspectives on large eddy simulation models for turbulent flows. *J. Math. Fluid Mech.*, 6(2):194–248, 2004.

145. J.-L. Guermond, J.T. Oden, and S. Prudhomme. An interpretation of the Navier-Stokes-alpha model as a frame-indifferent Leray regularization. *Phys. D*, 177(1-4):23–30, 2003.

146. M.D. Gunzburger. *Finite element methods for viscous incompressible flows.* Computer Science and Scientific Computing. Academic Press Inc., Boston, MA, 1989. A guide to theory, practice, and algorithms.

147. M.D. Gunzburger. *Perspectives in flow control and optimization.* Advances in Design and Control. Society for Industrial and Applied Mathematics (SIAM), Philadelphia, PA, 2003.

148. P. Hartman. *Ordinary differential equations,* volume 38 of *Classics in Applied Mathematics.* Society for Industrial and Applied Mathematics (SIAM), Philadelphia, PA, 2002. Corrected reprint of the second (1982) edition [Birkhäuser, Boston, MA; MR 83e:34002], With a foreword by Peter Bates.

149. N. Heitmann. Subgrid stabilization of time-dependent convection dominated diffusive transport. Technical report, University of Pittsburgh, 2002.

150. H. Helmholtz. Über die Theorie der Elektrodynamik. Erste Abhandlung. Über die Bewegungsgleichungen der Elektricität für Ruhende Leitende Körper. *J. Reine Angew. Math.,* 72:57–129, 1870.

151. J.O. Hinze. *Turbulence: An introduction to its mechanism and theory.* McGraw-Hill Series in Mechanical Engineering. McGraw-Hill Book Co., Inc., New York, 1959.

152. I.I. Hirschman and D.V. Widder. *The convolution transform.* Princeton University Press, Princeton, N.J., 1955.

153. Johan Hoffman. On duality-based a posteriori error estimation in various norms and linear functionals for large eddy simulation. *SIAM J. Sci. Comput.,* 26(1):178–195 (electronic), 2004.

154. J.L. Jr. Holloway. Smoothing and filtering of time series and space fields. In *Advances in Geophysics, Vol. 4,* pages 351–389. Academic Press, New York, 1958.

155. E. Hopf. Über die Anfangswertaufgabe für die hydrodynamischen Grundgleichungen. *Math. Nachr.,* 4:213–231, 1951.

156. K. Horiuti. Large eddy simulation of turbulent channel flow by one equation modeling. *J. Phys. Soc. Japan,* 54:2855–2865, 1985.

157. K. Horiuti. The role of the Bardina model in large eddy simulation of turbulent channel flow. *Phys. Fluids A,* 1(2):426–428, 1989.

158. L. Hörmander. *The analysis of linear partial differential operators. I.* Classics in Mathematics. Springer-Verlag, Berlin, 2003. Distribution theory and Fourier analysis, Reprint of the second (1990) edition.

159. T.J.R. Hughes. Multiscale phenomena: Green's functions, the Dirichlet-to-Neumann formulation, subgrid scale models, bubbles and the origins of stabilized methods. *Comput. Methods Appl. Mech. Engrg.,* 127(1-4):387–401, 1995.

160. T.J.R. Hughes, L. Mazzei, and K.E. Jansen. Large eddy simulation and the variational multiscale method. *Comput. Vis. Sci.,* 3:47–59, 2000.

161. T.J.R. Hughes, L. Mazzei, A. Oberai, and A. Wray. The multiscale formulation of large eddy simulation: Decay of homogeneous isotropic turbulence. *Phys. Fluids,* 13(2):505–512, 2001.

162. T.J.R. Hughes, A. Oberai, and L. Mazzei. Large eddy simulation of turbulent channel flows by the variational multiscale method. *Phys. Fluids,* 13(6):1784–1799, 2001.

163. E.C. Hylin and J.M. McDonough. Chaotic small-scale velocity fields as prospective models for unresolved turbulence in an additive decomposition of the Navier-Stokes equations. *Int. J. Fluid Mech. Res.*, 26(5-6):539–567, 1999.

164. T. Iliescu. *Large Eddy Simulation for Turbulent Flows*. PhD thesis, University of Pittsburgh, 2000.

165. T. Iliescu and P.F. Fischer. Large eddy simulation of turbulent channel flows by the rational large eddy simulation model. *Phys. Fluids*, 15(10):3036–3047, 2003.

166. T. Iliescu and P.F. Fischer. Backscatter in the rational LES model. *Comput. & Fluids*, 33(5-6):783–790, 2004.

167. T. Iliescu and P.F. Fischer. Large eddy simulation of turbulent channel flows by a scale-similarity eddy-viscosity model. Technical report, Virginia Tech, 2004.

168. T. Iliescu, V. John, and W.J. Layton. Convergence of finite element approximations of large eddy motion. *Numer. Methods Partial Differential Equations*, 18(6):689–710, 2002.

169. T. Iliescu, V. John, W.J. Layton, G. Matthies, and L. Tobiska. A numerical study of a class of LES models. *Int. J. Comput. Fluid Dyn.*, 17(1):75–85, 2003.

170. T. Iliescu and W.J. Layton. Approximating the larger eddies in fluid motion. III. The Boussinesq model for turbulent fluctuations. *An. Ştiinţ. Univ. Al. I. Cuza Iaşi. Mat. (N.S.)*, 44(2):245–261 (2000), 1998. Dedicated to Professor C. Corduneanu on the occasion of his 70th birthday.

171. M. Iovieno and D. Tordella. Variable scale filtered navier-stokes equations: a new procedure to deal with the associated commutation error. *Phys. fluids*, 15(7):1926–1936, 2003.

172. J. Jimenez. An overview of LES validation. In *A Selection of Test Cases for the Validation of Large Eddy Simulation of Turbulent Flows*. NATO AGARD, 1999.

173. V. John. The behaviour of the Rational LES model in a two-dimensional mixing layer problem, 2002. tech. report.

174. V. John. Slip with friction and penetration with resistance boundary conditions for the Navier-Stokes equations—numerical tests and aspects of the implementation. *J. Comput. Appl. Math.*, 147(2):287–300, 2002.

175. V. John. *Large eddy simulation of turbulent incompressible flows*, volume 34 of *Lecture Notes in Computational Science and Engineering*. Springer-Verlag, Berlin, 2004. Analytical and numerical results for a class of LES models.

176. V. John. An assessment of two models for the subgrid scale tensor in the rational LES model. *J. Comp. Appl. Math.*, 173:57–80, 2005.

177. V. John and W.J. Layton. Analysis of numerical errors in large eddy simulation. *SIAM J. Numer. Anal.*, 40(3):995–1020 (electronic), 2002.

178. V. John, W.J. Layton, and N. Sahin. Derivation and analysis of near wall models for channel and recirculating flows. *Comput. Math. Appl.*, 28:1135–1151, 2004.

179. T. Kato. On classical solutions of the two-dimensional nonstationary Euler equation. *Arch. Rational Mech. Anal.*, 25:188–200, 1967.

180. T. Kato. Nonstationary flows of viscous and ideal fluids in $\mathbf{R}^3$. *J. Functional Analysis*, 9:296–305, 1972.

181. T. Kato. Remarks on zero viscosity limit for nonstationary Navier-Stokes flows with boundary. In *Seminar on nonlinear partial differential equations*

(Berkeley, Calif., 1983), volume 2 of *Math. Sci. Res. Inst. Publ.*, pages 85–98. Springer, New York, 1984.

182. F.V. Katopodes, R.L. Street, and J.H. Ferziger. Subfilter-scale scalar transport for large-eddy simulation. In *14th Symposium on Boundary Layer and Turbulence*, pages 472–475. American Meteorological Society, 2000.

183. F.V. Katopodes, R.L. Street, and J.H. Ferziger. A theory for the subfilter-scale model in large-eddy simulation. Technical Report EFML Technical Report 2000-K1, Stanford University, 2000.

184. F. Katopodes Chow and R.L. Street. Modeling unresolved motions in LES of field-scale flows. In *15th Symposium on Boundary Layers and Turbulence*, pages 432–435, Wageningen, The Netherlands, 2002. American Meteorological Society.

185. M. Kaya. Existence of weak solutions for a scale similarity model of the motion of large eddies in turbulent flow. *J. Appl. Math.*, (9):429–446, 2003.

186. M. Kaya and W.J. Layton. On "verifiability" of models of the motion of large eddies in turbulent flows. *Differential Integral Equations*, 15(11):1395–1407, 2002.

187. S. Kaya. Numerical analysis of a subgrid eddy viscosity methods for higher Reynolds number flow problems. Technical Report TR-MATH 03-04, University of Pittsburgh, 2003.

188. S. Kaya and W.J. Layton. Subgrid-scale eddy viscosity methods are variational multiscale methods. Technical report, University of Pittsburgh, 2002.

189. J. Kim, P. Moin, and R. Moser. Turbulence statistics in fully developed channel flow at low Reynolds number. *J. Fluid Mech.*, 177:133, 1987.

190. A.A. Kiselev and O.A. Ladyzhenskaya. On the existence and uniqueness of the solution of the nonstationary problem for a viscous, incompressible fluid. *Izv. Akad. Nauk SSSR. Ser. Mat.*, 21:655–680, 1957.

191. A.N. Kolmogorov. The local structure of turbulence in incompressible viscous fluids for very large Reynolds number. *Dokl. Akad. Nauk SSR*, 30:9–13, 1941.

192. A.N. Kolmogorov and S.V. Fomīn. *Introductory real analysis*. Dover Publications Inc., New York, 1975. Translated from the second Russian edition and edited by Richard A. Silverman, Corrected reprinting.

193. R.H. Kraichnan. Eddy viscosity in two and three dimensions. *J. Atmos. Sci.*, 33:1521–1536, 1976.

194. G.J.M. Kuerten, B.J. Geurts, A.W. Vreman, and M. Germano. Dynamic inverse-modeling and its testing in large-eddy simulation of the mixing layer. *Phys. Fluids*, 11:3778–3785, 1999.

195. O.A. Ladyzhenskaya. New equations for the description of motion of viscous incompressible fluids and solvability in the large of boundary value problems for them. *Proc. Steklov Inst. Math.*, 102:95–118, 1967.

196. O.A. Ladyzhenskaya. Modifications of the Navier-Stokes equations for large gradients of the velocities. *Zap. Naučn. Sem. Leningrad. Otdel. Mat. Inst. Steklov. (LOMI)*, 7:126–154, 1968.

197. O.A. Ladyzhenskaya. *The mathematical theory of viscous incompressible flow*. Gordon and Breach Science Publishers, New York, 1969. Second English edition, revised and enlarged. Mathematics and its Applications, Vol. 2.

198. H. Lamb. *Hydrodynamics*. Cambridge Mathematical Library. Cambridge University Press, Cambridge, sixth edition, 1993. With a foreword by R.A. Caflisch.

199. L.D. Landau and E.M. Lifshitz. *Fluid Mechanics*, volume 6 of *Course of Theoretical Physics*. Pergamon Press, London, 1959.

200. B.E. Launder and D. Brian Spalding. *Lectures in Mathematical Models of Turbulence*. Academic Press, London, 1972.

201. W.J. Layton. On nonlinear subgrid scale models for viscous flow problems. *SIAM J. Sci. Computing*, 17:347–357, 1996.

202. W.J. Layton. Weak imposition of "no-slip" conditions in finite element methods. *Comput. Math. Appl.*, 38(5-6):129–142, 1999.

203. W.J. Layton. Approximating the larger eddies in fluid motion. V. Kinetic energy balance of scale similarity models. *Math. Comput. Modelling*, 31(8-9):1–7, 2000.

204. W.J. Layton. Analysis of a scale-similarity model of the motion of large eddies in turbulent flows. *J. Math. Anal. Appl.*, 264(2):546–559, 2001.

205. W.J. Layton. Advanced models for large eddy simulation. In H. Deconinck, editor, *Computational Fluid Dynamics-Multiscale Methods*. Von Karman Institute for Fluid Dynamics, Rhode-Saint-Genèse, Belgium, 2002.

206. W.J. Layton. A connection between subgrid scale eddy viscosity and mixed methods. *Appl. Math. Comput.*, 133:147–157, 2002.

207. W.J. Layton. Energy dissipation bounds for shear flows for a model in large eddy simulation. *Math. Comput. Modelling*, 35(13):1445–1451, 2002.

208. W.J. Layton and R. Lewandowski. Analysis of an eddy viscosity model for large eddy simulation of turbulent flows. *J. Math. Fluid Mech.*, 4(4):374–399, 2002.

209. W.J. Layton and R. Lewandowski. Analysis of the zeroth order scale similarity/deconvolution model for large eddy simulation of turbulence. Technical report, Univ. of Pittsburgh, 2003.

210. W.J. Layton and R. Lewandowski. A simple and stable scale-similarity model for large eddy simulation: energy balance and existence of weak solutions. *Appl. Math. Lett.*, 16(8):1205–1209, 2003.

211. W.J. Layton and R. Lewandowski. On a well posed turbulence model. *Discrete Contin. Dyn. Syst., Series B*, 2005. Accepted for publication.

212. A. Leonard. Energy cascade in large eddy simulations of turbulent fluid flows. *Adv. in Geophysics*, 18A:237–248, 1974.

213. J. Leray. Sur le mouvement d'un fluide visqueux emplissant l'espace. *Acta Math.*, 63:193–248, 1934.

214. M. Lesieur. Turbulence in Fluids. In *Fluid Mechanics and its Applications*, volume 40. Kluwer Academic Publishers, 1997.

215. M. Lesieur and R.S. Rogallo. Large-eddy simulation of passive scalar diffusion in isotropic turbulence. *Phys. Fluids A*, 1:718–722, 1989.

216. R. Lewandowski. Sur quelques problèmes mathématiques posés par l'océanographie. *C. R. Acad. Sci. Paris Sér. I Math.*, 320(5):567–572, 1995.

217. R. Lewandowski. *Analyse Mathématique et Océanographie*. "Recherches en Mathematiques Appliquees". Masson, 1997.

218. L. Lichtenstein. Über einige Existenzprobleme der Hydrodynamik homogener, unzusammedrückbarer, reibungsloser Flüssigkeiten und die helmholtzschen Wirbelsätze. *Math. Z.*, 23:89–154, 1925.

219. D.K. Lilly. The representation of small scale turbulence in numerical simulation experiments. In H.H. Goldstine, editor, *Proc. IBM Sci. Computing Symp. On Environmental Sciences*, pages 195–210, Yorktown Heights, NY, 1967.

220. D.K. Lilly. A proposed modification of the Germano subgrid-scale closure method. *Phys. Fluids A*, 4:633–635, 1992.

221. J.-L. Lions. *Quelque méthodes de résolution des problemès aux limitex non linéaires*. Dunod, Gauthier–Villars, Paris, 1969.

222. J.-L. Lions and E. Magenes. *Non-homogeneous boundary value problems and applications. Vol. I.* Springer-Verlag, New York, 1972. Translated from the French by P. Kenneth, Die Grundlehren der mathematischen Wissenschaften, Band 181.

223. S. Liu, C. Meneveau, and J. Katz. On the properties of similarity subgrid-scale models as deduced from measurements in a turbulent jet. *J. Fluid Mech.*, 275:83–119, 1994.

224. Y. Maday and A.T. Patera. Spectral element methods for the Navier-Stokes equations. In A.K. Noor, editor, *State of the Art Surveys in Computational Mechanics*, pages 71–143, New York, 1989. ASME.

225. Y. Maday, A.T. Patera, and E.M. Rønquist. An operator-integration-factor splitting method for time-dependent problems: application to incompressible fluid flow. *J. Sci. Comput.*, 5(4):310–337, 1990.

226. Y. Maday and E. Tadmor. Analysis of the spectral vanishing viscosity method for periodic conservation laws. *SIAM J. Numer. Anal.*, 26(4):854–870, 1989.

227. A.J. Majda and A.L. Bertozzi. *Vorticity and incompressible flow*, volume 27 of *Cambridge Texts in Applied Mathematics*. Cambridge University Press, Cambridge, 2002.

228. J. Málek, J. Nečas, M. Rokyta, and M. Růžička. *Weak and measure-valued solutions to evolutionary PDEs*, volume 13 of *Applied Mathematics and Mathematical Computation*. Chapman & Hall, London, 1996.

229. J. Málek, J. Nečas, and M. Růžička. On weak solutions to a class of non-Newtonian incompressible fluids in bounded three-dimensional domains. *Adv. Differential Equations*, 6:257–302, 2001.

230. C. Marchioro and M. Pulvirenti. *Mathematical theory of incompressible non-viscous fluids*, volume 96 of *Applied Mathematical Sciences*. Springer-Verlag, New York, 1994.

231. A.L. Marsden, O.V. Vasilyev, and P. Moin. Construction of commutative filters for LES on unstructured meshes. *J. Comput. Phys.*, 175:584—603, 2002.

232. J.E. Marsden and S. Shkoller. Global well-posedness for the Lagrangian averaged Navier-Stokes (LANS-α) equations on bounded domains. *R. Soc. Lond. Philos. Trans. Ser. A Math. Phys. Eng. Sci.*, 359(1784):1449–1468, 2001.

233. I. Marusic, G.J. Kunkel, and F. Porté-Agel. Experimental study of wall boundary conditions for large-eddy simulation. *J. Fluid Mech.*, 446:309–320, 2001.

234. J.C. Maxwell. On stresses in rarefied gases arising from inequalities of temperature. *Royal Society Phil. Trans.*, 170:249–256, 1879.

235. O. Métais and M. Lesieur. Spectral large-eddy simulation of isotropic and stably stratified turbulence. *J. Fluid Mech.*, 239:157–194, 1992.

236. K. Mikula and F. Sgallari. Computational Methods for nonlinear diffusion equations on image processing, 1998. tech. report.

237. G.J. Minty. Monotone (nonlinear) operators in Hilbert space. *Duke Math. J.*, 29:341–346, 1962.

238. C. Miranda. *Partial differential equations of elliptic type*. Second revised edition. Translated from the Italian by Zane C. Motteler. Ergebnisse der Mathematik und ihrer Grenzgebiete, Band 2. Springer-Verlag, New York, 1970.

239. B. Mohammadi and O. Pironneau. *Analysis of the K-Epsilon Turbulence Model*. John Wiley and Sons, 1994.

240. P. Moin and J. Kim. Numerical investigation of turbulent channel flow. *J. Fluid Mech.*, 117:341–377, 1982.

241. A.S. Monin and A.M. Yaglom. *Statistical Fluid Mechanics: Mechanics of Turbulence*. The MIT Press, Boston, 1981.

242. D.R. Moser, J. Kim, and N.N. Mansour. Direct numerical simulation of turbulent channel flow up to $Re_\tau = 590$. *Phys. Fluids*, 11:943–945, 1999.

243. S. Nägele and G. Wittum. Large-eddy simulation and multigrid methods. *Electron. Trans. Numer. Anal.*, 15:152–164 (electronic), 2003. Tenth Copper Mountain Conference on Multigrid Methods (Copper Mountain, CO, 2001).

244. C.L.M.H. Navier. Mémoire sur les lois du mouvement des fluides. *Mém. Acad. Sci. Inst. de France (2)*, 6:389–440, 1823.

245. J. Nečas. *Les méthodes directes en théorie des équations elliptiques*. Masson et Cie, Éditeurs, Paris, 1967.

246. F. Nicoud, J.S. Baggett, P. Moin, and W.H. Cabot. Large eddy simulation wall-modeling based on suboptimal control theory and linear stochastic estimation. *Phys. Fluids*, 13(10):2968–2984, 2001.

247. S.A. Orszag. Numerical simulation of incompressible flows within simple boundaries. I. Galerkin (spectral) representations. *Studies in Appl. Math.*, 50:293–327, 1971.

248. F. Pahlevani. *Sensitivity Analysis of Eddy Viscosity Models*. PhD thesis, University of Pittsburgh, 2004.

249. C. Parés. Existence, uniqueness and regularity of solution of the equations of a turbulence model for incompressible fluids. *Appl. Anal.*, 43(3-4):245–296, 1992.

250. C. Parés. Approximation de la solution des équations d'un modèle de turbulence par une méthode de Lagrange-Galerkin. *Rev. Mat. Apl.*, 15(2):63–124, 1994.

251. A.T. Patera. A spectral element method for fluid dynamics; laminar flow in a channel expansion. *J. Comput. Phys.*, 54:468–488, 1984.

252. R. Peyret. *Spectral methods for incompressible viscous flow*, volume 148 of *Applied Mathematical Sciences*. Springer-Verlag, New York, 2002.

253. U. Piomelli and E. Balaras. Wall-layer models for large-eddy simulations. In *Annual review of fluid mechanics, Vol. 34*, pages 349–374. Annual Reviews, Palo Alto, CA, 2002.

254. U. Piomelli, W.H. Cabot, P. Moin, and S. Lee. Subgrid-scale backscatter in turbulent and transitional flow. *Phys. Fluids*, 3(7):1766–1771, 1991.

255. U. Piomelli, J. Ferziger, P. Moin, and J. Kim. New approximate boundary conditions for large eddy simulation of wall-bounded flows. *Phys. Fluids A*, 1(6):1061–1068, 1989.

256. O. Pironneau. *Finite element methods for fluids*. John Wiley & Sons Ltd., Chichester, 1989. Translated from the French.

257. H. Poincaré. *Théorie des tourbillons*. Gauthier–Villars, Paris, 1893.

258. S.B. Pope. *Turbulent flows*. Cambridge University Press, Cambridge, 2000.

259. A. Pozzi. *Applications of Padé approximation theory in fluid dynamics*, volume 14 of *Series on Advances in Mathematics for Applied Sciences*. World Scientific Publishing Co. Inc., River Edge, NJ, 1994.

260. L. Prandtl. Bericht über untersuchungen zur ausgebildeten Turbulenzdie ausgebildete Turbulenz. *Z. Angew. Math. Mech.*, 5:136–139, 1925.

261. G. Prodi. Un teorema di unicità per le equazioni di Navier-Stokes. *Ann. Mat. Pura Appl. (4)*, 48:173–182, 1959.

262. O. Reynolds. On the dynamic theory of the incompressible viscous fluids and the determination of the criterion. *Philos. Trans. Roy. Soc. London Ser. A*, 186:123–164, 1895.

263. L.F. Richardson. *Weather Prediction by Numerical Process.* Cambridge University Press, Cambridge, 1922.

264. W. Rodi, J.H. Ferziger, M. Breuer, and M. Pourquié. Status of large-eddy simulation: results of a workshop. *ASME J. Fluid Engng.*, 119(2):248–262, 1997.

265. C. Ross Ethier and D.A. Steinman. Exact fully 3D Navier-Stokes solutions for benchmarking. *Internat. J. Numer. Methods Fluids*, 19:369–375, 1994.

266. W. Rudin. *Functional analysis.* International Series in Pure and Applied Mathematics. McGraw-Hill Inc., New York, second edition, 1991.

267. P. Sagaut. *Large eddy simulation for incompressible flows.* Scientific Computation. Springer-Verlag, Berlin, 2001. An introduction, With an introduction by Marcel Lesieur, Translated from the 1998 French original by the author.

268. P. Sagaut and T.H. Lê. Some investigations on the sensitivity of large-eddy simulation. In P.R. Voke L. Kleiser eds. J.P. Chollet, editor, *Direct and large-eddy simulation II*, pages 81–92. Kluwer Academic Publishers, 1997.

269. N. Sahin. *Derivation, Analysis and Testing of New Near Wall Models for Large Eddy Simulation.* PhD thesis, Department of Mathematics, Pittsburgh University, 2003.

270. V. Scheffer. Partial regularity of solutions to the Navier-Stokes equations. *Pacific J. Math.*, 66(2):535–552, 1976.

271. H. Schlichting. *Boundary layer theory.* McGraw-Hill Series in Mechanical Engineering. McGraw-Hill Book Co., Inc., New York, 1979.

272. U. Schumann. Subgrid scale model for finite difference simulations of turbulent flows in plane channels and annuli. *J. Comput. Phys.*, 18:376–404, 1975.

273. L. Schwartz. *Méthodes mathématiques pour les sciences physiques.* Enseignement des Sciences. Hermann, Paris, 1961.

274. J. Serrin. Mathematical principles of classical fluid mechanics. In *Handbuch der Physik (herausgegeben von S. Flügge), Bd. 8/1, Strömungsmechanik I (Mitherausgeber C. Truesdell)*, pages 125–263. Springer-Verlag, Berlin, 1959.

275. J. Serrin. The initial value problem for the Navier-Stokes equations. In *Nonlinear Problems (Proc. Sympos., Madison, Wis.*, pages 69–98. Univ. of Wisconsin Press, Madison, Wis., 1963.

276. K.B. Shah and J. Ferziger. A new non-eddy viscosity subgrid-scale model and its application to turbulent flow. In *CTR Annual Research Briefs.* Stanford University and NASA Ames Research Center, Stanford, CA, 1995.

277. J.S. Smagorinsky. General circulation experiments with the primitive equations. *Mon. Weather Review*, 91:99–164, 1963.

278. V.A. Solonnikov and V.E. Ščadilov. A certain boundary value problem for the stationary system of Navier-Stokes equations. *Trudy Mat. Inst. Steklov.*, 125:196–210, 235, 1973. Boundary value problems of mathematical physics, 8.

279. C.G. Speziale. Analytic methods for the development of reynolds stress closures in turbulence. *Ann. Rev. Fluid Mech.*, 23:107–157, 1991.

280. K.R. Sreenivasan. On the scaling of the turbulent energy-dissipation rate. *Phys. Fluids*, 27(5):1048–1051, 1984.

281. K.R. Sreenivasan. An update on the energy dissipation rate in isotropic turbulence. *Phys. Fluids*, 10(2):528–529, 1998.

282. L.G. Stanley and D.L. Stewart. *Design sensitivity analysis*, volume 25 of *Frontiers in Applied Mathematics*. Society for Industrial and Applied Mathematics (SIAM), Philadelphia, PA, 2002. Computational issues of sensitivity equation methods.

283. E.M. Stein. *Singular integrals and differentiability properties of functions*. Princeton Mathematical Series, No. 30. Princeton University Press, Princeton, N.J., 1970.

284. S. Stolz. *Large-eddy simulation of complex shear flows using an approximate deconvolution model*. Fortschritt-Bericht VDI Reihe 7, 2001.

285. S. Stolz and N.A Adams. An approximate deconvolution procedure for large-eddy simulation. *Phys. Fluids*, 11(7):1699–1701, 1999.

286. S. Stolz and N.A Adams. Large-eddy simulation of high-Reynolds-number supersonic boundary layers using the approximate deconvolution model and a rescaling and recycling technique. *Phys. Fluids*, 15(8):2398–2412, 2003.

287. S. Stolz, N.A. Adams, and L. Kleiser. Analysis of sub-grid scales and sub-grid scale modeling for shock-boundary-layer interaction. In S. Banerjee and J. Eaton, editors, *Turbulence and Shear Flow I*, pages 881–886. Begell House, 1999.

288. S. Stolz, N.A. Adams, and L. Kleiser. The approximate deconvolution procedure applied to turbulent channel flow. In P. Voke, N. D. Sandham, and L. Kleiser, editors, *Direct and Large-Eddy Simulation III*, pages 163–174. Kluwer, 1999.

289. S. Stolz, N.A. Adams, and L. Kleiser. An approximate deconvolution model for large-eddy simulation with application to incompressible wall-bounded flows. *Phys. Fluids*, 13(4):997–1015, 2001.

290. S. Stolz, N.A. Adams, and L. Kleiser. The approximate deconvolution model for large-eddy simulations of compressible flows and its application to shock-turbulent-boundary-layer interaction. *Phys. Fluids*, 13(10):2985–3001, 2001.

291. S. Stolz, N.A. Adams, and L. Kleiser. Discretization effects on approximate deconvolution modeling for LES. *Proc. Appl. Math. Mech.*, 1:282–283, 2002.

292. V. Šverák. On optimal shape design. *J. Math. Pures Appl. (9)*, 72(6):537–551, 1993.

293. A. Świerczewska. *Mathematical Analysis of Large Eddy Simulation of Turbulent Flows*. PhD thesis, Department of Mathematics, Darmstadt University of Technology, Germany, 2004.

294. R. Temam. On the Euler equations of incompressible perfect fluids. *J. Functional Analysis*, 20(1):32–43, 1975.

295. R. Temam. *Navier-Stokes equations. Theory and numerical analysis*. North-Holland Publishing Co., Amsterdam, 1977. Studies in Mathematics and its Applications, Vol. 2.

296. R. Temam. Behaviour at time $t = 0$ of the solutions of semilinear evolution equations. *J. Differential Equations*, 43(1):73–92, 1982.

297. R. Temam. *Navier-Stokes equations and nonlinear functional analysis*, volume 66 of *CBMS-NSF Regional Conference Series in Applied Mathematics*. Society for Industrial and Applied Mathematics (SIAM), Philadelphia, PA, second edition, 1995.

298. M.M. Vainberg. *Variational methods for the study of nonlinear operators.* With a chapter on Newton's method by L.V. Kantorovich and G.P. Akilov. Translated and supplemented by A. Feinstein from the 1956 Russian edition. Holden-Day Inc., San Francisco, Calif., 1964.

299. F. van der Bos and B.J. Geurts. Commutator-errors in the filtering approach to large-eddy simulation. Technical report, Dept. of Appl. Math., University of Twente, 2004.

300. F. van der Bos and B.J. Geurts. Dynamics of commutator-errors in LES with non-uniform filter-width. In R. Friederich and B.J. Geurts, editors, *Proceedings of DLES-5, ERCOFTAC Workshop Direct and Large-Eddy Simulation-V.* Kluwer Academic, 2004.

301. H. van der Ven. A family of large eddy simulation LES filters with nonuniform filter widths. *Phys. Fluids*, 7(5):1171–1172, 1995.

302. E.R. van Driest. On turbulent flow near a wall. *J. Aerospace Sci.*, 23:1007–1011, 1956.

303. O.V. Vasilyev and D.E. Goldstein. Local spectrum of commutation error in large eddy simulation. *Phys. Fluids*, 16(2):470–473, 2004.

304. O.V. Vasilyev, T.S. Lund, and P. Moin. A general class of commutative filters for LES in complex geometries. *J. Comput. Phys.*, 146(1):82–104, 1998.

305. M.I. Višik and A.V. Fursikov. *Mathematical problems of statistical hydromechanics.* Kluwer, Amsterdam, 1988.

306. J. von Neumann and R.D. Richtmyer. A method for the numerical calculation of hydrodynamic shocks. *J. Appl. Phys.*, 21:232–237, 1950.

307. B. Vreman. *Direct and large-eddy simulation of the compressible turbulent mixing layer.* PhD thesis, University of Twente, 1995.

308. B. Vreman, B.J. Geurts, and H. Kuerten. Realizability conditions for the turbulent stress tensor in large eddy simulation. *J. Fluid Mech.*, 278:351–362, 1994.

309. B. Vreman, B.J. Geurts, and H. Kuerten. Large-eddy simulation of the temporal mixing layer using the mixed clark model. *Theor. Comput. Fluid Dyn.*, 8:309–324, 1996.

310. B. Vreman, B.J. Geurts, and H. Kuerten. Large-eddy simulation of the turbulent mixing layer. *J. Fluid Mech.*, 339:357–390, 1997.

311. X. Wang. Time-averaged energy dissipation rate for shear driven flows in $\mathbf{R}^n$. *Phys. D*, 99(4):555–563, 1997.

312. H. Werner and H. Wengle. Large eddy simulation of turbulent flow around a cube in a plane channel. In F. Durst, R. Friedrich, B. Launder, U. Schumann, and J. Whitelaw, editors, *Selected Papers From the 8th Symp. on Turb. Shear Flows*, pages 155–168, New York, 1993. Springer.

313. H. Weyl. The method of orthogonal projection in potential theory. *Duke Math. J.*, 7:411–444, 1940.

314. S. Wiggins. *Introduction to applied nonlinear dynamical systems and chaos.* Springer-Verlag, New York, 1990.

315. G.S. Winckelmans, H. Jeanmart, and D. Carati. On the comparison of turbulence intensities from large-eddy simulation with those from experiment or direct numerical simulation. *Phys. Fluids*, 14:1809, 2002.

316. G.S. Winckelmans, A.A. Wray, O.V. Vasilyev, and H. Jeanmart. Explicit-filtering large-eddy simulations using the tensor-diffusivity model supplemented by a dynamic smagorinsky term. *Phys. Fluids*, 13:1385–1403, 2001.

317. W. Wolibner. Un théorème sur l'existence du mouvement plan d'un fluide parfait homogène incompressible, pendant un temps infiniment longue. *Math. Z.*, 37:698–726, 1933.

318. V.I. Yudovich. Non stationary flow of an ideal incompressible liquid. *Comput. Math. & Math. Phys.*, 3:1407–1456, 1963. Russian.

319. Y. Zang, R.L. Street, and J.R. Koseff. A dynamic mixed subgrid-scale model and its application to turbulent recirculating flows. *Phys. Fluids A*, 5(12):3186–3196, 1993.

320. E. Zeidler. *Nonlinear functional analysis and its applications. II/B.* Springer-Verlag, New York, 1990. Nonlinear monotone operators, Translated from the German by the author and Leo F. Boron.

Index

Scientific Computation

Scientific Computation

springeronline.com